An Economic History of
the English Garden

RODERICK FLOUD

An Economic History of the English Garden

ALLEN LANE
an imprint of
PENGUIN BOOKS

ALLEN LANE

UK | USA | Canada | Ireland | Australia
India | New Zealand | South Africa

Allen Lane is part of the Penguin Random House group of companies
whose addresses can be found at global.penguinrandomhouse.com

First published 2019
001

Copyright © Roderick Floud, 2019

The moral right of the author has been asserted

Set in 10.5/14 pt Sabon LT Std
Typeset by Jouve (UK), Milton Keynes
Printed and bound in Great Britain by Clays Ltd, Elcograf S.p.A.

A CIP catalogue record for this book is available from the British Library

ISBN: 978-0-241-23557-7

To my grandchildren,
Louisa, Alexander, Charlotte, Helena and James

Contents

List of Illustrations

1. Planting close to the City: the nurseries and market gardens of Southwark, south of London Bridge, in 1658. Detail from *An Exact Delineation of the Cities of London and Westminster and the Suburbs Thereof*, 1658, engraving by Richard Newcourt and William Faithorne / British Library, London, UK / © British Library Board. All Rights Reserved / Bridgeman Images.

2. The king's works: Charles II's canal in St James's Park, London, 1660. St James's Palace and Park / British Library, London, UK / © British Library Board. All Rights Reserved / Bridgeman Images.

3. A beautiful contrivance: the canal at Hall Barn, Buckinghamshire, in the eighteenth century. 'View of the Great Room etc. at Hall Barn near Beconsfield in Buckinghamshire', antique copper plate from P. Russell and Owen Price, *England Displayed* (London: 1769).

4. William III's view of the Thames: the Privy Garden at Hampton Court, Middlesex, 1702. Engraving from Ernest Law, *The History of Hampton Court Palace: Orange and Guelph Times* (London: George Bell, 1891). Private Collection / Bridgeman Images.

5. The cemetery as landscape garden: a design for Kensal Green Cemetery, London, 1832–3. © Museum of London.

6. Reflections in water: the canal and Archer Pavilion (1709–11) at Wrest Park, Bedfordshire. Photo by English Heritage / Heritage Images / Getty Images.

7. A man of taste: the Honourable Charles Hamilton (1704–86) at the age of twenty-eight, by Antonio David. Courtesy of Painshill Park Trust Ltd.

Graph of Changing Money Values, 1660–2020

Graph showing the modern equivalent of £1 spent at a date in the past, using a comparison with average earnings

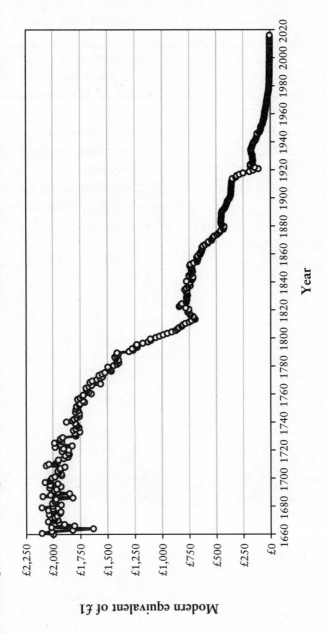

Source: Gregory Clark, 'What were the British Earnings and Prices Then? (New Series)', MeasuringWorth, 2019 (www.measuringworth.com/ukearncpi/)

An Economic History of
the English Garden

Introduction

I love visiting other people's gardens. It is a passion that I share with millions of others in this country who spend their weekends at the properties of the National Trust or at the myriad gardens open for charity. All of us admire the beauty of the landscapes, the quality of the planting, the serried ranks of vegetables; all of us secretly delight in discovering the odd weed or criticizing the colour schemes. In the great gardens we marvel at hidden grottoes and eye-catching temples, peaceful lakes and expansive vistas.

But as well as admiring all these things, I have a particular set of questions in mind when I tour a garden. A great deal of work is needed to create and look after a garden; I want to know how much it cost to make and to maintain. I want to know how many gardeners tend it, where the plants come from, what tools and techniques were used to create it and how they were invented. I want to be able to compare one garden with another in terms of the burden that it may have put on the family, who may have owned it for centuries. I want to know where their money came from and why they spent it on a garden. In short, I want to understand the economics of gardening and its history.

This book tries to answer these questions. It is, therefore, a new kind of garden history. Tens of thousands of books and articles have been written and published about the history of British gardens, and many more about those of other countries. But almost all of them focus on the design of gardens, or on their designers, on the trees, shrubs, flowers, fruit and vegetables that they contain, or on the temples, cascades, arbours, pergolas, statues or even the garden gnomes that adorn them. Only very rarely is the cost of making or maintaining a garden

3

ever mentioned. The prices of plants sold by nurseries, or imported from far-flung corners of the world, or the costs of employing a gardener, are sometimes discussed, but they are almost never put into context by comparing them with the price of other items at the time or by discussing the economic conditions of each period.[1]

Garden historians thus almost entirely ignore money.[2] But the subject of money in the garden is – oddly – ignored also by another group, the historians who study the economy in the past. This is despite the fact that, as this book aims to show, spending money on gardens has been one of the greatest, and certainly most conspicuous, forms of expenditure on luxury in England since the seventeenth century or earlier; it has employed hundreds of thousands of people at a time, given rise to a substantial foreign trade and created a whole industry of nurseries, garden centres and landscaping contractors whose current turnover exceeds £11 billion each year. So I hope that this book will help persuade economic historians and economists to recognize that gardening is an important British industry in its own right and has been one for hundreds of years.

In failing to appreciate the role of money in gardening, historians of all kinds have ignored, in addition, the importance of the impact of public spending, financed from taxes and borrowing, on the creation of gardens, both by the royal family and by those who in previous centuries occupied lucrative government posts. They have not appreciated the wealth that gardening brought to designers, seedsmen and nurserymen. Advances in the technology of gardening have led to wider improvements, such as in water and steam engineering, in central heating and in the construction of metal and glass buildings. The gardens both of the great estates and of the suburbs have changed the face of England. Gardens, in short, are far more important to our economy and society than even their greatest devotees have realized.

The book begins with the garden industry as it was in 1660 and what it has become today. The restoration of Charles II to the throne may not seem an obvious starting point for the history of an industry, but Charles spent large amounts of public money on gardens and, by that means and through his own personal interest, gave a crucial boost to gardening by the wealthy and the aristocratic. One of the results was a rapid growth of plant nurseries, particularly around London, which supplied an ever-increasing range of shrubs, trees and flowering plants to garden owners and garden designers throughout England. Their creations, which are documented in a range of books and pictures, show how large were both their ambitions and their purses. Quite how large can only be demonstrated by a full evaluation of how much was spent in the past. A proper translation from old to current values, by the method explained in Chapter 1, shows the huge sums that have been spent over the centuries, while modern evidence – for example, on our use of time – shows how important gardening is to us today.

Chapter 2 starts with the state and the monarchy that is its symbol. It describes the role of central government, which provided the funds with which successive monarchs, from Charles II onwards, created, maintained and enhanced the royal gardens and parks around London, including the world-famous Royal Botanic Gardens at Kew. Later, local government became important, building public parks with their floral displays and boating lakes throughout England. Where the monarchs trod, so the aristocracy followed and Chapter 3 describes the builders of the great gardens of Stuart, Georgian and Victorian England. It explores the sources of their wealth, the vast sums they spent to display their taste and discernment and how their conspicuous consumption,[3] in their gardens as well as their houses, helped to fuel the Industrial Revolution.

Chapter 4 turns from the customers to those who worked for them, did their bidding or influenced their taste. It considers the garden designers not just as artists but as businessmen, and a few business-women, who had a living to make and workers to employ and organize within the economic constraints and customs of their time. They worked closely with the nurserymen, discussed in Chapter 5, who provided the millions of plants which were and are required each year to meet the apparently insatiable demands of their customers for novelty as well as familiarity; they in turn worked with the plant collectors who scoured – and sometimes ravaged – other parts of the world in search of plants and varieties new to England. Last, there were the thousands – by the nineteenth century, hundreds of thousands – of working gardeners, who actually did the gardening in the great estates, in the suburban villas and in the public parks; their training, their careers and their rewards are discussed in Chapter 6.

All industries thrive on innovation and Chapter 7 describes several ways in which gardening took the lead in the technological changes that swept over England during the eighteenth century and beyond. It begins with water and steam engineering, where gardens developed techniques in construction and machinery that were later adopted by canal builders and machine makers. Then came the use of glass and iron in construction, pioneered in English greenhouses, together with the central heating systems that kept plants warm even while their owners shivered. Finally there was chemistry, where highly toxic substances were employed in a vain, and often destructive, attempt to wipe out insects and diseases.

Chapter 8 returns to the customers and to the huge expansion in the number and extent of English gardens and gardeners that came with the growth of the suburbs from early in the nineteenth century. Some of those gardens were used to grow fruit and vegetables, so Chapter 9 considers the vegetable garden within the context of the English diet and the state of nutrition of different groups of people within English society. Chapter 10 sums up the importance of gardening and its ramifications within the English economy both today and in the past.

Part of my aim in writing this book is to stimulate others to follow, to consider other aspects of the subject and to explore more historical records. There are over 1,500 gardens in Britain listed as being of historical interest and it is only possible to refer to a fraction of them. The archives in county record offices and stately homes are voluminous and I have therefore been forced to limit myself to England – although the gardens of Wales, Scotland and Ireland are equally beautiful and expensive – and to draw examples only from a small set of English counties.[4] I have had to leave, for future research, the economics of plant collecting and the recent development of the nursery industry. I have not been able to consider the impact of English gardens on the rest of Europe, nor on the rest of the world, much though I would like to be able to answer the question I have often been asked: was England, or Britain, unusual? I don't think that it was, but I look forward to others investigating the topic.

I

The English Garden in
1660 and 2020

'The World's a Garden; Pleasures are the Flowers'[1]

In 1664, Captain Leonard Gurle, a nurseryman who was later to become the king's gardener, received an order for sixty-five fruit trees. In pride of place were twenty different varieties of peach and five of nectarines, in which Gurle specialized, but there were also apricots, figs, plums and grape vines.[2] Gurle's nursery, where the young trees were growing, was not, as one might expect, deep in the English countryside, but in Shoreditch, only a few hundred feet outside the old walls of the City of London. It covered 12 acres of what is now the Brick Lane or Banglatown area of east London, with its south Asian restaurants and shops selling brightly coloured saris.

Peaches and nectarines were recent introductions to English orchards and the walls of kitchen gardens. Shakespeare does not mention 'peach' except as a colour, but some of the other plants that Gurle supplied were the luxuries that Queen Titania, in *A Midsummer Night's Dream*, ordered her fairies to give to her enchanted lover, Bottom, with his ass's head:

> Feed him with apricocks and dewberries,
> With purple grapes, green figs and mulberries.[3]

Gurle's trees were to be supplied, however, to a more mundane customer, William Alington, 3rd Baron Alington, for his new house, Horseheath Hall, in Cambridgeshire. Alington, whose family had owned land there since 1397, as well as manors in several other English counties, was rebuilding his mansion. He and his father had kept a low profile during the English Civil War of 1642–51 and the

7

following interregnum and had escaped the fines or confiscation of lands that affected other royalist aristocrats. Now, with the restoration of Charles II in 1660, Alington clearly felt confident enough to embark on a major building project, for which he engaged the aristocratic architect Sir Roger Pratt. It was 'on a grand scale with a 500 foot frontage, the most imposing in the country of that date';[4] in 1670 the diarist and garden writer John Evelyn dined there and remarked waspishly that Alington had 'newly built a house at great cost, little less than twenty thousand pounds . . . standing in a park with a sweet prospect and stately avenue, but water still defective. The house also has its infirmities.'[5] Evelyn says nothing else about the lake or the rest of the garden, of which little remains, but it seems to have been on an equally grand scale; there was a 'great terrace' between the house and a slightly sunken garden, with flanking walled areas for fruit and vegetables.[6] The stately avenue was over a mile long and the garden was divided into elaborate compartments.

The family's fortune rested on land and the rents from it, although Alington also held lucrative government posts and served as Constable of the Tower of London from 1679. Whatever its source, his wealth was enough to afford a very costly new house and garden. Gurle's trees for the orchard or kitchen garden represented a small fraction of that cost, at £8 and 3 shillings. However, each specimen of the most expensive varieties of peach, a Province, a Lion, a Violett Muscatt and a Persian Peach, cost 5 shillings, a large sum at the time. Gurle's customers were of the highest quality and he ended his career as the royal gardener at St James's, so Alington was clearly buying from the best or, at least, the most expensive.

THE GARDEN INDUSTRY

Alington and Gurle were part of the garden industry. We do not normally think of gardening as an industry; it is a hobby, a pastime, a search for beauty, even an obsession. But, as well as these, gardening is something on which we spend money: it employs people; it uses tools and machinery; it occupies land, from the smallest patio to the largest park; it constructs hedges and pergolas, temples, fountains and waterfalls. It is an unusual industry because many of its customers are also its workers, its designers and its entrepreneurs, but it is an industry none the less and one that has consumed great amounts of economic resources of all kinds – land, labour and money – for many centuries.

This book is about the myriad trades, professions, institutions, firms and people that have, over more than three centuries, interacted with their tens of millions of customers to create and maintain England's gardens. They have all acted within the society and economy of their time and their achievements can only be understood in that context. The book covers the gardens, parks and landscapes that were created for pleasure or to provide flowers, fruit and vegetables for personal consumption. It therefore includes the nurseries, such as Gurle's, which produced the plants for domestic gardens and parks, and all those who designed or provided the expertise, tools and machinery, but not – except as an aside when discussing kitchen gardens – what the English call 'market gardens' and the Americans 'truck gardens', growing vegetables, fruit or flowers for sale.

THE CHANGING VALUE OF MONEY – TODAY AND YESTERDAY

Since money underpins much of the discussion about gardens in this book, we have at the outset to confront a problem. Very few of us know how much prices have changed or what sums of money in the past meant to those who paid or received them. Lord Alington's mansion, the 'most imposing in the country' in 1670, seems very cheap at £20,000, but was the 5 shillings (£0.25) that he paid for a peach a little or a lot of money? And two centuries later, in the 1880s, was the £153,000 that Baron Ferdinand de Rothschild of the great European banking family spent on the gardens at Waddesdon, Buckinghamshire, an equally large sum in today's values? Let's take some other examples. The great garden designer Lancelot 'Capability' Brown became head gardener at Stowe, Buckinghamshire, in 1741 at the age of twenty-six; he was paid £25 a year, with £10 for housing. In 1826, Joseph Paxton – later designer of the Crystal Palace – was appointed head gardener at Chatsworth, Derbyshire; aged twenty-three, he was paid 25 shillings a week, or £65 a year, and given free lodging in a cottage that came with the job.

What do these numbers mean? How much would Alington's mansion, or his peach, cost in 2019, and what would the Waddesdon gardens cost if they were built today? Was Joseph Paxton really paid more than Brown? Was Brown well paid by today's standards? Are we spending more or less on our gardens than our ancestors did? Questions like this can only be answered if we translate money from the past into a common standard, such as its value today. But how? How do we take account of changes in the value of money and of all the other changes over three centuries in the things that we buy and the amounts we are paid?

A common method is to look at the prices of consumer goods such as bread and milk, to calculate the average change in those prices, and to apply that change to a salary or the cost of a building in the past; this method uses the retail price index (RPI) or consumer price index (CPI). Since consumer prices have increased by 128 times since 1741, we would multiply Brown's salary of £25 by 128 and conclude that he was paid, in today's values, £3,200. But this figure doesn't make sense. It suggests that the head gardener of the greatest garden in England in the 1740s was paid for a year the equivalent of someone working for twelve weeks at the current national minimum wage of an unskilled labourer.

Comparing consumer prices works well over short periods; that's why we use them to make yearly adjustments to pensions or other benefits. But it is a misleading method for comparing prices over long periods; the things on which we spend money have changed in nature and quality – white-van couriers have replaced the horse-drawn carters of the eighteenth century and we send emails instead of letters. Foreign trade means that bread, for example, is much cheaper, because of grain imports from Canada and the United States. We also spend much less of our income on manufactured goods and food than we did two centuries ago and much more on services such as entertainment or restaurant meals. Finally, we use much more machinery than we used to. We live our lives and spend our money quite differently; few of us spend half our income on food and half of that on bread, as was common until the twentieth century. All these changes make the use of RPI or CPI inappropriate as a means of comparison over long periods.

A better method, particularly for examining salaries or wages or the objects built or made by the people who were paid these wages, is to compare their pay with changes in average earnings over the centuries. In 1700 the average worker earned £12 and 8 shillings a year; in 2015 he or she received £25,609. This large increase in the amount of money at our disposal has occurred mainly because of inflation in the twentieth century – caused by a number of different factors, including wars and the prices of commodities such as gold or petroleum. But, also, each worker – often aided by machinery – now produces much more than he or she used to; the agricultural labourer can plough far more land in a day with a tractor than his ancestor could with a horse, the metal-worker can produce far more screws or tools than could the Georgian village blacksmith. This has meant that wages have risen and our average standard of living has improved; our houses are of much higher quality, we drive cars rather than walking or riding a horse, we have more nutritious and much more varied food.

All this improvement is reflected in average earnings, which take account of changing values of money and the improvement in our welfare over the centuries. So in this book we translate the average annual wage of £12 and 8 shillings in 1700 to £25,609 today. By 1750 the annual average had risen to £14 and 1 shilling, but we still translate that to the average today of £25,609. If someone was paid twice as much as the average wage in 1800, the translation is twice the average today. This also makes sense if we are comparing expenditure on gardens over the centuries, since most of the cost of making and maintaining gardens is used in paying wages for labourers and gardeners. So, if a garden project employed twenty-five men in 1850 and cost roughly twenty-five times average earnings in that year, we would say that the modern equivalent of that project would be one costing twenty-five times average earnings today. To put it another way, if something cost two-thirds of average earnings in 1800, we say that its equivalent is two-thirds of average earnings today.*

* Equating proportions of average earnings in this way may somewhat overstate the value in today's money of work done in the past. The garden labourer of today can

Economic historians have estimated what average earnings were in each year from 1270 onwards, so we can calculate the modern value of any sum of money in each year since then; all the data are conveniently displayed on the MeasuringWorth website and the conversions are done automatically there. That website shows and explains other ways of making the translations, but in this book I use the index of labour earnings.* The value of £1 at each date since 1660, according to the change in average earnings, is shown in the graph on page xi.

On this basis, Capability Brown was paid – in modern values – £45,580 a year, or £63,810 if one includes the value of the house that came with the job; it isn't a princely sum, but it is reasonable for Lord Cobham's trusted gardener, who was in charge of a number of under-gardeners at Stowe. Using this method, we can also answer the question of whether Paxton was paid more than Brown. It turns out that Paxton was getting, in modern terms, £47,120 together with his cottage, so his basic salary was about £2,000 more than Brown's. Both of them were earning quite a bit more than the UK average. We can use the same method to measure how much the gardens at Waddesdon cost, since so much of it comprised the wages of labourers digging out vast quantities of earth: the £153,000 that Baron Ferdinand spent translates to £68.8 million in today's values. It's a huge sum but he was, after all, one of the richest men in Europe and even this amount did not greatly dent his fortune. Finally, Lord Alington's mansion and garden of £20,000 cost him the equivalent of £33 million, which helps to explain why John Evelyn thought they were so grand, while each of the peach trees that he bought from Gurle cost him £400 in modern values, an indication of their rarity at the time.

---

shift far more earth, using a mechanical digger, than his predecessor could with his spade and wheelbarrow. As an economist would put it, capital in the form of machinery has been substituted for labour and the desired result has been to reduce the overall cost. However, much gardening remained labour intensive for almost all the period covered by this book, so the use of average earnings is likely to give the most reliable results.

* In this book, all the translations in values from one period to the present were calculated in June 2019. As more data become available and are added to the index of labour earnings database, some of the translations after that date may differ slightly from those given here. For the sake of consistency, all are related to average British earnings in 2015.[7]

In case this expenditure on historical gardens may seem too large to believe, we can also consider whether these translations – using fractions or multiples of average earnings – give reasonable results, by comparing them with expenditure on gardens today. Capability Brown was paid £21,538 by the 4th Duke of Marlborough for his work at Blenheim Palace in Oxfordshire between 1764 and 1774, which equates via the earnings index to about £34.4 million today. It's a huge sum and it may not seem believable that so much could be spent. Most of the money went on the 150-acre lake; no one has built anything like it recently, but the Marlborough estate is spending £6 million in 2020 simply to clear the silt that has accumulated in only a relatively small part of the lake that Brown created. At Alnwick Castle in Northumberland, at least £42 million was spent between 2000 and 2006 on the restoration of part of the garden, while the Royal Horticultural Society began work in 2018 to create its new garden in Salford at an estimated cost of £35 million. Many gardens in the past contained elaborate buildings, few more dramatic than the pagoda at Kew; its latest restoration in 2018 has cost £5 million.[8] The terrace in front of the great house at Cliveden has just cost £6 million to restore. Expenditure on the Royal Parks in London is currently about £27 million each year, which is somewhat less than the equivalent that a succession of monarchs, from Charles II onwards, through the eighteenth century, were spending every year to create and maintain the same parks. In contrast, a peach tree that cost Alington the equivalent of £400 would today cost about £30, illustrating the fact that buying goods has become much cheaper in relation to average earnings.

In other words, the sums of money spent on gardens, when translated into modern values, seem to be huge because that is exactly what they were and still are. It is why current annual UK expenditure in nurseries and garden centres and on landscape contractors is over £11.4 billion, without including the amount we spend on gardeners, or the value of all our own labour or the cost of all the land that we use. For centuries we have been spending, and now continue to spend, far more on our gardens than almost anyone realizes.

It is both confusing and clumsy to burden the reader with repeated sentences such as: 'Capability Brown was paid £25 a year, which in modern values is £45,580.' So this book adopts a unique approach for a work of history and gives the modern value followed by the original sum in brackets and italics: for example, 'Capability Brown was paid £45,580 *(£25).*' The original values are quoted in pounds, shillings and pence – each pound consisting of 20 shillings (20s) and each shilling of 12 pence (12d) – which were the British currency until 1971. Finally, since measures of length and area are still mired in confusion between imperial measures and their metric equivalents, the original feet, yards or acres have been used: 1 foot is 0.3 metres, 1 yard is 0.9 metres and 1 acre is 0.4 hectares.

Armed with these tools, what can we say about the garden industry in 1660?

GARDENING IN 1660

An industry supplies what its customers demand and can pay for. By 1660, the English garden industry was already sufficiently large and well developed to grow vast numbers of trees and other plants and to organize their transportation to wherever they were wanted. It could supply the labour and the expertise to dig lakes, construct terraces, plant hedges and avenues; it was soon to construct greenhouses and to use them to nurture an increasing number of exotic plants, initially from North America. It had sent out its gardeners, such as John Tradescant the Younger, in the years before the English Civil War to seek out new plants and seeds in America and to arrange their transportation – with great difficulty – back to England. John Harvey, historian of the nursery trade, places 'the beginnings of the garden trade as we think of it in the middle of the 17th century',[9] but its antecedents go back centuries, even millennia.

We know that there were gardens at the villas of Roman England,[10] for instance, and probably next to houses even during what used to be called the 'Dark Ages'. In medieval England, gardens are documented in the records of monasteries and royal palaces. The accounts of Winchester College in the late fourteenth and early fifteenth centuries record numerous payments for seeds, seedlings and cuttings, together with the wages of gardeners and the purchase of tools and measuring cords and lines.[11] If there were purchases, there must have been sales. In other words, there was a nascent garden industry in the sense, at least, that someone was saving seed and cultivating plants that were then made available for sale. By the sixteenth century, there was clearly a thriving overseas trade in plants. It is documented, for instance, in the accounts of the royal gardens, which show that Henry VIII paid £101,100 *(£20)* in 1547 to Sir Jehan Le Leu to bring 'trees and sets of sundry kinds out of the realm of France'.[12] Sir Jehan, who is credited with introducing the apricot to England in 1542, was known as *confector viridariorum* (literally, 'garden maker'), or royal garden designer, and was granted an annuity for life in 1538 of £78,500 *(£13 6s 8d)*; he worked at Whitehall and Hampton Court, and he was not alone, as there were a number of other royal gardens and gardeners.[13]

It was not only kings who had gardens. C. Paul Christianson has unearthed the records of five other large London gardens, along the River Thames, including those of the grandest of bishops, of Canterbury, Winchester and London, together with the garden at Chelsea of Sir Thomas More. According to More, in his *Utopia*, published in 1516, the Utopians – citizens of his ideal society, which was modelled on London – were

> very fond of these gardens of theirs. They raise vines, fruits, herbs and flowers, so thrifty and flourishing that I have never seen any gardens more productive or elegant than theirs. They keep interested in gardening, partly because they delight in it, and also because of the competition between different streets which challenge one another to produce the best gardens. Certainly you will find nothing else in the whole city more useful or more pleasant to the citizens. And for this reason, the city's founder seems to have made gardens the primary object of his consideration.[14]

The Thames of More's time in early sixteenth-century London was bordered by gardens.

Archives recording royal gardens and other gardens in London document the purchase of seeds, plants and trees, tools and pots. In other words, there is plenty of evidence of a developing garden industry in London and, probably, in other urban centres, small though they then were. The gardeners of the big houses might also have been in business on their own account and we know that many of them were contracted to maintain gardens, employing their own labour force and probably supplying their own plants and seeds.[15] While the range of plants was very limited compared with what is available today or with what became available later, in the eighteenth or nineteenth century, it was still wide enough to provide colour and variety, both in flower gardens, herb and rose gardens and in the kitchen garden. Fruit trees seem to have been particularly valued.

This is part of the background to the mature garden industry that emerged in the second half of the seventeenth century. But there are other underlying factors. One is the state of the English economy, which has affected gardens and garden-making through the centuries.

Gardens are, in the main, a form of luxury. Flower gardens or landscaped parks are not essential to life, even if a part – though usually only a small part – of our diet is supplied by a vegetable garden. Therefore money spent on gardens comes from whatever is left after we have provided for our basic needs for food, clothing and housing, together with the energy required for work. For much of English history there was, for most people, very little or nothing to spare in terms of unused income, but there are signs that, from the seventeenth century onwards, the surplus available after providing for bare necessities was growing. Despite the disruption and heavy loss of life in the English Civil War, the period from 1650 to 1700 saw the productivity of the economy – that is, the total economic output divided by the number of people in the country – growing more rapidly than during any other fifty-year period between 1270 and 1870. It was founded on the expansion of industry, particularly textiles, coal and iron, the concurrent development of internal and foreign trade and commercial activities, and on an increased role for the state in regulating the economy and the financial system.

By 1690, England's productivity surpassed that of its closest rival, Holland. What makes this period particularly unusual is that there was very little increase in the population overall, although the proportion to be found in London and a few other urban centres was growing. This minimal population growth is in marked contrast to the period of the so-called Industrial Revolution, from 1780 to 1840 and beyond, when the increase in goods and services produced was nearly matched by a similar increase in the number of people.[16] All this may be a surprise to people brought up on the traditional account of an agricultural revolution of the eighteenth century and an industrial revolution to follow it, but we now know that the economy had been growing and changing since much earlier on.

In simple terms, people at the end of the seventeenth century had a bit more money to spend. As their probate inventories compiled at their death show, they spent it on clothes, furniture and more comfortable houses. They spent it in an increasing number of shops; some employed more servants and people also paid for better transport, for more services, such as shipping and carriage of goods by land, and a developing legal and banking system. Although study of this aspect of consumption has been neglected, people also spent it on gardens. Some had a great deal more money and were able and willing to spend part of it on very elaborate and costly garden schemes. The royal family – funded by taxes levied on the rest of the population – led the way, but they were soon followed and emulated in both the town and the country by the rest of the aristocracy, such as Lord Alington, and by the growing middle class.

These were the demands that Leonard Gurle and an increasing number of nurserymen were successfully meeting. Gurle had been preceded during the early 1600s by three leading London growers of fruit trees: Banbury of Tothill Street, Westminster; Warner of Southwark; and Pointer of Twickenham. There was also the florist – then the term for a flower grower – Ralph Tuggie of Westminster, whose specialism was carnations. There were the Tradescants, father and son, royal gardeners and botanical collectors, whose cabinet of curiosities was to become the foundation of Oxford's Ashmolean Museum. There was the Millen family of the Old Street nursery, just outside the City of London, who were known for their gooseberries. There were an increasing number of nurseries in Southwark, across London Bridge, as shown in plate 1, as well as in other areas along the Thames and in Shoreditch, to the north-east. Seedsmen could be found all over London. Thus Captain Gurle, who had founded his nursery in 1643, faced increasing competition, particularly – towards the end of his life, in the early 1680s – from the great Brompton Park Nursery.[17]

By then, nurserymen were branching out into the design and construction of gardens and landscapes. By then, too, the nurseries of London were occupying hundreds of acres; as every gardener knows, one can grow a very large number of plants on an acre. Among them, by this time, were thousands of forest trees, a whole range of shrubs and, of course, the prized fruit trees in which Gurle specialized. His peaches and nectarines, his apricots and figs, the eight orange trees 'in boxes of ye biggest sort' and seventeen 'of the next sort' – valued altogether at £97,750 (£49 10s) in his probate inventory of 1685 – were luxuries, to be afforded only by the few. They were desired as items that denoted wealth and social status. They were scarce, to be fought over, even to bankrupt some of those who desired so desperately to possess them.[18]

GARDENING TODAY

In many ways, gardening in England today has changed little since 1660. The hand tools that we use are still much the same. We buy plants and seeds from nurseries and seedsmen whose predecessors, like Gurle, were well established by then. Garden designers have existed for all that time. Gardening is still mainly governed by the seasons. We still love admiring or criticizing other people's gardens, as past visitors to the Tradescants' garden and cabinet of curiosities or to numerous stately homes must have done too.

The main differences are ones of scale. There are millions more gardens and in them thousands more different species of plants, originating from all over the world. Nurseries in Stuart England probably contained millions of plants, as did those in France and the Netherlands, but now English gardens have access to hundreds of millions, either propagated locally or imported throughout the year. A few greenhouses have been replaced by millions and are supplemented by polytunnels and conservatories. Garden visitors are now numbered in the tens of millions. How did this increase in scale occur?

England's population is now about ten times larger than it was three and a half centuries ago. Moreover, each person today, on average, consumes at least eighteen times more by way of resources – after taking account of changes in prices – than his or her equivalent in 1660. These are the two most important facts in understanding the growth of English gardens and the garden industry. It is economic growth, partly driven by population growth, that has produced a surplus of income over expenditure on the necessities of life; part of that surplus is spent on luxury consumption such as gardening.

But actually those two facts understate the impact of economic growth in several ways. One example is our use of time. Until very recently – in fact only until the past century or more recently – production in the economy, and the wages that were paid, depended on long hours of work. Most of our ancestors worked for twelve hours a day, six days a week, with only short meal-breaks. They had little spare time and also little energy left to enjoy it. Today, we have far more leisure time and we spend a good deal of it on gardening. On average across the whole population of the UK, we spend ten minutes per day, or over an hour a week, on that pastime, men slightly more than women; this may not seem much, but in 2014 – things may have changed a little since then – it was as much or more than the time we spent on other hobbies and forms of entertainment, other than watching television. It was the same as the number of minutes spent on computer and video games and other computing, for example, and much more than is spent on any form of sport. However, there is great variation across age groups: men aged fifteen to twenty-four spend only one minute on gardening in an average day, while those aged sixty-five and over spend twenty-six minutes. This statistic is important because there are now more older people who live longer in retirement and hence have more time for gardening.

We are not alone in this use of our time. The English, or indeed the British, may believe that they are exceptional in their devotion to gardening, but actually, and despite the fact that our temperate climate is particularly suitable, we spend less time on it than all but two – Finland and Spain – of the fifteen countries who participated in a recent 'Harmonised European Time Use Survey'. Top comes Bulgaria, with thirty minutes per day, closely followed by Slovenia with twenty-eight minutes.[19] We also spend less money on our gardens than do people in many other European countries, less than in France, Italy or the Netherlands and much less than in Austria, Sweden,

Denmark, Norway and Luxembourg.[20] We even seem to do less gardening than men and women in the United States, although the data there are collected in a different way.[21]

As well as time, space is a factor. Part of our country's economic growth has come from making the land more productive; we have also been able to spend some of our income on importing food from abroad. This has made it possible for us to devote more and more of our land to building houses and to creating gardens around them. Particularly since the garden city movement at the end of the nineteenth century and the growth of the suburbs in the twentieth, we've been prepared, collectively and individually, to pay to use land to make gardens. Today, even though land prices have risen so rapidly in recent years, we still devote 15 per cent of the land area of our towns and cities to parks and gardens, most of it taken up by the individual front and back gardens that we tend so lovingly. This is changing, and the average new house has a smaller garden than it would have had a century ago, but there has been some compensation in a rise in the use of pots, patios and window boxes to cram more plants into smaller places.

In many areas of life, increased incomes and overall wealth have meant that we have employed other people to do things that we used to do ourselves. We no longer make our own clothes and we often go to restaurants or buy take-away meals rather than cooking ourselves. Gardening bucks this trend. In past centuries, every upper-class and most middle-class households would have employed a gardener, but today only 5 per cent of us still do so.* Garden machinery has helped, just as the vacuum cleaner, washing machine and electric cooker have made housework easier. So too have the nurseries and garden centres, with their ranks of container-grown plants and seedlings, which make it unnecessary for most of us to raise plants from seed.

For the 95 per cent of us who don't employ a gardener, gardening is something that we do ourselves. Surveys tell us this and what we think about it. First, and perhaps most importantly, we enjoy it. The Horticultural Trades Association published a survey in 2006 that found that 55 per cent of respondents said that they enjoyed gardening a lot or a

---

* This figure probably does not include the occasional employment of landscape contractors, such as the army of workers paving over so many front gardens today.

little. Enjoyment rose to 70 per cent among individuals aged fifty-five to sixty-four, from 25 per cent among those aged fifteen to seventeen. Other surveys have found that the older you are, and the richer you are, the more gardening you are likely to do. Six out of ten of those interviewed thought that spending money on plants was a good investment, and eight out of ten thought the garden was important for relaxing and entertaining; seven out of ten took pride in their gardens, although about the same proportion were always looking for ways to reduce their garden workload.[22]

Even though we do so much garden work ourselves, we seem rarely if ever to keep a record – as I did while writing this book – of how much time and money we spend on our gardens; probably some of us dread what we would discover. I was amazed to find that my wife and I had spent £7,500 in a year on a half-acre garden in Buckinghamshire. Oxford Economics has recently calculated that in 2017 we spent £5.6 billion in the nurseries and garden centres of the UK, together with £6.8 billion on landscape services – designing, building, planting and maintaining the UK's green spaces. In addition, £1.35 billion worth of ornamental plants were produced and sold in the UK and a further £1.2 billion was spent on imported plants, although about two-thirds of those were cut flowers.[23] These figures do not include the cost of innumerable television programmes, books and magazines on the subject. Together, gardening is a big industry. Creating gardens can be particularly expensive; it is said that the average cost of a show garden at the annual Chelsea Flower Show of the Royal Horticultural Society, which will be on display for only five days, is now over £1 million. But garden maintenance is expensive too, particularly as few gardeners can resist making constant improvements. As the writer John Claudius Loudon put it in 1822:

> A man whose garden is his own for ever, or for a considerable length of time, whether that garden be surrounded by a fence of a few hundred feet, or a park-wall of ten or twelve miles, will always be effecting some change in arrangement, or in culture, favourable to trade and to artists.[24]

Much of this book is about the trade and the artists who have made up the garden industry since 1660. But it is also about their interaction with us, their customers – we who love, create and maintain gardens and who supplement the professional expertise of the gardener, nurseryman or designer with our own skill and artistry, as well as our sometimes fickle likes and dislikes. Demand and supply in gardening produces constant change; we alter our gardens far more frequently than we modify our houses. That is what produced the garden industry in the seventeenth century and still animates it today. This book celebrates an industry that is oddly unrecognized in our history, certainly in comparison to cotton, iron or the railways, but which has nevertheless changed the face of England not once but many times.

2

Gardens and the State

'To see our Prince his matchless force employ'[1]

On 29 May 1660, his thirtieth birthday, Charles II returned to London from his exile abroad to take up the throne, eleven years after his father, Charles I, had been executed outside the Banqueting House in Whitehall. The new king had plenty to do. He had to re-establish the monarchy, appoint his ministers and courtiers, call a new parliament and cement relationships with the Anglican Church, which was still suspicious of his Catholic mother; not least, he had to pursue and put to death the men who had killed his father. Yet despite all these tasks, he still had time before the end of 1660 to begin work on another of his major priorities: renovating the royal gardens. In St James's Park, only a few hundred yards from Whitehall, he made use of unemployed soldiers to dig a huge rectangular lake, 850 yards long by 42 yards wide, which he called a 'canal';* plate 2 shows the results of their work. It linked together a series of ponds that had been neglected under Cromwell's Protectorate. The next step for Charles was to restock the menagerie and aviaries that had been established by his grandfather, James I; soon he was able to enjoy himself feeding the ducks on his new lake. Gardens were important to Charles and indeed to most of his successors down to today.

* A French word, which has now come – in both languages – to mean a waterway, but in the late seventeenth century also meant a rectangular pond or lake. Charles II's canal was constructed at about the same time as the canal at the Château de Vaux-le-Vicomte, but predated both the great French Canal du Midi and Louis XIV's Grand Canal at Versailles.

Ducks were the least exotic of the attractions. As early as 1663, there were elk, deer, antelope, 'goates from Guinea' and monkeys, together with cranes, storks, peacocks, pelicans – still a favourite with visitors – partridges and even a cassowary.[2] By 1670 there were parrots, a golden eagle, a vulture and two eagle owls, and by 1682 thirty ostriches, the gift of Moulay Ismail ibn Sharif, the Sultan of Morocco. John Evelyn recommended that enclosures and aviaries should be provided in gardens; he had one of his own at Sayes Court in Deptford.[3]

As it was soon opened to the public, Charles was often seen walking there, mingling with passers-by in a way that would be inconceivable in today's security-conscious times. In the winter of 1662, he and Queen Henrietta Maria were observed watching the 'strange and wonderful dexterity of the sliders [skaters] on the new canal, after the manner of the Hollanders'.[4] John Evelyn and Samuel Pepys record such royal perambulations, but the park became known for less innocent pleasures, indulged in by both Charles and his subjects, and Evelyn also censoriously describes, on 1 March 1671, how he walked with the king through the park

> to [Nell Gwyn's] garden, where I both saw and heard a very familiar discourse between ... [Charles] and Mrs Nelly [Nell Gwyn] ... she looking out of her garden at the top of the wall and ... [Charles] standing on the green walk under it. I was heartily sorry at this scene. Thence the King walked to the Duchess of Cleveland, another lady of pleasure, and curse of our nation.

Much less elevated ladies of pleasure, and their activities in the park, were described in graphic detail in the 1661 poem 'A Ramble in St James's Park' by the courtier John Wilmot, 2nd Earl of Rochester.

St James's Park saw the first royal 'improvement' after the Restoration. But by the end of the seventeenth century, equally ambitious or even larger works – under Charles I, James II and William and Mary – had been carried out at Hampton Court, Greenwich, Newmarket and, abortively, Winchester, while the next century saw more – at Kensington Palace, Hyde Park, Windsor Castle and Windsor Great Park (including Virginia Water), Frogmore House, Carlton House, Richmond and Kew. In the nineteenth century came Frogmore Kitchen

Garden and works at Osborne House, Balmoral and Sandringham, in the twentieth at Highgrove. Each royal generation tried to outdo the one before.

So kings, queens and princes each did their 'matchless force employ'. But it wasn't with their own money. That came from the government of the day and, ultimately, from all the nation as taxpayers. Nor was building and maintaining royal gardens the only role that the state played in English gardening. During the late nineteenth century, public parks multiplied throughout the country, usually through the initiative of local government. In the twentieth, new towns were surrounded with parks, lakes and woodland; colleges and universities laid out gardens. Public events such as the Great Exhibition of 1851, the Festival of Britain in 1951 and the London Olympics of 2012 called for landscaping and planting.

Together, these projects make up a very large – this and the next chapter aim to show how large – public investment in English gardening, stimulating the creation of a garden industry and underpinning its growth. Royal patronage and enthusiasm led the way, creating and sustaining a fashion that was emulated not only by the aristocracy – who piled into gardening right across the country after the Restoration – but by the growing middle classes, the more prosperous working class and eventually the poor. The government responded and applied its 'matchless force' to the task. But the decision, taken by generation after generation, to spend billions of pounds of public money on gardens and parks, required not only a wide consensus that this was a sensible thing to do but also an economy that could afford to do it. How was that possible?

The answer lies in the increase in our population and the size of the economy. More people has meant more workers and more consumers – along with more taxpayers – but we have also become better at producing and selling; we have invested in machinery, developed new technology, trained and educated individuals and traded throughout the world.

This change, which has transformed the lives of successive generations, has not been a sudden one, however. England has never experienced economic growth as rapid as in today's China or India, but it has been growing for much longer. Very broadly, our economy grew slowly between 1650 and 1750,* a bit more rapidly from 1750 to 1825 and a bit faster again from then until today – for this last long period at roughly 2 per cent each year, although there have been ups and downs, booms, recessions and slumps. At first, we were accompanied or sur-passed by Holland and France, which then faltered; Holland suffered from a lack of coal and France had a revolution. Britain carried on at the front of the pack until we were challenged by the United States and Germany at the end of the nineteenth century.

We have decided, collectively, to allocate part of the proceeds of the past three centuries of economic growth to gardens, both to our own and to those created and maintained by the state. Government expenditure, financed by taxation and by borrowing, has had a bad press recently. People talk of the 'burden of taxation' and of govern-ment expenditure 'crowding out' private investment. Spending by the state is seen as inherently less efficient and less desirable than spend-ing by individuals or companies. Some want to reduce the role of government in all areas.

However, successive governments in the late seventeenth century were essential in re-establishing the rule of law and the property rights that underpin contracts and commerce. In 1694 they set up the Bank of England, which gradually established confidence in the cur-rency and provided – after the crisis of the South Sea Bubble – a stable way for the government to borrow from the public. This, at the least, greatly reduced centuries of conflict over money between king and Parliament, one of the main causes of the Civil War that cost King Charles his head. Governments defended the country and attacked its enemies.† Governments made rules, for example about the building and running of canals and railways. They set taxes, which affected individuals and companies in many ways; the excise duties favoured by eighteenth-century governments were levied on a bewildering range of transactions, from employing servants to distilling gin or manufacturing glass, and made a difference to both production and consumption.

* However, in the first half of this period, output per head – productivity – rose quite rapidly, which would have allowed for a similar improvement in living standards.
† By one count, Britain was involved in as many as twenty-three wars during the eighteenth century and thirty-four in the nineteenth.

Governments also spend money; this pays wages and stimulates the development of new methods and products. Most public expenditure in the eighteenth and nineteenth centuries was on the army and navy, which meant putting public money into the shipbuilding and armaments industries.* That had a knock-on effect on other parts of the economy. It helped to develop coal-mining and iron-making. Naval dockyards and army and militia garrisons also injected cash into Plymouth, Portsmouth and many other towns, which in turn stimulated growth in local amenities, such as food and clothes shops, pubs and plant nurseries.

London, as usual, did best of all. There, governments built and refurbished palaces, parks and gardens to uphold the dignity of the royal family. The money went to builders and contractors, artists, furniture makers, sculptors, plant nurserymen, landscape designers and gardeners. Much more was spent, in London and then throughout the country, both by central and local government, in the nineteenth century. This chapter is about the impact of all this money on the English garden industry.

ROYAL GARDENS — PUBLIC FUNDS

In 1701, the Lords of the Treasury were worried about their king, William III. They were not concerned that he was asserting too much royal power, as his (and his wife's) executed grandfather, Charles I, had done. Nor were they anxious about his religion, the problem that had led to William's uncle, James II, losing his throne. The Act of Settlement of 1688 and the Bill of Rights of 1689 had dealt with both issues, even if the repercussions had to be worked out throughout the next century. No, the problem with William was the amount he was spending on 'works' – palaces and gardens – particularly at Hampton Court and Kensington.

Both his predecessors had been big spenders. Charles II followed

---

* As governments borrowed more, partially to finance wars, the payment of interest on the national debt accounted for larger and larger proportions of government expenditure; military always dwarfed civil expenditure.

up his canal in St James's Park with building and garden work at Whitehall, Newmarket and Windsor Castle; one of his other early efforts was to build a court to play the ball game pell-mell (a precursor to croquet), as Pepys recorded in his diary in April 1661 and on other occasions.[5] The name of the fashionable London street Pall Mall still recalls its location. James II continued the tradition at Whitehall, St James's and Hampton Court. William III and his wife Mary II, who had already constructed two large gardens in Holland before securing the British throne, lost no time; in 1689, the year after they arrived – as Charles II had done before them – they were improving Kensington House and its grounds and building greenhouses at Hampton Court. Much more work followed.

In response to an inquiry from the Treasury, the Office of Works – which had some responsibility for the palaces, parks and gardens[6] – estimated that Charles II had spent about £39 million (£20,000) per annum on them during his twenty-five-year reign (1660–85). James II had spent about £63 million (£30,000) in each year of his brief reign (1685–8). But William (with Mary, until her death in 1694) had spent £92 million (£45,000) each year and by 1701 his spending seemed to be increasing.[7] So the four monarchs had between them spent nearly £2.4 billion in modern values between 1660 and 1701.*

We don't know exactly how much of this was spent on parks and gardens, but we can make an educated guess. The financial accounts of a number of great country houses suggest that at least one-third of the cost of a new house and garden went on the latter. This seems to be confirmed by expenditure later in the eighteenth century, when the Office of Works produced detailed accounts; between 1761 and 1776, George III spent about £22 million (£13,500) each year on his gardens, about one-third of the annual expenditure of the Office of Works on buildings and gardens.[8] So it is not unreasonable to think that total royal spending on parks and gardens may have been between £800 million and £1 billion in the last forty years of the seventeenth century.

Some of the money went, for example, on Kensington Palace and its gardens as a principal royal residence. William III, who suffered from

* The modern equivalents are computed from the mid-year of each reign.

asthma, urgently needed to find an alternative to living with the smoke pollution and damp atmosphere of the Palace of Whitehall. He and Queen Mary took a liking in 1689 to Hampton Court Palace and renovations there began almost immediately, but it was too far from London – either by road or river – for the court and ministers who needed to be near Parliament and the Law Courts. Kensington – less polluted than Whitehall – appealed as the village was on one of the routes to Hampton Court, and William and Mary chose to buy Nottingham House, on the western edge of Hyde Park, for £39 million (£20,000). It needed renovation and extension and Sir Christopher Wren, architect of St Paul's Cathedral and Surveyor-General of the King's Works, was soon at work with Nicholas Hawksmoor, architect of several renowned churches in and around the City of London. Work was sufficiently advanced that the king and queen were able to move in on Christmas Eve 1689.

While Wren and Hawksmoor worked on the house, which became Kensington Palace, the Earl of Portland – Hans Willem Bentinck – newly appointed as Superintendent of the Royal Gardens, set to work with George London on the grounds. Initial landscaping and laying out of gravel paths was followed, between 1690 and 1696, by the planting of huge numbers of trees and other plants, most probably supplied by London's own large nursery close by at Brompton Park. There was a menagerie with 'curious wild fowl', tortoises, 'tygers' and snails.[9] There may have been a pause after the death of Mary in 1694, as there was at Hampton Court, but London was again at work extending the garden in 1701.

Queen Anne succeeded William III in 1702, vowing to restrain expenditure on the royal gardens; her restraint didn't last long and her new gardener, Henry Wise, was soon at work at Hampton Court and Kensington. 'Thousands of bulbs and shaped hollies' were planted at Kensington in 1702–3, no doubt also supplied from Brompton Park Nursery, where Wise was a partner. Hawksmoor and Sir John Vanbrugh, later architect of Blenheim Palace, built the orangery, an essential attribute of the formal gardens of the time since citrus fruit and other tender plants, which played a major part in garden designs, could not survive English winters. A wilderness – a carefully controlled woodland area – and a sunken garden followed, constructed by a labour force of at least a hundred men, together with a 100-acre paddock known as a zoological garden, stocked with antelope. A plan of 1713–14 shows a huge formal garden with box hedges and elaborate, maze-like paths.[10]

Anne's successor, George I (r. 1714–27), spent between £12 million and £16 million in modern values each year between 1715 and 1717 and at least this in succeeding years. Some of it went on a vast eastward extension of the Kensington Palace garden, improving the menagerie to house tigers and other exotic animals, and enclosing yet more acres of Hyde Park; the result was the largest pleasure garden in England, of 170 acres. George I started work on the 'Great Bason' – today called the Round Pond – and the other waterworks that ultimately became the Serpentine, a project of Queen Caroline, the wife of George II (r. 1727–60), although she got rid of the menagerie.[11] In the taste of the time, a huge lawn replaced the formal gardens of earlier years. The public, or at least those considered sufficiently genteel, were admitted, in a gesture to those who were paying for all this magnificence. Expenditure seems to have continued at this rate or more for the rest of the eighteenth century and probably, although the records are less complete, thereafter. This is actually quite modest by the standards of today's Royal Parks – now, despite their name, almost all open to the public – whose annual combined expenditure is £27 million.[12]

Three questions arise here. First, where did the money come from? Second, what did it buy? And third – to be left until we've looked at public parks and new towns in the next section – what did all this state aid do for the English garden industry?

Answering the first question is easy. The money came from the British taxpayer. It was allocated to the royal family and its gardens from the 'civil list', the name given to the funds issued by the Lords of the Treasury to cover the expenses of running the civil government, including the upkeep of the royal household, as well as, for example, tax collection, the legal system and the costs of ambassadors posted abroad. The other, always much larger, part of government expenditure was on the army and navy. Civil and military spending was paid for by taxes and, increasingly, by borrowing from the British public.

The political and financial settlement that made this possible was the outcome of the 'Glorious Revolution' that brought William and Mary to the throne in 1688. Eighteenth-century Britain had some of the trappings of a democracy, with an elected House of Commons, but it was in essence an oligarchy. Parliament and government were controlled by a small number of very wealthy families, working with the monarchy; their money came mainly from their land, but individuals from some of these families also had positions at court and became generals, admirals and bishops, all lucrative appointments. Local government was also in their grip, as was law enforcement, particularly at the local level. Only a small proportion of the population – men only, of course – could vote either in national or local elections. This oligarchy had been profoundly shaken by the events of the Civil War and the establishment by Cromwell of the Protectorate. The conflict between king and Parliament had pitted aristocratic family against aristocratic family, father against son, and had, like most civil wars, led to very heavy loss of life: about 85,000 died in combat and 100,000 more from war-related diseases, from a total population in England at that time of about 5 million.* The Restoration promised stability, but then the succession of a Catholic, James II, to the throne in 1685 threatened to spark further conflict.

The settlement of 1688 – whereby William and Mary came to the throne and James was exiled – stabilized the country. Although there were foreign wars and the Jacobite rebellions of 1715 and 1745, these common dangers cemented the alliance of the monarchy with the aristocratic oligarchy; they had a shared interest in avoiding further internecine conflict. Within the oligarchy, there were two political parties – Whigs and Tories – who vied for favour and a place in government but, as Britain's economy continued to grow, the aristocracy and the king felt increasingly secure. In particular, the country seemed to have solved the problem that had plagued all European countries for centuries: how to raise sufficient funds to finance the monarchy, the government and foreign wars.

The solution was twofold: the government borrowed money, principally from its own citizens, occasionally from those of other countries, and imposed a bewildering range of taxes on consumption

* War-related mortality in Ireland and Scotland was much higher.

and trade to pay its expenses, including interest on the money that had been borrowed. The national debt – the total amount owed by the state – rose at a rate that would today give the Treasury, or the International Monetary Fund, collective heart failure. It was £6.2 billion in modern values in 1691, £139 billion by 1750 and – partly because of the wars with France – £483 billion in 1800. This was only possible because the British public – who had lent most of the money to the Bank of England, which managed the debt – trusted that the government would pay interest on the loans. Interest from government bonds or consols (short for 'consolidated annuities', a means by which the government borrowed), or 'the funds', as the debt was known, soon came to be relied upon by large swathes of upper- and middle-class society – including men such as Capability Brown – to provide a regular and reliable income. They were benefiting from a growing economy and so had money to invest; the funds were safer than any other form of savings and investment in an age when there was no risk-free way of investing in trade or industry and when there were frequent bank failures.

Most of the rise in the national debt came from the need to pay for wars. The cost of the civil list – paying for the royal household, their luxuries, their palaces and gardens – rose only gradually. It was a price that the oligarchy were prepared to pay for the stability that William and Mary, Anne and the Hanoverian monarchy had brought to the country. Indeed, the oligarchs themselves paid very little, since the excise duties and other indirect taxes that paid the interest were largely borne by the rest of the population; the oligarchy saw to it that taxes on the land that they owned fell and there was no income tax until 1799.

So, to answer the second question posed above, the money was spent on larger and larger royal palaces, parks and gardens. Although most of the public who paid for them were never allowed to see them, they excited little concern. It was an age when broadsheets and cartoons viciously attacked the royal family, but nothing very unpleasant seems to have been said about their gardens as they put up and took down buildings, erected pagodas and other forms of chinoiserie and dug lakes. Even the Dowager Princess of Wales, Augusta, who was mercilessly lampooned in the 1750s for her supposed relationship with Lord Bute, wasn't attacked for her gardening, although her daughter-in-law, Queen Charlotte, did not entirely escape ridicule for her works at Kew. To be fair, the British royal gardens were never as lavish or ostentatious as those of their continental cousins and rivals; they were also, with the possible exception in the seventeenth century of Hampton Court Palace, hardly more so than the gardens of the rest of the oligarchy.

What did the state get in return? First, the dignity of the monarchy – an important aspect of the maintenance of an oligarchy – was upheld. None of the British palace gardens was as enormous or elaborate as the original 20,000 acres of Versailles; Hampton Court was only 27 acres in 1689, although William III increased it to 46 acres by 1696.[13] William and Mary thought the palace and park of Honselaarsdijk and Het Loo in Holland at least equal to their English gardens. George I and George II, who were each brought up at Herrenhausen in Hanover, both considered it superior. But the royal gardens as they developed during the eighteenth century were at least respectable in their size and design, even if the Royal Gardens Inquiry of the 1830s concluded that none of them then really matched the dignity desired of the English monarchy. Kew, the Royal Botanic Gardens, acquired special status, but mainly after it ceased to be a private royal garden in 1840 and became the foremost botanical garden in the world.

The state made some beautiful gardens, even if many of them long remained closed to the public. Possibly the greatest formal garden ever created in England was that at Hampton Court, Cardinal Wolsey's palace on the Thames fifteen miles west of London. Seized by Henry VIII from his disgraced Lord Chancellor in 1529, it was a royal residence from then until 1737.[14] While its gardens were always notable, they were developed further, particularly under Charles II, James II, William and Mary, and Anne, though they stood still after that. In the 1660s, Charles II commissioned André and Gabriel Mollet, who had recently completed the canal for him in St James's Park, to design and build one at Hampton Court; the result was the Long Water, 105 feet wide and 3,800 feet long, bordered by 758 Dutch elms, and the semicircular basin that fronted the palace.[15] The water came from a diverted river, although it never properly supplied the fountains that were an essential feature of formal gardens of the period.

More changes followed; Charles laid out an entirely new garden, the Wilderness. William and Mary constructed, on another front of the palace, the largest parterre (a formal garden, usually next to a house, with planted beds separated by paths and sometimes ponds with fountains) built in Britain in the seventeenth century. Long avenues were laid out, thousands of trees planted; three greenhouses, each 55 feet long, were constructed for Queen Mary's collection of rare and exotic plants. After a pause following the death of Mary in 1694, work carried on; the accounts show that Henry Wise, the royal gardener, carried out work costing £18.9 million (£9,384) in 1700–1701 and a further £6.8 million (£3,407) in 1701–2. By 1703, despite her resolution to economize, Queen Anne was remodelling the Great Parterre, probably because the fountains had again failed to work. Later in her reign, she altered the gardens north and east of the palace, ordered extensive tree-planting in Bushy Park and built the famous maze; since the nineteenth century, the maze has been the palace's greatest tourist attraction.

Hampton Court was just one of a series of royal residences that were renovated and, sometimes, abandoned as fashions changed, the royal family decided that they liked another house or garden better, or the growth of London made it sensible to relinquish land for housing. Successive generations of monarchs, their spouses and children, supervised garden improvements or even worked in them. During the eighteenth century, George I and George II modified the gardens of Kensington Palace, while George III and his wife Charlotte, an accomplished botanist, altered the gardens of his parents at Richmond and Kew at the behest of Capability Brown. During the nineteenth, Victoria and Albert superintended the creation of the largest kitchen garden in Britain, at Frogmore on the Windsor estate, improved the grounds of Windsor and Balmoral and created their own personal paradise at Osborne on the Isle of Wight; their son Edward VII built notable rockworks at Sandringham. Edward VIII, when Prince of Wales, commanded his guests, probably including a succession of mistresses, to plant with him at Fort Belvedere; most recently, Prince Charles has transformed his garden at Highgrove.

A particular royal garden, destroyed in the mid 1820s, was once hailed as the epitome of gardening taste. Frederick, Prince of Wales and son of George II, was estranged from his father for many years and a focus for the parliamentary opposition to the king. He worked in the 1730s with his wife Augusta to build the garden of Carlton House, north of St James's Park in central London. Its designer was the protégé of Lord Burlington, William Kent, who had been responsible for a trendsetting garden – still to be seen – at Rousham in Oxfordshire. Documented in Frederick's household accounts, Carlton House is an eighteenth-century example of today's 'instant' gardening. A year spent preparing the 9 acres of grounds cost over £1.3 million (£710). Frederick couldn't wait for trees and shrubs to grow and they were transplanted wholesale, at a cost of £2.2 million (£1,239); there were over 15,000 trees, including a 25-foot tulip tree and an 18-foot Virginia black walnut, together with thousands of shrubs, bulbs and flowering plants.[16] The ground was levelled and turfed, an aviary (clearly a necessity, as the prince had another at Cliveden, which he rented from 1837 onwards) and bath-house were built, statues installed and, in February 1735, Thomas Fowke delivered two pumping engines, 654 feet of leather forcing pipes, 10 feet of suction pipes and '22 pair' of brass screws, all 'for the use of HRH the Prince of Wales for Watering his Gardens' and costing £203,400 (£113 10s). Altogether, the garden cost at least £10.4 million (£5,781), spent in less than three years.

Lavish spending was a feature of many of the royal gardens. The royal family, like all garden owners, liked to show off. Part of a Roman temple, transported from Leptis Magna (in modern-day Libya), became a picturesque ruin at Fort Belvedere on Virginia Water, newly constructed in the 1750s. The Duke of Cumberland, who had built the fort as a summer house, kept a 50-ton ship on the lake there to entertain his friends with mock naval battles.[17] From the 1740s onwards there were elaborate menageries and aviaries at Kew and Richmond, containing zebra, elephants and a range of birds; they were joined in the 1790s by kangaroos from the new colony of Australia.[18] The basin for the fountain at Frogmore Kitchen Garden was 30 feet in diameter and made of polished Peterhead granite, like those in Trafalgar Square.[19] The hothouses were 840 feet long and there were 1,665 feet of pits devoted to asparagus, cucumbers, melons, pineapples and grapevines. Perhaps the record for extravagance is held by William III. In 1701, he visited Hampton Court Palace and noticed that he could not see the River Thames from his first-floor rooms; he immediately ordered that the level of the Privy Garden should be lowered by 8 feet. This was done in the winter of 1700–1701 and the garden was fully planted, pipework was installed for the fountains and statues placed on their plinths. Then, in June 1701, William visited again and discovered that he could still not see the Thames; he insisted that the whole garden should be lowered again. All the newly installed plants, grass, paths, pipes, fountains and statues had to be removed and 3 feet of soil taken from an area of 5 acres before all the plants and equipment were put back. Henry Wise, the royal gardener, employed hundreds of labourers on the project and, remarkably, it was all done by November. It cost £2.8 million *(£1,426 4s 4d)* on top of the millions spent on the original work. But the king had his view of the river. The garden can still be seen, as in plate 4, restored in the 1990s to its original splendour at a cost of between £4 million and £5 million.

No one can doubt that the royal parks and gardens – most of them now open to the public, who paid for them over the centuries – are a major ornament of London and the envy of many other capital cities. If, at a conservative estimate, their annual cost was about £15 million in modern values, they have cost more than £5 billion over the course of 350 years. Cheap at the price, many will say; less than the cost of two aircraft carriers. But there was much more to come, as the fashion for building public parks spread across England and local took over from central government in the funding of gardening.

PEOPLE'S PARKS: PUBLIC SPACES FOR THE LIVING AND THE DEAD

During the second quarter of the nineteenth century, the average height of Englishmen began to fall.[20] Like the canaries that warned of poisonous gas in coal mines, it was a sign of danger. Height is a very sensitive indicator of the welfare of the population and it has normally risen in line with the growth in people's incomes and the economy in general. But in the 1830s and 1840s in Britain, several other European countries and the USA, that relationship was broken; the economy seemed to be growing, but the average height was faltering. The reason was the poor living conditions in towns and cities.

As the economy grew from early in the eighteenth century and the Industrial Revolution gathered pace, England's population rose and was increasingly concentrated in towns and cities. By 1800, England was the most urbanized country on earth and the process accelerated in the new century. It was entirely unplanned, the result of a myriad individual decisions; millions of people, men and women, needed jobs and places to live, and employers and builders, in their thousands, tried to satisfy their needs. As there was no cheap and efficient public transport, jobs and houses had to be near each other. The result was more and more people crammed together and in close proximity to the amenities that supported their lives but also contaminated them: abattoirs, dairies and their cows, stables generating tons of horse manure, tanneries, soap boilers, dye works – all constantly churning out smoke, noxious smells and other pollutants. There was little local government and what there was could not cope: waste poured into rivers; water sources were polluted and caused cholera to spread; overcrowded houses were fire traps; graveyards were overflowing with putrid corpses, if they had not been purloined by the body-snatchers. Central government either did not care or was helpless, not least because local politicians, such as those in the City of London, resisted any interference with what they called their 'liberties' and their right to pay as little as possible in taxes.

By the 1830s and 1840s, the situation had become intolerable. In 1833, a Factory Commissioner (a government commissioner inquiring into the state of British factories) visited Manchester and was 'struck by the lowness of stature, the leanness and paleness which present themselves so commonly to the eye'. In the 1840s, the great sanitary reformer Edwin Chadwick argued that 'noxious physical agencies' were producing a population 'having a perpetual tendency to moral as well as physical deterioration'.[21] It was not only in the large towns that chaos reigned: in 1842 the authorities administering the system of poor relief that existed in various forms from the reign of Elizabeth I until the establishment of the welfare state after the Second World War were told that the small town of Windsor, nestling below the royal castle, was 'the worst beyond all comparison. From the gas-works at the end of George-street a double line of open, black and stagnant ditches extends to Clewer-lane. From these ditches an intolerable stench is perpetually rising.'[22] Famously, in 1858, Members of Parliament fled from the 'Great Stink' of the River Thames, into which flowed all the human waste and industrial effluent of London. Mourners didn't dare to accompany their loved ones to the graveside, for fear of what they would see and smell in the overflowing churchyards. No wonder that disease was thought to be spread by miasmas.

Nothing could be more different from the towns and cities than the landscape parks of aristocratic England. Yet it was to that model that social reformers turned, in the hope that light, air and space could alleviate the horrors of urban existence. In the process, they changed the meaning of 'park' from the deer parks of medieval England or the rolling landscapes of Capability Brown to the more regimented, flower-bedecked spaces that still adorn our towns and cities and which continue to be loved by their inhabitants. They, together with the splendid town halls, libraries, concert halls and, later, swimming pools are the most visible remnants of the late nineteenth-century activism of local government that has been increasingly stifled by centralized control from Whitehall in the twenty-first century.

As ever, concern for the general health of the population was motivated by various factors: some were philanthropic, others utilitarian attempts to improve the productivity of British workers. Some were self-interested: several early parks, from Regent's Park onwards, were part of schemes for building luxury villas and aimed to make a profit. On 21 February 1833, the House of Commons established a Select Committee to consider 'the best means of securing Open Spaces in the Vicinity of populous towns, as Public Walks and Places of Exercise, calculated to promote the Health and Comfort of the Inhabitants'.[23] Its chairman was Robert Slaney, MP, a wealthy Shropshire landowner whose overriding concern in a long parliamentary career as a Liberal was 'improving conditions for the lower classes, especially in the industrial towns'.[24] The committee found that 'as respects those employed in the three great Manufactures of the Kingdom, Cotton, Woollen and Hardware, creating annually an immense Property, no provision

has been made to afford them the means of healthy exercise or cheer-
ful amusement, with their families, on their Holidays or days of
rest'.[25] With the usual mix of anxiety and condescension they concluded that open places 'reserved for the amusement (under due
regulations to preserve order) of the humbler classes, would assist to
wean them from low and debasing pleasures ... drinking-houses,
dog fights and boxing matches'.[26]

Although the committee did not actually recommend that central
government funds should be used, a succession of bills in Parliament
produced, in 1841, a grant of £7.2 million *(£10,000)*, which was used
between 1842 and 1856 to subsidize parks in ten English, Scottish
and Irish towns. The largest sums went to Manchester and Bradford,
but those for the north of England were dwarfed – a familiar story –
by the money that was separately spent on London, a total of £339.2
million in modern values as compared to £7.2 million for the whole
of the rest of the United Kingdom.[27]

One of the new London parks, Primrose Hill, still today a much-
loved open space, north of Regent's Park and surrounded by expensive
houses, had actually been the proposed site of a landscaping solution
to another of London's problems, the overflowing graveyards.
Although the problem had been identified as early as the 1720s, nothing had been done either by the Anglican Church, which owned the
graveyards and received the fees for burials, or by the local authority,
the City of London Corporation. But by the 1820s, about 40,000
people needed to be buried in London each year. The solution was
left to private initiative in the shape of George Frederick Carden, who
combined the roles of lawyer and philanthropist, and was the prime
mover in what became the General Cemetery Company. Carden,
who was inspired by the Parisian Père Lachaise cemetery, argued that
a private burial ground was needed: it would provide vaults and
graves secure from body-snatchers,[28] it would remove dangers to
public health and it would create a fittingly peaceful and beautiful
landscape. Carden was enthusiastically supported, in the *Morning
Advertiser*, by the garden writer and designer John Claudius Loudon,
who suggested that London should be surrounded by several burial
grounds that could be made 'at no expenses [sic] whatever' into
botanic gardens.[29] They could also serve as 'breathing places' for

urban populations.[30] Primrose Hill was soon abandoned as a site and Kensal Green, in inner west London, was chosen to house what quickly became a fashionable cemetery, particularly after the burial there in 1843 of Queen Victoria's uncle, the Duke of Sussex – although he had actually wanted to donate his body to be dissected.

Kensal Green, which cost about £155 million *(£200,000)* to establish but soon made a profit after its opening in 1833, was not the first private cemetery in Britain. Bunhill Fields had been established in the City of London, for the burial of dissenters, in the 1660s. The first true garden cemeteries were laid out in Norwich in 1819 and in Manchester in 1820; from 1825, Liverpool had two – one nonconformist, the other Anglican – making use of abandoned quarries.[31] But Kensal Green deserves its place in the history of gardening because it was the first to be laid out as a landscape, on the models established by Capability Brown and Humphry Repton and systematized by Loudon, who in 1843 published the first – and apparently still the only – treatise on the subject: *On the Laying Out, Planting, and Managing of Cemeteries; and on the Improvement of Churchyards*. Plate 5 shows an early design, not fully carried out in practice. Kensal Green was laid out, like Regent's Park, with a circular carriage drive and avenues of trees, while the plants alone – supplied by Hugh Ronalds, a celebrated Brentford nurseryman – cost £232,500 *(£300)*.[32] Other large commercial cemeteries – Highgate, Norwood and Tower Hamlets, among many – soon followed, part of the 'Magnificent Seven' established around London by 1841.

The greatest boost to cemetery-building followed the cholera epidemics of the 1840s, which were initially attributed – wrongly, as we now know – to miasmas in the air. A series of Burial Acts were passed and consolidated into one in 1857, stimulating local government, in the shape of Burial Boards, to build cemeteries across England; nearly a hundred competitions for their design were held in the 1850s.[33] Municipal cemeteries were established everywhere, planned by such eminent designers as Joseph Paxton, Thomas Mawson, Henry Milner and Edward White; they became a lucrative source of commissions and sales for the burgeoning garden industry. Bradford's Undercliffe Cemetery cost £8.1 million *(£12,000)* to lay out in 1854 and the City of London Cemetery at Ilford £29.4 million *(£45,450)* in 1858.[34] The commercial Brookwood, still the largest cemetery in Europe, occupied more than 350 acres of Surrey heathland, partly financed by selling off building plots on surrounding areas; it was famous for having two railway stations, to which bodies and mourners were conveyed – by first, second or third class – from a special terminus at Waterloo Station in London.[35]

Landscapes for the living also flourished. The story of activist local government in the late nineteenth century – spearheaded and symbolized by Joseph Chamberlain in Birmingham – is a complex one, but two landmarks in a convoluted progress were the Municipal Corporations Act of 1835, which reformed a corrupt and inefficient system of local government, and the Public Health Act of 1848. The latter tried to give as much responsibility as possible to the reformed councils, while encouraging them to follow the precepts of a government body, the General Board of Health, led initially by Edwin Chadwick. Unfortunately, the main concerns of the small shopkeepers who made up most of the electorate were to spend as little money as possible and to resist central government. *The Economist*, for example, saw the Public Health Act as a much worse evil than bad smells.[36] It was only with the Second Reform Act of 1867 and the Municipal Franchise and Assessed Rate Act of 1869 that the vote was extended to about 60 per cent of working-class men. They were willing to spend money and, crucially, to borrow against the security of the rates (local taxes). One result was the flourishing of public parks.

Pleasure gardens had been a feature of eighteenth-century London and other British cities. Some, such as Vauxhall in London, initially enjoyed royal patronage; most underwent a cycle of becoming the epitome of high fashion before losing their novel appeal and degenerating into haunts of depravity. Some were quite long-lasting: the 'Vauxhall' at Dudston Hall in Birmingham was opened in 1746 and survived until 1850.[37] They were the creation of private entrepreneurs and, in their heyday, were elaborate and expensive, the scene of concerts and fireworks to which high society flocked. They were one of the few places where young unmarried men and women could meet. But entrance was costly and limited, at least in theory, to the respectable or at least monied classes. Pleasure gardens did not meet the need for open space for the masses.

Public parks as we now know them were first laid out in the 1830s, but they were initially privately financed by various means. The first of many to bear the name 'Victoria' was opened in Bath in October 1830. Later renamed Royal Victoria Park, as it is still known, it was private, though it used what had been common land, but was open for free to the walking public, or at least those of 'a decent appearance and good behaviour'. People on horseback or in carriages were expected to pay a subscription and donations were also sought and, sometimes, received.[38] Other parks were funded by private philanthropists; the first of these may have been the Derby Arboretum, laid out by John Claudius Loudon and opened in 1840 on the basis of a gift from Joseph Strutt, a local industrialist, although it too charged a small fee on five days of the week. Another business model was that of Birkenhead, the park that was to be the inspiration for Frederick Law Olmsted's Central Park in New York. The town, across the Mersey from Liverpool, was essentially new, with suburban housing but also a shipyard and engineering works. Marshy land was drained and the park laid out with terraces, hills, rockeries and lakes. Considered to be the first park established at public expense, it contained a 'Roman' boathouse, a pagoda and a covered wooden bridge in the Swiss style. Like the aristocratic gardens on which it was modelled, the land was shaped and planted to give successive vistas to visitors. The park was laid out by Joseph Paxton and financed, to the tune of £50.2 million (£70,000), by reserving nearly half the land for building plots.[39] A similar model was that of Lord Vernon in Stockport and Henry Bolckow in Middlesbrough,[40] who each donated the parkland but kept back the surrounding area to build villas. There seem to have been over twenty parks, financed by different means, in England and Scotland by 1860.

From then on, it was the newly reformed and enfranchised local authorities that took the lead. Inspired by Joseph Chamberlain's 'gas and water socialism' in Birmingham, town councils everywhere started, invested in or took over the supply of essential services such as electricity and tramways, as well as gas and water, financing the new municipal enterprises from their profits, from the rates (local taxes) and, above all, by borrowing. They borrowed also to provide other essentials of a modern town: schools, workhouses, roads, hospitals and, not least, commons and parks. Between 1883 and 1903, £144 billion (£400 *million*) was borrowed, representing about 40 per cent of the total national debt, while total local government expenditure was about half of all public spending.[41] Just as the royal parks of the eighteenth century were built and maintained by higher and higher levels of government borrowing, so too were the people's parks of the nineteenth.

Public parks differed, of course, both in size and in how they were adapted to the topography of the plot allocated to them. But there were common features. Many included lakes, usually laid out on the serpentine principles established in the eighteenth century to imitate rivers. That gave the opportunity for ornamental bridges. Parks were often surrounded by shelter belts of trees, to give the illusion that they were in open countryside. A bandstand was essential. Flower-beds were increasingly florid, as the mid-nineteenth-century fashion for carpet-bedding swept across the country. Older generations today remember with affection the elaborate floral clocks. There were drinking fountains – Angela Burdett-Coutts donated one to Victoria Park in Hackney that was in neo-Gothic cum Moorish style and cost £4.4 million (£6,000). Martial themes were introduced. Hull's East Park gloried in an imitation of the Khyber Pass, scene of many Victorian battles in the 'Great Game' being played out between Britain and Russia in Central and Southern Asia. Scarborough staged – indeed it still does – battles between miniature naval ships, the equivalent of the 'naumachia' held on eighteenth-century estates. Russian cannon seized in the Crimea were displayed around the country. Rock gardens and ornamental rockwork were common as alpine flowers became fashionable. Artificial, Pulhamite, stone made it possible to devise cascades, rocky streams and picturesque bridges. Battersea Park had a Pulhamite waterfall and cave, as later did the royal gardens at Sandringham. Ramsgate in Kent had its elaborate Madeira Walk with falls, pools and tunnels, while Folkestone, to the south, had an artificial zigzag path down its cliff.[42] There was a particular, although contentious, growth in sports grounds, stimulated by the fears about the physical state of the population that have gripped the press and governments at intervals since the late nineteenth century.

The construction of parks was mostly financed from loans, on which interest had to be paid, but maintenance was a direct charge on the rates. Like the great estates, public parks needed large numbers of gardeners. Their greenhouses were filled with the thousands of annual plants required by bedding schemes. Lakes needed to be kept free of silt and weed. Sports facilities, increasingly popular towards the end of the nineteenth century, needed groundsmen and machinery to maintain them. All this was extremely expensive, rising from over £600 million a year by the time of the First World War to over £4.3 billion a year by 1980.*

Recent history is, for garden and park visitors, a much less happy one.[43] Providing open spaces was – unlike education, libraries or social services – never a statutory duty imposed on local authorities by law. When governments in the last quarter of the twentieth century and especially since 2010 started to try to cut public expenditure, they looked to local government; to councillors desperate to meet spending limits, parks were an easy target. Park keepers, gardeners and groundsmen – all skilled professionals – were laid off, with the inevitable consequence that maintenance suffered and vandalism increased. In the interests of 'efficiency', park management was merged with a miscellany of other 'recreational services' and maintenance was put out to open tender; profit-making companies swept in with gangs of under-trained workers who could not care properly for the shrubs and trees and were only to be seen occasionally in each park. Groups of volunteers, devoted to their local open space, struggled to fill the gaps. Councils began to look for ways of making money from their parks – there were hard-fought battles over plans to use public land for commercial theme parks – and some even spied an opportunity to kill two birds with one stone and build houses on them. After all, as

---

* In 1884, when statistical records begin, local authorities were already spending on parks and gardens £63.2 million *(£140,000)* a year from the rates on maintenance and another £45.1 million *(£100,000)* on capital projects financed by loans. By 1890 the total was over £250 million a year in modern values and by 1900 over £400 million; by the time of the First World War, it was over £600 million a year. There was a check in the immediate post-war period, a time of slow recovery in the economy, but by 1925 the growth in spending had returned, with over £900 million spent then and £1.5 billion on the eve of the Second World War. The war produced another check, but again spending resumed and, by 1980, after particularly rapid growth in the 1970s, the total was over £4.3 billion a year – 2.5 per cent of local government expenditure.

one councillor is alleged to have said: 'All people want to do in parks is sit in them.'

Actually, as an unlikely saviour – the Heritage Lottery Fund – has found, over 57 per cent of the British population use their local park at least once a month and, for those with young children, it rises to 90 per cent.[44] The fund is financed by Britain's gamblers – perhaps the least well-off sector of the public – and has been spending millions each year on conserving and restoring historic landscapes, public parks among them.[45] Perhaps most importantly, it has imposed conditions on its grants to provide that councils will properly maintain these landscapes in the future. We shall see.

The state and local government also created a large number of gardens that were attached to Poor Law institutions and lunatic asylums. Some of the latter were laid out on a large scale, imitating private estates, and were equipped with farms and parkland as well as kitchen and pleasure gardens. An important feature, as Sarah Rutherford has shown, is that the inmates were employed as unpaid labour.[46] In Poor Law institutions this was part of the punitive regime introduced by the New Poor Law from 1834, while in the huge lunatic asylums built in the Victorian period the purpose was at least in part therapeutic. However, since few people were cured sufficiently to emerge from the asylums, the result was a continuing supply of free labour, producing fruit, vegetables and farm produce on a scale that contributed significantly to covering the costs of the asylum. The State Criminal Lunatic Asylum at Broadmoor, for instance, produced 'articles of provision' to the value of £317,000 (£509 10s) from its farm and 14-acre kitchen garden in 1865.[47] Poorhouse gardens were typically much smaller, but since every group of parishes across the country had to have a workhouse, the total acreage of their gardens must have been substantial.

Another kind of garden has been encouraged and protected by the state since the 1830s – the allotment. Stimulated by high rural unemployment and the prospect of violent discontent, no less than three acts of Parliament were passed in 1832 to allow parishes to enclose land for use by agricultural labourers.[48] Numbers of allotments grew rapidly, first in areas of particular rural poverty but then much more generally; by 1914, when local authorities had a duty, not just the power, to provide 'adequate provision', there were as many as 600,000 plots.[49] Both world wars gave a huge boost to the allotment movement, with numbers of plots rising to about 1.5 million; since 1945 there have been waves of interest, but there are probably still at least 300,000 in use, a total of about 18,500 acres. With subsidized rents and intensive cultivation, they make a contribution to the diets of the people who work them; how much they provide is discussed in Chapter 9.

The state also played a role in transforming large parts of England into suburbs, millions of houses each with its own garden. Most of the suburbs were built by private developers and builders, described in Chapter 8, but central and local governments initially permitted and then, under the influence of the garden city movement, encouraged through regulations what has been characterized by some hostile observers as 'urban sprawl'. After the Second World War, a succession of new towns were created, with much greater attention to civic design. But both suburbs and new towns were composed of detached or, more usually, semi-detached houses with front and back gardens that peer pressure, or the strictures of local officials on council housing estates, demanded should be maintained; the weekend in the garden was the result for millions of middle- and working-class households. As Anthony Alexander puts it: 'The New Towns were founded on the idea that life consisted of a steady job and weekends of ideal bliss tending to one's garden.'[50]

350 YEARS OF STATE SUPPORT

From palaces to new towns, from royal parks to floral clocks, from cemeteries to suburbs, the English state in its various forms has therefore been spending money on gardens for more than 350 years. And that doesn't take account of the money that was spent on their private gardens by courtiers or army officers, colonial administrators and government contractors, much of it the fruits of public expenditure. It's difficult to add it all up, but a conservative estimate would be at least a billion pounds, in modern values, in each decade since the seventeenth century, with much more from the middle of the nineteenth century onwards, when local authorities were spending billions annually on parks. It must have had some economic impact, but what did it achieve?

The money was not a subsidy to the garden industry. Sir John Vanbrugh, architect of Blenheim Palace and Castle Howard, was Comptroller of the Office of Works from 1702 to 1726 and Surveyor of Gardens and Waters from 1715 to 1726. He was once challenged by the Lords of the Treasury about the cost of plants for the royal gardens and, after an investigation, assured them that normal commercial prices had been paid. The royal gardeners, it is true, from George London and Henry Wise in the seventeenth century, through Capability Brown in the eighteenth to William Townsend Aiton at Kew in the nineteenth, had lucrative contracts, on which they were able to make large profits.[51] Some of this would have gone to underpin their own businesses: London and Wise ran the Brompton Park Nursery and Brown had his many aristocratic clients. But the royal gardening contracts were not out of line with those given to the other tradesmen who served the royal household. Nor, despite rumours of backhanders, are there examples of excessive payments made by later local authorities.

However, the expenditure, much of it spent with nurseries from Leonard Gurle's and Brompton Park onwards, was so large and so regular that it must have contributed to 'economies of scale' in the garden industry – the lower unit-costs of doing something 'in a large way of business'. This benefited every customer. The spending underpinned, for example, the plant-hunting that brought plants to the royal gardens and, later, to the Royal Botanic Gardens at Kew. In the late eighteenth and early nineteenth centuries, every naval ship on a long voyage carried a botanist – paid for by the navy or by British botanical gardens – whose job was to dash ashore when the ship docked at a foreign port and seek out new plants to be brought back, often with great difficulty, to Kew and then cultivated for sale by nurseries. The royal gardens, and Kew in particular, have had a long tradition of training gardeners who would then work in the industry; the Royal Gardens Inquiry in the 1830s bemoaned the fact that the gardens were then in such a poor state that this duty had been neglected. But Kew has certainly made up for this since then and the parks departments of local authorities took up the role in later years, though sadly no longer. University and college departments of botany, landscape architecture and horticulture have been supported. Landscape designers, from Thomas Mawson and Edward Milner in the nineteenth century to the teams who set out the new towns, have been employed on large and sometimes long-term contracts.

Why? What made it possible for the state to support gardening in so many different ways? The billions that were borrowed or acquired through taxation could have been otherwise spent. They could have funded more battleships, or provided more palaces, or reduced poverty, or simply have been left in people's pockets. There are many reasons why, instead, the money was spent on gardens, but perhaps the central one is that relatively small sums of money – compared to state spending as a whole – spread over many years, produced great beauty in a medium that was, and remains, accessible to and enjoyed by everyone.

The state in its various forms – from Charles II to the municipal gardeners who tended the floral clocks – set an example. In economists' jargon, it primed the pump; but it then continued to subsidize the garden industry for centuries. It bought plants from the nascent nurseries, spent vast sums on landscaping and reshaping the English countryside, trained botanists and gardeners and fostered the collection of thousands of plants from all over the world. The state's example was then followed by myriads of individuals, men, women and children, creating their own gardens, spending money in nurseries and garden centres and visiting the thousands of gardens, great and small, across the country that open their gates each week or each year; this created the garden industry as it is today. That is the story now to be told.

3

The Great Gardens

'Laying out with taste your ground'[1]

There are few garden views in England equal to that of Thomas Archer's baroque pavilion reflected in the Long Water, the centre of the garden of Wrest Park in Bedfordshire, as is shown in plate 6. The rectangular 'canal', from the late 1680s, with its surrounding woodland garden, takes one's eye across the terrace in front of the great mansion. Its shallow, still waters give the reflections loved by garden designers and owners. On either side, rides and paths criss-cross the woods and each turn brings a new vista of statues, monuments, antique altars, bridges and columns. Winding around them all is a sinuous river. There is also a huge walled kitchen garden. It looks the quintessential English great garden, all of a piece – yet it isn't.*

Wrest is the ancestral home of the de Grey family, earls and dukes of Kent, and what makes it special is that, although each part of the garden comes from a different era, it blends into one harmonious whole, the result of three centuries of changing fashion, experimentation and

---

* What is a 'great' garden? One way to define it is to think of the estate of which it formed a part. Wrest, and gardens like it, were at the centre of estates of 10,000 acres or more; there were probably between three hundred and four hundred of them in eighteenth-century England.[2] The focus of this chapter is on the greatest of the gardens that were created between 1660 and 1800, normally as a centrepiece of a very large landed estate. However, such enormous, complex and beautiful gardens have continued to be created ever since. In the nineteenth century the Rothschilds, at the end of that century Lord Leverhulme, in the twentieth century the Getty family and, very recently, Lord and Lady Heseltine – among many others – have designed, built or restored gardens on a great scale, with many acres of pleasure gardens, lakes, ornamental buildings and often an arboretum.

modification. Each of the de Greys embraced new garden styles by adding them to the garden, not by sweeping away the work of their ancestors. So Wrest reveals to us the formal garden of the seventeenth century, the informality and winding waterways of Capability Brown as well as the renewed formalism of the nineteenth century.

Wrest, which lies in undulating countryside fifty miles north of London, was celebrated even in the 1630s; adopting the typically florid style of the time, Thomas Carew wrote in 'To my friend G.N. from Wrest' how

> Here steep'd in balmie dew, the pregnant earth
> Sends forth her teeming womb a flowrie birth . . .[3]

Flowers must always have been grown at Wrest, but it was after the Restoration in 1660 that water and woodland began to dominate. The Long Water was created and round it, in the first decades of the eighteenth century, the next generation of de Greys laid out the terraced parterre, seen in an engraving of 1704, which shows the huge scale of this and many other gardens of the time. Then came the great woodland garden, in its turn surrounded by more rectangular canals. Then, fifty years later, these straight canals were curved by Capability Brown for yet another generation of the de Greys, to make the waterways look like a river – a favourite Brown device. Finally, in the early nineteenth century, an amateur architect de Grey rebuilt the house on a slightly different site and set out its formal flowerbeds and paths. Now, English Heritage are restoring it in all its complexity.[4]

Wrest offers today, therefore, a beautiful lesson in garden history. But it is more than that, because the story of the de Grey family and their garden helps to shed light on the motives and aspirations, as well as the resources, of the men and women who created it and the many other great gardens of the seventeenth and eighteenth centuries. The gardens were created by a partnership – albeit an unequal one – between very rich people and the garden designers who did their bidding but also shaped their employers' tastes. This chapter is illuminated by those men and women who paid for the gardens, by the society in which they lived, by their wealth and their fashions, their marriages, their inheritances and their land and titles.

Thousands of large gardens were created, altered or 'made over' between 1660 and 1800, but about two hundred of the greatest, and the individuals who created them, emerge from two sources. The first is *Britannia Illustrata: or Views of Several of the Queen's Palaces. As also of the Principal Seats of the Nobility and Gentry of Great Britain* by Johannes Kip and Leonard Knyff; it was published in London from 1707 and contains eighty bird's-eye views of parks, houses and gardens. The second is the account book of Capability Brown, now owned by the Royal Horticultural Society and kept in the Lindley Library, which lists 125 of his clients during the period of his greatest success between 1761 and 1783. Together, these two books enable us to gain an insight into the elite of British society and the gardens they made.

Wrest Park, with its owners, represents a microcosm of that elite and what they built. It appears in *Britannia Illustrata* as the seat of the Right Honourable Henry, 12th Earl of Kent (1671–1740). His family could trace its ancestry back to the time of William the Conqueror and had several times married minor royalty. Its power and wealth had fluctuated over the centuries and was at a relatively low point when the 10th Earl inherited the estate in 1643. He set about retrieving the family fortunes by a method that runs through the history of the great gardens, marriage to an heiress. But he did so twice, first to Mary, the daughter of one of the richest merchants and financiers of the time, and second to Amabella (1607–98), the daughter of a wealthy lawyer.

In the next generation, another Mary, the wife of Anthony, the 11th Earl, had large landholdings and the title of baroness in her own right, while Henry, the 12th Earl, married first Jemima, who had inherited £41.6 million *(£20,000)* in 1697, and then, second, Sophia Bentinck, daughter of the Earl of Portland – favourite of William III and Superintendent of the Royal Gardens – the richest non-royal nobleman in Europe. Finally, in a gender reversal, Henry's granddaughter, another Jemima (1722–97) – who held the title of Marchioness Grey in her own right – married Philip Yorke (1720–90), son of the extremely wealthy Lord Chancellor of the time.

The first phase of the Wrest gardens – the Long Water, the terrace in front of the house and the wilderness – was the work of Amabella and her son and daughter-in-law. The cost of digging out and water-proofing the Long Water is not known, but between 1685 and 1701 the bills for terracing, building walls, planting trees and shrubs and installing ornamental gates came to £6.8 million *(£3,349)*. The iron gates cost £926,000 *(£455)*; 51,500 bricks were needed to build a wall; 'a pack of thread to sett out ye Wilderness Quarter' cost £203 *(2s)*; 'beer for the workmen' occurs regularly. So Amabella and Mary's fortunes were put to good use.[5]

But Henry Grey, the 12th Earl, had another source of income. He has had a bad press from contemporaries and historians, described as 'very insignificant', 'a *yes* and *no* hireling to the Court for forty years' and ridiculed for, of all things, his body odour – his nickname was 'Bug' or, in one account, 'His Stinkingness'.[6] But it is successive court and government appointments that resonate, like aristocratic marriages, through the history of the great gardens. They transferred what are – to modern eyes – inconceivable sums of money from the public purse to Grey and other aristocrats like him.

In 1704, Henry Grey became Lord Chamberlain, in charge of part of the royal household. Appointments like this were political and Grey's seems to have been made because any other appointment might have offended someone. The alternative explanation, rumoured at the time, was that he had bribed Sarah, Duchess of Marlborough, the confidante of Queen Anne, with £19.5 million *(£10,000)* by pretending to lose at cards – an example of eighteenth-century money laundering on a grand scale. Why would he, allegedly, pay such a huge sum?

The answer is that Grey, as Lord Chamberlain, was paid £2.3 million *(£1,200)* a year, together with plate worth £781,000 *(£400)* and other fees. So his six uninspiring years in the office brought him nearly £14 million in modern values. He then traded it in for the title of Duke of Kent – if the story about Sarah Churchill is true, the dukedom may have cost him £6 million.[7] But his court appointments continued: gentleman of the bedchamber, Constable of Windsor Castle, Lord Steward of the Household, Lord Keeper of the Privy Seal. The last two were the most lucrative, but the total he received between 1714 and 1719 was at least £15.7 million.

Thomas Archer's pavilion and the great woodland garden at Wrest were one product of all these newly acquired riches, together with the removal of the walls around the vast parterre that Henry's father had erected at such enormous expense. Henry worked with his head gardener, John Duell, and rated him so highly that he had his portrait painted. Later in the century, as English gardens became less formal, informality was introduced to Wrest and this was completed by Henry Grey's granddaughter, Jemima, Marchioness Grey. It was Jemima and her husband Philip Yorke who invited Capability Brown to Wrest in 1758 to add the sinuous curves of the 'river' that winds round the garden. In 1773–80 they recorded the annual upkeep of the gardens as £479,000 *(£300)*,[8] noting that this had been the case for many years, though this huge sum probably did not include the cost of wages. Their grandson Thomas, 2nd Earl De Grey (1781–1859), completed the garden in the early nineteenth century and built the new house at Wrest as it can be seen today, in an imposing French style that matched the grandeur of its surroundings.

Henry Grey and his family laid out their grounds 'with taste'. Taste was an important concept in eighteenth-century Britain, though a slippery one, as the philosopher David Hume found when he tried to define it in 1739. But Grey, like many other aristocrats of the period, would have wished to be seen as a man 'of delicate taste . . . easily to be distinguished in society, by the soundness of [his] understanding and the superiority of [his] faculties above the rest of mankind'.[9] Such sensibility was developed by a classical education and, particularly, by the 'Grand Tour' through France to Italy that Henry Grey, and many other young men of the British nobility, undertook after leaving university. It exposed them to the art, architecture and sculpture of antiquity and of the Renaissance, to music and the theatre and, sometimes, to amorous adventure. Important steps on the Grand Tour were Versailles, with its fabulous palace and gardens, the journey through the grandeur of the Alps, Florence, with the gardens of the Medici, and finally the landscapes of Tuscany and Rome. If they were particularly adventurous, they might continue to Athens and Constantinople. The young aristocrats were expected to return home with paintings of themselves in classical – or sometimes oriental – settings and dress, with collections of antiquities but, above all, with an appreciation of art and architecture, literature and poetry that would equip them with what was needed to demonstrate sensibility and taste in their houses and gardens.[10] Lord Burlington, Horace Walpole, Alexander Pope, Charles Hamilton and many others saw a display of taste in gardening, like that in the other arts, as central to their place in polite society. Grey made a public demonstration of it in laying out Wrest.

SPENDING

The cost of making, remaking and maintaining the great gardens was enormous. At Wrest the bill for alterations between 1685 and 1701 totalled, as we saw above, £6.8 million *(£3,349)*. One way of assessing the cost of such a large undertaking is to compare it with other major infrastructure projects, particularly those that – like the gardens – required moving large amounts of earth. The biggest such projects of the time were the river navigation works, the precursors to the canal-building of the eighteenth century. One of the greatest of these was the Aire and Calder Navigation, carried out between 1699 and 1704 to link Leeds and Wakefield with the River Ouse, a distance of about forty miles; its main use was to carry coal. The work included the construction of twelve locks on the Aire, each 59 feet long by 26 feet wide, together with four on the Calder, and it cost about £19.2 million *(£10,000)*. So the works at Wrest Park cost one-third of what has been described as the 'most important enterprise of this generation and type'.[11]

But that £6.8 million *(£3,349)* at Wrest did not include earlier works, such as the Long Water, nearly 60 feet wide by 600 feet long, which together must have cost at least as much again. So to create the garden shown in *Britannia Illustrata* may have cost around £15 million in modern values. But that was before Henry Grey, or his granddaughter Jemima, got going. Henry's works included the great woodland garden; the pavilion by Thomas Archer of 1712, itself one of the grandest garden buildings in Britain; a greenhouse of 1735–7 by Batty Langley (1696–1751), modelled on a French *trianon* (villa) and stocked with orange trees; several different lakes and numerous statues and monuments – it would be reasonable to guess at a figure of at least another £20 million in modern values. Capability Brown's surviving account book does not cover the work at Wrest, but the average cost of his contracts for garden works at similar houses between 1761 and 1783 is £5.1 million.[12] So the total capital spending at Wrest may well have approached £40 million between 1660 and 1760, or £400,000 on average each year, excluding the regular yearly maintenance and labour costs.

But Wrest was only one garden and not even the grandest; many of the other parks and gardens shown in *Britannia Illustrata* were even larger, with wooded avenues stretching for miles in every direction. Some of the biggest created after its publication, in the first half of the eighteenth century, such as Blenheim, Stourhead and Stowe, dwarfed even them. There were many more smaller, but still expensive, creations such as Rousham and Hartwell. Then came Capability Brown, with receipts from his clients during his career of close to £1 billion in modern terms.* But even the large gardens shown in *Britannia Illustrata* and in Brown's accounts are just the tip of the iceberg; there were many other gardens created by many other landscape designers. It has been calculated, for example, that Brown was responsible for only 5 per cent of the landscape improvements of the eighteenth century.[13] David Jacques believes that there were about three hundred country seats established or modified between 1660 and 1735: 'Almost all the politically powerful had fine gardens, and men of less elevated origins who were raised to greater heights took immense pains to acquire them.'[14] This remained true for the rest of the eighteenth century, when many more gardens were established.

A hundred years after Brown's time, a *Return of Owners of Land* was compiled in 1873.[15] By that time, much land had been taken up by the growing cities. But a reliable estimate based on the *Return* is that there were about five thousand English country houses, each linked to an estate of at least 1,000 acres; all of them would have had the full panoply of pleasure gardens, hothouses, stables and kitchen gardens required to support the lifestyle of a landed gentleman.[16] Their owners would certainly not all have spent as much on garden works as did the de Greys, but even if on average they spent an eighth as much – £50,000 each year – this would still amount to capital expenditure of £250 million each year. This is, to return to a comparison with the building of the canals and river navigations, about

---

* Brown's business is described in Chapter 4.

half the total cost of the forty-one miles of the Bridgewater Canal, or the ninety-four miles of the Trent and Mersey Canal. But the building of a canal was a one-off expense, unlike the costs of garden works, which were incurred year after year. The annual sum may be less than two-tenths of 1 per cent of gross national product at the time, but that is still a very sizeable amount to spend on gardens. And it doesn't include the cost of the gardeners.

1660 TO 2020: LUXURIOUS PLANTS

Part, though certainly not most, of it was spent on plants. Garden fashions come and go. One of the fashions that has ebbed and flowed over time is that for planting English trees, shrubs or flowers and rejecting foreign imports. It is difficult to do, for there are actually remarkably few 'endemic' English plants – only forty-eight species, in fact – that reached Britain before the land bridge joining it to the rest of the European continent was submerged about eight thousand years ago. Almost all of them are varieties of *Euphrasia* or eyebright, *Limonium* or sea-lavender, and *Sorbus*, the whitebeam, rowan or mountain ash.[17] The vast majority of the trees, shrubs and flowers that ornament our gardens are not strictly of native origin, but have been imported by human hands or as the result of seeds blown by the wind – or carried by birds – across the English Channel.

Plants have been deliberately imported for millennia: the grapevine by the Romans, rosemary in the fourteenth century, and firs, pines and plane trees by the end of the sixteenth. John Harvey dates an acceleration in the process to the reign of Henry VIII,[18] but it was in the succeeding century, as European exploration pushed further and further into other parts of the world, that the flow of new species really began, growing to a flood in the eighteenth and nineteenth centuries. Its story has often been told, perhaps most eloquently, as far as the eighteenth century is concerned, in Andrea Wulf's *The Brother Gardeners: Botany, Empire and the Birth of an Obsession*.[19] Wulf emphasizes the cooperative aspects of the gardening trade, but there was a darker side: cut-throat competition for novelties; the poaching of skilled gardeners from one nursery to another; the pillaging of orchid fields; the obsessive search for new species that drove men to their deaths in North American rivers.*

Novelty is expensive. One of the earliest prices recorded is for a peach tree that was sold in 1614 by Arnold Banbury of Tothill Street, Westminster; it cost £640 *(5s)*, while a single musk rose sold to Sir Henry Slingsby of Moor Monkton, York – over two hundred miles from London – cost just over £250 *(2s)*. As we saw in Chapter 1, Leonard Gurle's peaches and nectarines cost much the same. When, in the next century, major importation of plants and seeds from North America and other parts of the world began, and nurseries competed to be the first to grow them in England, prices were sky-high. In 1738 John Telford, of York, one of the first nurseries outside London, sold a Benjamin tree (*Lindera benzoin*), a native of the south-eastern states, for £223 *(2s 6d)*. A cassine (*Maurocenia capensis*) from the Eastern Cape of South Africa sold for the same sum. In 1775, Telford sold a rhododendron, either from the Appalachian Mountains or from Asia, for £1,142 *(15s)*. A Cape jasmine (*Gardenia jasminoides*) cost £773 *(10s 6d)* from William and John Perfect of Pontefract in 1777. Shortly afterwards, in 1788, another York nurseryman, William Thompson, sold a swamp magnolia (*Magnolia virginiana*), from the southern American states, for £736 *(10s 6d)*.

Novelty combined with size could produce eye-watering prices. The most expensive tree in a database of about five thousand plant prices was sold in 1734 by Robert Furber of London to Frederick, Prince of Wales, for his new garden at Carlton House; it was a 25-foot tulip tree (*Liriodendron tulipifera*), related to the magnolia, from North America, and cost £38,120 *(£21)*, while a smaller tulip tree cost £7,624 *(£4 4s)* and a large horizontal cypress (*Cupressus sempervirens var. horizontalis*) went for £9,530 *(£5 5s)*.

* It is impossible to study all the tens of thousands of plants that have appeared in nursery catalogues since the first printed copies were published in the eighteenth century: not only is there a forbidding number of different species and varieties, but they have been sold at different ages and stages of growth; botanical and common names appear but often change. But by concentrating on trees and hardy shrubs, it has been possible to amass a database of about five thousand prices, from 1614 to the present day, and to use this to try to gain insight into how the garden industry has evolved and how it works.

In the next half-century, between 1800 and 1849, the list of expensive plants is dominated by azaleas and rhododendrons; Mackie of Norwich sold a *Rhododendron arboreum* for £694 *(18s)* in 1833. Prices for new introductions remained high in the late nineteenth century. An *Athrotaxis laxifolia*, an evergreen coniferous tree from Tasmania, sold in 1867 for £908 *(£1 11s 6d)* and a *Cephalotaxus harringtonia var. drupacea*, or plum yew, from Japan, for £3,026 *(£5 5s)*, while a Japanese larch (*Larix kaempferi*) and a chestnut (*Castanea chrysophylla*) from the western United States each went for £6,051 *(£10 10s)*.

Different areas of the world were in fashion at different times, as aristocrats and, then, nurseries, sent their men out to find the choicest specimens. But the novelty value typically lasted only as long as it took for English nurserymen to nurture the plants that were sent back, acclimatize them to English conditions and produce them in larger and larger quantities. The smaller tulip tree for which the Prince of Wales paid £7,624 in 1734 was selling in modern values for £220 in 1777, £58 in 1833, £50 in 1880 and £33 in 1892; you can now buy one at your local garden centre for about £5, though larger and older specimens can still cost quite a lot more.

This reduction in the price of something new is very common. The cotton and metal goods that were the mainstay of the Industrial Revolution dropped in price, sometimes precipitously, in the same way. That was the result, in their case, of greater efficiency of manufacture due to the introduction of the spinning jenny and early machine tools; garden plants benefited from the gentler processes of propagation, seed sowing and general nurture, all conducted by nurseries on a larger and larger scale, leading eventually to the mass production of tens or hundreds of millions of plants that occurs today.

Novel plant varieties such as these must have made good profits, at least initially, for those who found them and then those who grew them in England. John Bartram (1699–1777) came to supply much of the English aristocracy with plants and seeds from North America in the early to mid eighteenth century. From 1733, he was commissioned by the London merchant Peter Collinson (1694–1768) to collect and supply seeds, nuts and cones; in 1735, for example, Collinson supplied Lord Petre, a mad-keen collector, with two boxes of seeds from Bartram, containing 500 swamp laurel cones, 2,000 swamp Spanish acorns, 1,500 tulip tree cones, 2 pecks of red cedar cones, 1 peck of dogwood berries and several thousand others, worth a total of £35,000 (£18 13s 3d).[20] Bartram and Collinson set up a successful business; Bartram packed his specimens and seeds into boxes and Collinson sold each of them for around £9,250 (£5 5s) to a growing list of patrons. Their most successful year was 1752, when twenty-nine boxes were sold.[21] The fact that Bartram's specimens were so rapidly diffused throughout England and that the species he had collected were soon being grown by a number of provincial, as well as London, nurseries, demonstrates the scale of the market at the time.

However, the price lists also show the importance of another source of demand for nurseries: enormous numbers of forest trees. In 1664, the diarist John Evelyn, concerned about the supply of timber for constructing naval ships, published *Sylva, or A Discourse of Forest-Trees and the Propagation of Timber in His Majesty's Dominions*; it is still in print. He was a strong advocate of the planting of trees for this purpose; the English aristocracy took him at his word – partly because they saw an opportunity to make money – and the great gardens of succeeding centuries often incorporated large areas of woodland. The nurseries responded to this fashion and price lists soon show trees being sold in quantities of a hundred or more. By 1754, William Joyce of Gateshead was supplying lime, elm, beech, hornbeam, horse chestnut, walnut, poplar, plane, oak, larch and fir by the hundred, many of them in five or six different sizes.[22] The archives of stately homes show the purchase of thousands of 'quicks' or 'quicksets', hedging plants. Telford's in York in 1775 quoted prices for beech, fir, holly, oak and many others by the thousand, as did the Perfects in Pontefract; Mackie in Norwich and Barron in Derby did the same in the next century. The market continued to be huge; the engineer and armaments maker Lord Armstrong bought and had planted 7 million trees at his estate, Cragside, in Northumberland between 1869 and 1882.

Not only do the price data suggest the large size of the market, they also demonstrate that there was a demand for a huge variety of plants, far more than today. Even in the eighteenth century this was remarkable, with over 359 species of herbaceous plants, most of them with several varieties, and 1,000 types of hyacinth, shown in lists published in 1775 and 1776.[23] The range continued to expand in the nineteenth century, when nursery catalogues typically contained many more varieties than can be bought today. Jefferies sold eleven varieties of *Thuja*, a cedar from North America, each in forty-two different sizes, while most modern nurseries seem to sell four varieties at most. As a result, the nineteenth-century catalogues preserved in the Lindley Library of the Royal Horticultural Society are weighty tomes, even though they contain far fewer illustrations or photographs than their modern counterparts.

The plants and the catalogues reflect the changes in fashion that have swept across English gardens and the garden industry since 1660. Driven by competition to create striking effects through landscaping and planting, the evolution of gardening styles has been often described, since it forms the basis of the current discipline of garden history; for that reason, the story is not told again in this book. But the changes in fashions that have occurred – from the formal geometric designs of the late seventeenth century to the naturalistic landscape gardens of the eighteenth, to the more rugged, picturesque styles, the return to formality, then back to naturalism and the cottage garden movement, then the avant-garde designs of the late twentieth century – are all characteristic of luxury consumption. So too are the transient obsessions with particular plants, from the 'greens' that filled the early greenhouses* to the North American plants, pineapples, auriculas and, in the nineteenth century, rhododendrons and azaleas, orchids, bamboo, palms, ornamental grasses, ferns and many others. Fashion succeeds fashion, in gardening as in clothing, architecture, furniture or interior decoration, and just as clothes are discarded when they become unfashionable, so too were, and still are, garden designs, garden features and garden plants. It is fashion that drives people to buy plants, to employ garden designers, to ornament their land with lakes, temples, conservatories and even gnomes. And it is fashion that drives culture wars, in which particular styles and their advocates are set against each other in bitter conflict. For every person who extols

---

* These were not the leafy green vegetables that children are told to 'eat up, because they're good for you', but evergreen trees and shrubs, often introductions from the Mediterranean, such as oranges, lemons, myrtle, oleander, *Pyracantha*, which could not withstand the English winter and were moved from October until March into greenhouses.[24]

the landscaping style of Capability Brown as England's greatest contribution to European culture, there is another who laments the destruction of the Jacobean formal garden. Such conflict is what fuels the garden industry.

INEQUALITY

Wrest, like the other great gardens of its time, is the beautiful product of an extremely unequal society. Its creation cannot be understood without exploring that inequality and the vast wealth of those at the top of the tree who made the gardens. In 1696, the disparity of incomes within English society was set out in a 'social table' that listed the different classes of society and their average income. It was produced by Gregory King (1648–1712), who was Europe's, and probably the world's, first social scientist, although he was also, in his job as Lancaster Herald (an English officer of arms at the College of Arms in London), a relic of a bygone age of chivalry. King was working in the midst of a ferment of intellectual curiosity about science, the economy and society; many of his friends – men such as Christopher Wren, John Aubrey and Robert Hooke – were leading members of the world's first and greatest scientific association, the Royal Society, founded in 1660 by Charles II.

The society that King described was one in which a small group of very rich people led lives very different from those of the rest of the population. Two hundred noble families each received an average income of £11.9 million *(£6,060)*, 400 times that of a labouring family, while vagrants and paupers received even less. This inequality is much greater than in Britain or the USA today; as an example, the chief executive of a major British company today gets, on average, 145 times the salary of the average worker in his – almost all of them are men – company.[25] There are certainly societies today – such as South Africa and Brazil – that are even more unequal than late seventeenth-century England; in such societies, as in England in the past, large numbers of men, women and children were and are barely able to survive. Even though England was then one of the richest countries in the world, the poor looked different from the rich, not just because of their clothing but because the lack of food stunted their growth – they were short and very thin; it was literally true that the upper classes could look down on the lower, so great was the disparity in height. The rich had great mansions, filled with goods from around the world – Chinese wallpaper, Roman antiquities, French furniture and Italian marble floors; the poor owned very little and their hovels had floors of mud.

The elite who possessed such wealth and the political power that went with it was very small. Even if one adds to the nobility all the families of baronets, knights and esquires – the gentry – it was a ruling class made up of only about 1 per cent of English people. This is unusual, for the equivalent groups in other European countries were larger, but members of the English upper class jealously guarded their exclusivity and their status and protested when it was threatened. They resented the Dutchmen given peerages by William and Mary, the men from Hanover elevated under the Georges and, above all, they resented the Scots. They intermarried, except when the need for money drove them to heiresses with large dowries, and thus kept their class small and distinct. It remained, later social tables demonstrate, as small – in proportion to the population – in 1801 as it had been in 1696. Meanwhile, its share of national income and wealth actually increased and its political power was undiminished.

This was the elite that created the great gardens of the seventeenth and eighteenth centuries. *Britannia Illustrata* shows us that in 1707 half the gardens pictured there were owned by nobles. They were led by Queen Anne, five dukes, two marquesses and fifteen earls. Although the book makes no mention of the fact, two of the gardens were created, or at least greatly modified, by a woman important in the history of gardening. From a gardening family herself, Mary, Duchess of Beaufort (1630–1715), altered the gardens of Badminton, near Bath; its landscaped park still stretches to the horizon, but in the late seventeenth century more than a dozen avenues extended for miles beyond one of the most elaborate formal gardens of its day, with one-quarter devoted to an intricately patterned near-maze of tall box hedges. Meanwhile, the duchess created another vast formal garden at Beaufort House in Chelsea, described by one observer as one of the best in Europe, with a very wide range of plants, which she catalogued.[26]

Of the owners of the rest of the gardens, baronets predominated, with knights and a few gentlemen making up the rest. Sixty years later, nothing had changed. It was said at the time that Capability Brown's clients included half the House of Lords; this was an exaggeration, but the 125 that we know about certainly included – as his account book shows – the king (George III), 7 dukes, 26 earls, 20 lesser peers, 19 knights or baronets, 2 generals and a judge. These men – they were all men – had the resources to pay for Brown's work. Clive of India spent £49.9 million *(£31,612)*, the 4th Duke of Marlborough £35.3 million *(£21,538)*, Lord Palmerston £33.4 million *(£21,150)* and the Earl of Bute £30.6 million *(£19,409)*. But where did the money come from?

SOURCES OF WEALTH

The British are, today, very reticent about money. It is considered impolite to boast about how much money you have or, even more so, to ask someone else how much they earn or own. Income tax records are state secrets for ever, although oddly the value of someone's wealth at their death is made public.

This reticence was not so marked in past centuries. The marriage market among the nobility and gentry was one reason; it was extremely important to know the value of the estate of a prospective husband or the dowry of a prospective wife. It did not take long for the news of Mr Darcy's £10,000 a year – £6.9 million today – to spread around Longbourn in Jane Austen's *Pride and Prejudice* (1813). Another reason was that heirs often wanted to borrow on their expectations of inheritance. Moreover, in a society that relied on credit and trust, everyone, from a shopkeeper waiting for his money to an industrialist looking for a partner, needed to know how much someone was 'good for'.

Although people were relatively open about their income and wealth in the eighteenth century, it is difficult, on the whole, to work out exactly how wealthy individual members of the aristocracy and gentry were, or what were the sources of their wealth.* However, the

* One major problem is that, until 1898, the value of land and property such as houses was not included in valuations for probate at the time of death. Wills are

biographies of the men who showed off their gardens in *Britannia Illustrata* at the turn of the eighteenth century, or who employed Capability Brown between the 1750s and 1780s, are very illuminating.[27] They show that the immense wealth that made the great gardens came from just a few sources.

First there was, of course, land; almost all wealthy individuals owned a lot of it and some of those who created gardens came from families who had possessed thousands of acres for several centuries. These families intermarried, hence estates and the wealth that they generated were exchanged over the centuries. But inherited land was not always enough. It was bolstered, secondly, by what the radical William Cobbett (1763–1835) was later to call 'Old Corruption', the system of placeholders and sinecures operating around government and the royal court. Thirdly, money came from the armed services, particularly the army, where officer commissions could be bought and came with lavish salaries and other perquisites, while the navy offered the possibility of massive prize money from captured enemy ships. Other sources included income from slavery in the West Indies and trade in the East Indies; marriage to an heiress from outside the nobility; and, more rarely, income from other forms of trade. All these are the sources of wealth that lay behind the great gardens and which will now be explored in more detail.

Land

At the core of landed society, as its name implies, was the agricultural estate. By the late seventeenth century, England was far from the 'merrie England' of labourers or small farmers exercising their common rights and cultivating their strips of land in the open fields; an

---

sometimes available but can often be misleading, since they usually provide only details of specific bequests, while the residue of the estate, which may have been large, is not given a value. Both problems afflict, for example, estimates of the wealth of Capability Brown; his will contains specific bequests of £14 million *(£10,000)* and this has led some historians to conclude that he was not very wealthy. But the residue of the estate is unknown and neither the value of his Fenstanton estate, which he had bought in 1767 for £20.3 million *(£13,000)*, nor that of his other estate in Lincolnshire, is included.

active land market had seen to that. Some land had been inherited by noble families from generation to generation, even in some cases from the time of the Norman Conquest in 1066, but much more had been bought and consolidated into large estates. Henry VIII's dissolution of the monasteries in the sixteenth century had seen a huge land grab of a quarter of the cultivated land in England. Much of it went to the friends of the king and of his ministers, such as Thomas Cromwell. Meanwhile, beginning in the sixteenth but accelerating through the seventeenth and eighteenth centuries, landlords used their wealth and political power to consolidate their landholdings and to extinguish common and customary rights over land, the process known as 'enclosure'. Tenant farming continued, for the landowners could not physically farm vast acreages themselves, but they controlled the land, benefited from rising rents and, increasingly, exploited the mineral resources that lay beneath their territory.

The biographies of the great garden owners abound with accounts of individuals owning thousands of acres, in both Britain and Ireland; often, noble families had several estates – acquired by inheritance or advantageous marriage – scattered across the country, with a country house and garden attached to each. The 'Architect Earl', Lord Burlington (1694–1753), known as the 'Apollo of the Arts' for his role in bringing Palladian architecture to Britain, owned Burlington House in London, now the home of the Royal Academy of Arts, and other 'seats', at Chiswick to the west of London and Londesborough in Yorkshire – all three of which had appeared in *Britannia Illustrata* – together with estates in Ireland. His grandfather, reputed to be the richest man in Ireland, had owned land yielding over £60 million (*£30,000*) a year, while Burlington's marriage brought more estates throughout the north of England. With essentially unlimited funds, he proceeded to spend millions on houses and gardens. At Chiswick, which then consisted of close to 150 acres along the River Thames, his grandfather's formal garden, the serried ranks of trees illustrated by Kip and Knyff, was entirely remodelled. Burlington built an exquisite Palladian villa and a garden to match, laid out with the aid of William Kent, whom he had first encountered on his Grand Tour in 1714–15, on the model of ancient Roman gardens, sweeping away much of the formalism of the early eighteenth century. A straight canal became a serpentine river and a patte d'oie (goose foot) was the centrepiece of a series of alleys and walks through different garden compartments. Burlington also spent more than £3 million *(£1,600)* in the 1730s to improve Londesborough, which had been laid out in the late seventeenth century by the scientist Robert Hooke (1635–1703), rebuilding a large kitchen garden to include a stream and cascade, fed from one of at least four lakes in the park and gardens. Avenues of walnut and Turkey oak radiated outwards into the distance and Burlington's improvements included the demolition of a village that stood in the way.[28] Nearly a century later, his descendant, the 6th Duke of Devonshire (1790–1858), spent further enormous sums on Chiswick House and its gardens, for which in the 1820s he acquired an elephant and later kangaroos, elks, emus, an Arctic fox, Indian bulls, goats and a llama.[29]

During the eighteenth century, all the land owned by such men became increasingly valuable. Rents rose and with them the income and overall wealth of the families. The cause was simple: the English population was growing – from 5.2 million in 1700 to 8.7 million in 1801 – and it was increasingly difficult to produce enough food to feed the additional mouths. People found jobs outside agriculture, in craft production, in internal and foreign trade, in the service industries, from maidservants to teachers and lawyers, and in the mines and manufactures, which, gradually, transformed the economy through the process known as the Industrial Revolution. So they had money and they needed food.

This greater demand for food confronted what was essentially a fixed amount of land, with the inevitable result that land became more valuable. Rents rose and farmers, along with their landowners, tried to grow more from each acre of land; this was the process known as the 'Agricultural Revolution', symbolized by improving landlords such as 'Turnip' Townshend and Coke of Holkham. New crops were introduced, such as the turnip and the potato, and great efforts went into breeding larger and more productive animals. All these changes were very successful: food production rose fourfold and the growing urban population was fed. The substantial investment in land and farming, by both landlords and farmers, deserved a fair return, but the increase in rents – an indication of the underlying value of the land that was rented – was correspondingly large: average rents rose fourfold between 1700 and 1800.[30] The increase was about twice that of prices in general, so landlords were enjoying good times – particularly at the period when Capability Brown was in greatest demand.

Some of the resulting wealth was spent on gardens. In 1780, for example, the 2nd Earl of Ashburnham (1724–1812), who employed Brown at Ashburnham Place, received an annual rental income of £4.2 million *(£2,880)* from his 8,019 acres in Sussex (and more money came from the iron industry); by 1820 this had risen to £4.9 million *(£6,479)*. He had another 3,117 acres in Suffolk. John Spencer, 1st Earl Spencer (1734–83), inherited the great family estates, including Althorp – later the childhood home of Diana, Princess of Wales – in 1746 when he was eleven, but was also left Wimbledon Park by Sarah, Duchess of Marlborough, on the unusual condition that he never accepted any office or pension from the crown. He lavished money on a house in Park Lane and spent more on riding with the Pytchley Hunt and his stables than on the day-to-day running of his many houses. He was also known as an excessive gambler and left his son massive debts. But he and Brown transformed Wimbledon Park into something 'perhaps as beautiful as anything near London';[31] we can't judge, as much of the park is now covered by the All-England Tennis Club, although Brown's 30-acre lake survives.

Income sometimes came from estates in Ireland. Henry Temple, 2nd Viscount Palmerston (1739–1802) – an Irish title that meant he was not a member of the House of Lords, so his political career in England was in the House of Commons – had an income from Sligo, which reached £19 million *(£12,000)* a year from the work of landless labourers on an estate that he described as 'the most dreary waste I ever yet beheld'.[32] Broadlands, his estate of 2,500 acres in Hampshire, was much more to his taste, however, and he remodelled it from 1767 onwards with Brown and his son-in-law Henry Holland Junior;* he was one of Brown's largest clients, spending £33.4 million *(£21,150)* to create the Palladian mansion and park where the Queen

---

* There were three Henry Hollands: Henry Senior (1712–85), a builder and friend of Brown's; his son, Henry Junior (1745–1806), a successful architect and Brown's son-in-law; his son Henry (1775–1855).

and Prince Philip, and later Prince Charles and Princess Diana, spent their honeymoon. Brown didn't have to create an apparent river near the house, as he did in other places, as the grounds of Broadlands run down to the River Test, so his main legacy consists of his trademark open lawns and grassland, punctuated by clumps of trees, particularly in the distance, drawing the eye towards far views of the New Forest. In the foreground were water meadows, prized for the reflections that they created with the slow movement of the water.[33]

Inextricably intertwined with land came marriage and inheritance. Several of Brown's clients owed their wealth to successive inheritances, often accompanied by lucrative alliances through marriage. Hugh Percy, 1st Duke of Northumberland (1712–86), was an example. 'His genealogical luck was phenomenal';[34] through his marriage to Lady Elizabeth Seymour, Baroness Percy, Hugh took the name of Percy in 1740 and, nine years later, after some manoeuvring, the title of Earl of Northumberland. Exploiting land, marriage and coal raised his total income to £82 million *(£50,000)* a year. Appointed Lord Lieutenant of Ireland, he and his 'junketaceous' wife, as Horace Walpole called her,[35] entertained lavishly. Further manoeuvring brought his dukedom. He could well afford his gardens. The Brown account book shows total expenditure of £3.2 million *(£1,950)* at Syon House in west London – like Burlington's Chiswick, an estate alongside the River Thames of around 150 acres; Brown's work, some of it still to be seen, included constructing two lakes on a former course of the Thames, a large area of woodland and a botanical garden. A sum of £4.6 million *(£2,690)* had already been spent at Alnwick Castle in Northumberland to make 'the old naked castle . . . the noblest seat I think in the kingdom'.[36] Brown's work at Alnwick included damming a stream to create an apparent river. Percy's marriage, although apparently happy, was not enough: he fathered several illegitimate children, including James Smithson, whose wealth, inherited from his mother, founded the Smithsonian Institution in Washington, DC. But it was the current Duchess of Northumberland, wife of one of his legitimate descendants, who recently created the Alnwick Garden within the walls of the old kitchen garden, supposedly at a cost of at least £42 million.

Childlessness could sometimes play havoc with even the largest landed fortune, particularly when, as in the case of the Ferrers family, of Staunton Harold in Leicestershire, it was combined with insanity. Robert Shirley, 1st Earl Ferrers, created around 1680 the great garden that appears in *Britannia Illustrata*; it has been attributed, though without much evidence, to the designer and nurseryman George London. The house, church and outbuildings are surrounded by ponds with elaborate fountains, lawns, formal gardens and a huge rectangular canal; there are fishponds, planted woodland and an orangery. It was said in 1711 that the garden was 'well watered with fountains and canals, very good aviaries and a decoy and a great many exotic waterfowls'.[37] Ferrers's two eldest sons died, so the third brother inherited, only to die unmarried and having been kept in confinement owing to mental illness. His nephew and successor, the 4th Earl, Laurence Shirley (1720–60), became notorious as a domestic abuser and murderer. He imprisoned his wife until her release was ordered by the court of King's Bench; she had to obtain a divorce, by act of Parliament, on the grounds of his cruelty. He lived at Staunton Harold with his mistress but became paranoid, seeing secret plots all around him. A key suspect was his land steward, John Johnson, the administrator of the estates. Ferrers shot him in a locked room at Staunton Harold, was then caught by a gang of colliers and handed over to be tried by his peers, as was his right, in Westminster Hall. He pleaded periodic insanity but was convicted and sentenced to death in 1760, even though two of his brothers gave evidence of their own lunacy in an attempt to save him. Ferrers was hanged at Tyburn, now Marble Arch, in London; William Hickey described the scene: 'His lordship was conveyed to Tyburn in his own landau, dressed in a superb suit of white and silver, being the clothes in which he was married, his reason for wearing which was that they had been his first step towards ruin, and should attend his exit.'[38] The garden, with all his estates, was seized by the crown, although restored to the family in 1763 and remodelled with the obligatory lakes, which still survive.

Later in the eighteenth century, as the Industrial Revolution gathered pace, more and more landowners exploited their land by becoming entrepreneurs. One of the most successful examples was Francis Egerton, 3rd Duke of Bridgewater (b. 1736): not content with an annual income from his estates, in twelve counties, of £50 million (*£30,000*), he built from 1760 the Bridgewater Canal, forty-one miles long, to take coal from his mines to Manchester. It was a gamble – the only British canal ever built by one individual – but it paid off; by his death in 1803, his income was said to be over £76 million *(£80,000)*. Bridgewater, possibly the richest nobleman in England, used his profits to build more canals, to buy paintings, including commissions for J. M. W. Turner, for his house at Ashridge, in what is now Hertfordshire, and to employ Brown to create there the spectacular 'Golden Valley', a natural feature embellished by Brown and edged with woodland. His family thought he was uncouth – he swore, didn't wash regularly and withdrew from and despised polite society – but he was probably Britain's greatest aristocratic entrepreneur. His house is now one of the country's leading management schools and some of his park has been turned into a golf course.

'Old Corruption'

But breathtaking although the income from land often was, it never seems to have been enough for those who owned it. The landowners who created the great gardens also sought lucrative positions at the royal court and the sinecures allotted by different governments to their supporters; this was Cobbett's 'Old Corruption', checked in the 1780s but not fully abolished until the Victorian era. In its heyday it provided extraordinary sums – in modern terms – to those who profited from it.

'Old Corruption' represented the nexus between money and power: money bought power – political, social and even military – and that in turn brought more money. It was a remarkably open system – it was actually called 'the system' by those who benefited from it – by modern standards; the salary of posts at court and the cost of purchasing promotion among the ranks of army officers, for example, were widely published. So was the cost of purchasing a seat in Parliament. Posts and honours were openly traded, as Henry Grey, the 12th Earl of Kent, did to obtain the post of Lord Chamberlain in 1704. Jobs at court, or in the household of the Prince of Wales, were particularly prized for their prestige and for the salaries that went with them, paid from the civil list provided to the king by the government.

At the end of the seventeenth century, many posts in the royal household were personal appointments by the sovereign – although of course he or she would be lobbied or bribed by friends and supporters to grant them. But as what we would now recognize as a governmental system developed, with political parties at least loosely defined, even posts close to the king and queen came to be in the gift of the government, given or withheld in return for support in Parliament. The money to pay for the system came, like the money that paid for the royal gardens, from the taxpayer; it was part of what is now called 'public expenditure'.

Thirteen of the gardens in *Britannia Illustrata* were owned or created by men who held a position at court, and this does not take account of positions in the gift of the government that did not require royal warrants.[39] Nothing had changed later in the eighteenth century: thirty-six of Capability Brown's clients, listed in the Lindley Library account book, held court office at some point during their lives.

To modern eyes, a bewildering aspect of the system was the pursuit of honours and titles. The king, and powerful political patrons, were importuned for aristocratic titles; sometimes – as had been routine in the seventeenth century – they were actively bought and sold. Others were given as the result of gaining, or sometimes losing, political office. Some men seem to have devoted their lives to 'the unremitting pursuit of advancement',[40] as has been said of Wills Hill (b. 1718), who paid Brown £1.8 million *(£1,200)* for work at Hill Park in Kent, now also a golf course; by his death in 1793, he had secured a barony, two viscountcies, two earldoms and a marquessate. He was, in this respect alone, the equal of the Duke of Wellington and twice as honoured as Admiral Lord Nelson.

Some wanted titles, others money. New Park, in Surrey on the edge of Richmond Park, was designed by George London, the royal gardener, after 1692 within one of Charles I's deer parks; *Britannia Illustrata* shows it to have been 'an exquisite masterpiece of the terraced formal garden'.[41] Large parterres and formal walks, with ponds and fountains, around the house were surrounded by huge areas of wilderness and woodland, 'a fine wood so interspers'd with Vistos & little innumerable private dark walks',[42] and, in the distance, an artificial mound from which the whole park could be viewed. It also provided a view of the distant St Paul's Cathedral in the City of London.[43] The woods prefigure the similar woodland at Castle Howard in Yorkshire. Laurence Hyde, 1st Earl of Rochester (1642–1711), who paid for New Park, was said – though by an enemy – to have received £39 million (£20,000) in perquisites as Master of the Robes in the 1670s; he went on to be First Lord of the Treasury and later Lord Treasurer under Charles II, apparently proving a successful administrator, before being dismissed in 1687 but consoled with an annual pension of £8.4 million (£4,000) and grants of land valued at £42 million (£20,000). Sadly, only the ghost of his garden remains.[44]

The Waldegrave family of Navestock in Essex, to take a slightly later example, were poor by the standards of many aristocrats – their estates under George I were worth under £5.6 million (£3,000) a year – but they made up for it with court offices. James Waldegrave, 2nd Earl Waldegrave (b. 1715), was a lord of the bedchamber, on £1.8 million (£1,000) a year, and from 1751 Lord Warden of the Stannaries (the tin mines of Cornwall), an utter sinecure, demanding no work at all, worth £803,000 (£450) a year. He then became governor to the young Prince of Wales, which brought him into conflict with the Earl of Bute (1713–92), who won the post of groom of the stole to the prince but lost the propaganda battle that Waldegrave fomented by alleging an illicit liaison between Bute and the prince's mother, Augusta. Waldegrave got the lucrative Tellership of the Exchequer on £11.5 million (£7,000) a year; his estates produced only £3.8 million (£2,300) a year, but the tellership meant that he could spend £7.6 million (£4,550) on Brown's work at Navestock and still leave each of his three daughters £13.6 million (£8,000) on his death in 1763. The title was inherited by Waldegrave's brother John, the 3rd Earl, and then by a succession of army and navy officers, one of whom demolished the house and gardens in 1811, the park being turned into farmland; only the lakes survive.

Some posts were total sinecures: Wills Hill was appointed joint postmaster-general in 1766, receiving £4.8 million (£2,900) a year for 'occasional light duties'.[45] Others – such as posts as governor of disease-ridden islands in the West Indies – could be carried out by deputies paid a fraction of the salary. Often, the offices were spread around the family. Sir Marmaduke Wyvill (c.1666–1722), of Constable Burton Hall in Yorkshire, became a commissioner of excise under Queen Anne, at a salary of £1.5 million *(£800)*, while his daughter and three sons snaffled posts worth £3 million *(£1,800)* between them.[46] The house was rebuilt by a later Marmaduke and only remnants of the formal and woodland gardens remain.

Court places were not always total sinecures, but they certainly depended more on royal patronage than on qualifications. Caesar Hawkins (b. 1711), who spent £782,000 *(£500)* on Brown's services at Kelston Park in Somerset in 1767–8, was the son of a surgeon and, in 1737, had become surgeon to the Prince of Wales. Ten years later he was made serjeant-surgeon to George II, on £718,000 *(£400)* a year, and continued in that role under George III until his death in 1786, when his son succeeded him as baronet. It was said that his baronetcy 'clearly rewarded his royal practice, not his intellectual or scholarly attainments',[47] which were minimal, but he was said to have made £1.8 million *(£1,000)* a year from phlebotomy – blood-letting – alone.

Some people took their public service seriously and carried out their duties conscientiously. Naval officers, in contrast to those in the army, were increasingly expected to have relevant skills, in particular to know how to sail a ship. Some political posts were removed from aristocrats on the grounds of total incompetence or lack of inter-est. But the system as a whole did, nevertheless, transfer large sums of public money to political and personal cronies within the tiny elite. And much of it was spent on gardens.

Perhaps the greatest creation that can be directly attributed to Old Corruption is not displayed in Kip and Knyff, nor is it a creation of Capability Brown. It is Painshill, a marvel of garden design on the outskirts of south-west London. Charles Hamilton (1704–86), who made it, was the fourteenth child, and youngest son of nine, of James, 6th Earl of Abercorn, an Irish peer; despite his aristocratic hauteur, displayed in plate 7, little family money came to him. But the public purse stepped in. Powerful friends got him, in 1738, the post of Clerk of the Household in the retinue of Frederick, Prince of Wales, which paid £1.8 million *(£1,000)* a year, later rising to £2.3 million *(£1,250)* – although he seems to have been required to work only every other month, checking and certifying the household accounts. With a loan from another friend, the banker Henry Hoare – himself the creator of

Stourhead – he began to create the beautiful and innovative Painshill gardens, part of which is displayed in plate 8. The 14-acre lake lies 15 feet higher than the adjoining River Mole, so Hamilton had to experiment with a series of different pumps to fill the serpentine centrepiece of his creation. It is matched by tree-lined 'Alpine' valleys, a vineyard, flower gardens on an 'Elysian plain', temples, a 'Chinese' bridge and a hermit's hut that is reputed to have housed, briefly, a real (paid) hermit. Most amazing of all is an entirely artificial grotto, on an island, lined with a variety of sparkling minerals, while wooden pillars embedded with crystals form stalactites. It cost about £14 million *(£8,000)*. A gardener was paid to conceal himself and let fall a cascade of water when guests entered. More money went on classical sculptures and there is even a 'ruined abbey' built to conceal a failed venture in brick-making.

Charles Hamilton, like Henry Grey, had developed his passion for gardening and his good taste on the Grand Tour. Renowned for his sensibility, he was invited to advise on other gardens. He was an amateur designer of genius, but his political and financial career was much less successful. In 1746, he objected to the fact that Prince Frederick had taken his sister, Lady Jane Hamilton, as his mistress; he also lost sympathy with the prince's wayward political manoeuvring and relinquished his position as Clerk of the Household. However, he had already been appointed in 1743 by Henry Pelham, the prime minister, to be 'Receiver-General of the King's Minorcan Revenues', with a salary of £2.1 million *(£1,200)* a year; the duties of the post, which Hamilton held for fourteen years, were carried out by a deputy. Unfortunately, the money ran out, his hopes of a substantial legacy from his mother were dashed and in 1756 he wrote to another wealthy friend, Henry Fox, that his income was only £1.4 million *(£800)* a year. Fox obliged in 1757 by appointing his friend as Deputy Paymaster-General, with another salary of £2.1 million *(£1,200)* a year, which continued at least until 1766. But by 1773 the combination of increasing debts and reduced income meant that Hamilton had to sell Painshill and retire to Bath, although not from gardening; he proceeded to create a 9-acre garden in the centre of that fashionable city.[48]

Few who have seen the beauties of Painshill today would begrudge the money, although the middle and working classes of eighteenth-century Britain, who paid for it, might have felt differently. The fact remains that Hamilton was paid well over £60 million *(£37,600)* from public funds for doing very little. 'Old Corruption' indeed. As with the royal gardens described in the last chapter, it was a means by which the taxes of the many were used to fund the pleasures of the few, with the incidental benefit of the creation of the modern garden industry.

Army and Navy

Posts in the army and navy were another way by which public money found its way into gardens. Britain was often at war and this brought rich pickings for officers and, particularly, for those who supplied food, clothing and armaments to the British forces or were responsible for paying their wages. Throughout the eighteenth century and well into the nineteenth, the purchase of a commission as an officer in the armed forces was an accepted part of British public life. In 1765 the prices for buying rank in each branch of the army were fixed and published. In the cavalry, for example, it cost £83,500 *(£50)* to progress from the rank of cornet, the lowest grade of commissioned officer at the time, to that of lieutenant; £1 million *(£600)* was needed to progress from captain to major. A captain-lieutenancy in the foot regiments in 1772 cost £1.5 million *(£950)*, in the dragoons £3.2 million *(£2,100)* and in the cavalry £3.8 million *(£2,450)*. These were the prices paid to the army authorities, but the commissions were also treated as property and could be bought and sold at another price by private agreement.

People were prepared to lay out such large sums simply because officers' pay was very high and the cost of a commission could soon be recouped. A colonel in the infantry, for example, was paid, at the end of the eighteenth century, £436,000 *(£411 15s)* a year, a colonel in the cavalry £636,000 *(£600 17s)*.[49] There were substantial expenses and living costs, of course, but a man could still expect to make a profit on the purchase price some years later. There were also numerous other benefits – such as the opportunity to delay the payment of wages to your troops and pocket the interest – and finally a generous pension. In the navy, in addition, there were also opportunities for earning prize money – through the capture of foreign ships and their cargoes.

Generals, colonels and admirals were important in the history of gardening.[50] Blenheim Palace, named after the Battle of Blenheim in 1704 and the gift of a grateful nation to the victorious general, the 1st Duke of Marlborough, featured one of the most important and most expensive of early eighteenth-century gardens. It was initially the creation of the architect Sir John Vanbrugh and the gardener Henry Wise, whose great parterre – now grassed over as a cricket ground – was flanked by bastions in imitation of military fortifications. Grand avenues march in all directions, particularly from the front of the house over Vanbrugh's three-storey, now partly submerged, bridge to the tall column that marks Marlborough's victories. A later makeover by Capability Brown turned Vanbrugh's rectangular canal into the huge lake that matches the vast scale of the house, later the birthplace of Sir Winston Churchill. Close by, the garden at Rousham, north of Oxford, created initially by Colonel Robert Cottrell-Dormer (d. 1737) and then by his brother James (1679–1741), working with Charles Bridgeman and then William Kent, is still a delight – its Arcadian designs forming one of the best examples of the transition in the 1730s and 1740s between the formal and the English landscape garden. One wanders through sylvan glades, happening upon classical statues and tinkling waterfalls, with glimpses of the River Cherwell winding below. Some of the gardens clearly reflected the military origins of their owners and others were actually built by soldiers; it has even been suggested that the ha-ha – the ditch that replaced an obtrusive hedge in many of Brown's creations – owes its origin to early trench warfare.

Richard Temple, 1st Viscount Cobham (1675–1749), retired in 1713 from his position as colonel of the Princess Anne of Denmark's Regiment of Dragoons to create his garden at Stowe in Buckinghamshire, still one of the grandest in England, after being cashiered, for political reasons, and stripped, temporarily, of his military posts; he later went back to the army. The whole garden carries a political message, of the importance of the defence of English liberties, even if some of the characters who are said to embody this may surprise us today.

The army or navy was often the destination of second or third sons. Augustus Keppel (1725–86) was the second son of the 2nd Earl of

Albemarle and entered the navy in 1735 at the age of ten, rising to become colonel of marines, rear-admiral and finally First Lord of the Admiralty. Along the way he acquired £42.2 million *(£25,000)* in prize money at the capture of Havana in 1762, together with court office as a groom of the bedchamber – not bad for the younger son of a spendthrift father. He was able to spend £2.3 million *(£1,460)* on employing Brown at Elveden from 1765 to 1769, though little of what was created now remains. Keppel's career ended in confusion and accusations by him and against him of treason, although he and his antagonist, an admiral serving under him, Sir Hugh Palliser, were both acquitted by court martial.

One didn't have to be a general, or even a soldier, to make money from the army in the eighteenth century. John Calcraft (1726–72), whose estate at Leeds Abbey in Kent was refurbished by Brown in 1771–2 for £3.1 million *(£2,000)*, gained his wealth as a paymaster, as 'deputy commissary of musters' and as an agent for half the regiments in the army; 'this effectively made him quasi-banker and contractor for the forces', as well as the private agent for colonels, 'their wives and children, their mistresses and bastards'.[51] By the time he retired in 1764 – to become a Member of Parliament – this had made him £849 million *(£500,000)* even after the bribes that he had to pay. This ranks him, in the pursuit of dubious or ill-gotten gains, with Clive of India and other East India Company nabobs. He used some of the money in liaisons with two actresses, in supporting numerous illegitimate children, in large purchases of property and, of course, in making landscape gardens at Leeds Abbey, which have not survived.

East and West Indies

However much could be made from court office or army rank at home in Britain, it was dwarfed by proceeds from the colonies – including those from slavery – and service in the East India Company or in the other chartered companies that monopolized much of Britain's trade from the seventeenth to the nineteenth centuries.

Two great houses with Brownian gardens, Dodington and Harewood, represent slave money. The Codringtons had large estates in Gloucestershire, where they employed Capability Brown at Dodington

House near Bristol, but their real wealth came from West Indian slavery. Most people who read about the principled campaign to abolish slavery do not realize that in 1833, when Parliament abolished slavery in British possessions, £15.4 billion *(£20 million)* was paid in compensation to the slave-owners for their loss of 'property'. The 'property', their slaves, received nothing. The Codringtons as a family, then still active slave-owners, were given £31.6 million *(£41,001)*, since they owned 4,618 slaves.[52] Dodington's park of – at one stage – 700 acres, was praised in September 1766 when a visitor, Mrs Boscawen, enthused: 'Mr Brown having been before Us, and finding great Capabilities of Hills and Vales, shade and Water has dispos'd of the whole in a scene which greatly excited our Admiration.' There were two large lakes, one of them of about 50 acres, with a rotunda on an island.[53]

One of the greatest monuments to British slavery – and to Capability Brown's work in its gardens and 30-acre lake – is Harewood House in West Yorkshire, known in the eighteenth century as Gawthorpe. Henry Lascelles (1690–1753) went to Barbados in 1711 or 1712 and soon married the daughter of a Barbados merchant and slave-trader. He held the contract from the victualling commissioners for the provisioning of British troops in Jamaica, Barbados and the Leeward Islands from at least 1734 until 1747. He was also contractor with the Sick and Hurt Board to supply sick and wounded seamen on the same stations. These contracts proved lucrative, Lascelles acknowledging that the 'victualling in my time was a Branch of business which through good management (I reckon) I chiefly made my fortune by'.[54] He also became collector of customs in Barbados and was accused of corruption and fraud, although he was not convicted of the charges, which were probably politically motivated.

Having returned to England, Lascelles created a slaving syndicate of merchants shipping slaves from the Guinea coast. He invested in West Indian and East Indian loans and became a director of the East India Company.[55] His eldest son, Edwin (1713–95), inherited the Harewood estates and began building Harewood House in 1759. He spent £10.4 million *(£6,203)* there on work directed by Brown. It was not one of Brown's greatest successes, as the lake leaked: 'the water ran out half as fast as it came in', as the Harewood steward wrote to Edwin Lascelles.[56] Another surveyor had to be called in to repair it.

Money cannot have been a problem. In 1786, Edwin Lascelles added twenty-two working plantations and 1,947 slaves to his holdings, worth £420 million (£293,000). When the family's slaves were emancipated in 1833, Henry Lascelles, 2nd Earl of Harewood (1767–1841), the son of Edwin's cousin Edward, received £20.3 million (£26,309) in compensation for 9,737 slaves, almost all in Barbados. Slavery had provided 'the income upon which the transformation of his family into one of the principal aristocratic dynasties of Yorkshire was founded'.[57] It also helped to create one of the greatest of Yorkshire gardens; the huge terrace in front of the house is a later creation, but it gives a splendid view of Brown's lake – now successfully holding its water – which extends around a peninsula whose trees conceal a large walled kitchen garden. There is a wooded valley with cascades and cliffs, as well as areas of sweeping lawns and parkland.

Claremont, in Surrey, to the south-west of London, shows that even slave wealth was dwarfed by the rewards of service with the East India Company. Robert Clive (1725–74), 'Clive of India', came from the Shropshire gentry, one of a large family, and began his career as an impoverished clerk with the East India Company in 1742. War with the French led to his transfer to the military arm of the company and, in 1749, the position of commissary for the supply of provisions to their troops; this, with its opportunities for skimming off commissions, was the foundation of his wealth. By 1753, aged only twenty-eight, he had amassed £70 million (£40,000), which he invested in diamonds. After some time in England, he returned to Madras as deputy governor and led the expedition to rescue the British in Calcutta (now Kolkata), many of whom died in 1756 as prisoners-of-war in the notorious 'Black Hole'. The establishment of British-backed rule in Bengal was the outcome and British leaders received lavish 'presents' from the new ruler.[58]

When he returned to England in 1760, Clive bought estates and a London house.[59] He returned to Madras as governor and speculated in East India stock, benefiting from inside knowledge, though, on the principle of 'set a thief to catch a thief', promising to reform the administration and reduce the opportunities for corruption. Even so, he returned to England worth at least £668 million (£400,000); he bought more estates, including Claremont in Surrey. It cost him £41.7 million (£25,000) and he then spent over £50 million (£31,612) on a new house and gardens there, designed by Brown and Henry Holland Junior. Important features of the gardens were an aviary and a menagerie, which entailed raising the height of the park walls to keep the animals confined – deer, cranes, nylghai and antelope among them and, best known, a zebra.[60] Before Clive could live in his new mansion, he died, either by an accidental drug overdose or suicide.[61] He had suffered long periods of illness and pain, relieved by opium.

The Claremont estate was already an important landscape garden, the creation of Sir John Vanbrugh, who lived there for a time, and of Charles Bridgeman, who worked with Henry Wise and later himself became Royal Gardener; he made the lake – later transformed from its round shape to a more sinuous outline by William Kent – and turf amphitheatre that still delight visitors. The formality of Bridgeman's garden, with grand avenues of trees praised by the garden writer Stephen Switzer, was softened by William Kent, and further changes took place when Clive employed Brown as architect of a new house – he knocked down Vanbrugh's – and to make further changes to the landscape. The spectacular hilly and wooded site is the result, therefore, of the work of three of England's greatest landscape architects and a great deal of dubious money.[62]

Patshull in Staffordshire contains one of Brown's largest lakes, the Great Pool of 65 acres, which survives, though as part of a golf course. George Pigot, 1st Baron Pigot (1719–77), was the son of the clerk to the stables of the Prince of Wales and the dressmaker to Caroline, Princess of Wales. Like Clive, he began his career as a clerk in Madras and, by 1755, had become its governor. War with the French brought him fame and a great deal of money, including a notoriously large diamond and a substantial pension of £7.4 million (£4,500) from Muhammad Ali Khan, the nawab of the Carnatic. He bought Patshull Hall for £134 million (£80,000) and consulted Brown about the gardens in the 1760s, though he does not seem to have paid him for the advice. His contemporaries thought him ruthless, domineering and a considerable sexual predator, who never married but fathered ten children by six women. Apparently running out of money, he returned to India in 1775 but fell out with his colleagues and was deposed and put under house arrest; he soon died from excessive exposure to the sun while gardening.

It was a contemporary and associate of Clive, Francis Sykes (1730–1804), youngest son of a yeoman farmer, who between 1751 and 1769 turned a clerkship in the East India Company into 'one of the largest contemporary fortunes to come out of India',[63] estimated to be between £410 million *(£250,000)* and £820 million *(£500,000)*. Part of it was used to buy the Basildon estate of 2,500 acres, near Reading in Berkshire, and it was there that he employed Brown, but paid him only £80,000 *(£52 10s)*: 'For my journey there and for plans of the Kitchen Garden and Stoves'.[64] Little sign of Brown can be seen, however, in what is still a splendid eighteenth-century landscape park. Sykes was vilified by contemporaries as a typical nabob, 'accumulating wealth by dubious means in private trade',[65] but he was probably more honest than Richard Barwell (1741–1804), another of Brown's clients, who was renowned for 'his trust in the powers of bribery'.[66] He made possibly £640 million *(£400,000)* in twenty years in India – 'engaged in corrupt dealings in timber and salt, and in multifarious financial transactions'.[67] A disappointed blackmailer exposed some of his amorous exploits in a pamphlet of 1780: *The Intrigues of a Nabob: or Bengal the Fittest Soil for Lust*. Barwell spent £164 million *(£102,500)* of his ill-gotten gains in buying Stansted House in Hampshire, to which he brought a sixteen-year-old American wife, having discarded the beautiful mistress by whom he had several children. His gambling and extravagance led to great debts and his estates were sold soon after his death. It is said that Brown improved the grounds of his house but little remains and no payments are recorded. Perhaps that was the last of Barwell's frauds.

Marrying an Heiress

If more money was needed – and it usually was – there was always marriage to a rich woman. Even if the British aristocracy jealously guarded its titles and privileges, the desire to preserve the purity of 'blue blood' did not extend to rejecting heiresses from inferior social classes. One such marriage created the wealthiest landed estate in Britain today, though little remains of the original house and garden of Sir Thomas Grosvenor (b. 1655), at Eaton Hall, Cheshire, which is illustrated in *Britannia Illustrata*. That book shows it with a highly formal, symmetrical, design leading from a terrace with semicircular summer houses through a parterre to a wilderness and then, via a rectangular canal lined with trees, to a distant prospect of Beeston Castle on the horizon. This 'eye-catcher' was unusual at the time, although it became a feature of many eighteenth-century gardens, which were equipped with artificial follies in the shape of towers, castles, temples and Turkish tents set on any convenient hill. Grosvenor married Mary Davies in 1677, the daughter of a London scrivener (someone employed to write or copy letters or legal documents) who owned estates in Chelsea and what is now Park Lane; Mary first decided to become a Catholic, against her husband's wishes, and then, after his death in 1700, married the brother of her chaplain. The Grosvenor family, alarmed at the prospect of losing the estates, had the marriage dissolved and Mary declared a lunatic. This ensured that the Grosvenor estates, the foundation of the wealth of the Duke of Westminster, now the richest landowner in Britain, remained in the family.

This was not the only marriage between different classes to go wrong. George Booth, 2nd Earl of Warrington (1675–1758) and owner of Dunham Massey in Cheshire, now the property of the National Trust, married the daughter of a merchant of the City of London in 1702. She brought a fortune of £46 million *(£24,000)* but 'Some few years after my lady had consign'd up her whole fortune to pay my lord's debts, they quarrelled, and lived in the same house as absolute strangers to each other at bed and board';[68] Booth devoted himself to writing pamphlets in favour of divorce. His huge moated house, built around a central courtyard, was surrounded by formal gardens, though little remains of them except for a mound – possibly the remains of a medieval castle – which was originally topped by a gazebo. Later landscaping produced pools, woodland and avenues.[69]

Since women often died in childbirth or soon afterwards, it was sometimes possible to acquire more than one fortune through marriage. Sir John Pakington (1671–1727), of Westwood in Worcestershire – whose grandfather of the same name had been an ardent Royalist in the Civil War – first married, in 1691, Frances Parker, who had a portion of £7.8 million *(£4,000)*. When she died in 1697, he married the daughter and heiress of Sir Herbert Perrott, acquiring a 'good £1,100 a year' (£2.2 million today), although he still found it useful to have a secret pension of £1.6 million *(£800)* a year from the Irish establishment.[70] Westwood, of which much survives, was built as a banqueting house and then became a hunting lodge; *Britannia Illustrata* shows a hunt in progress. The house is unusually tall, so that the guests could watch events from the upper floors or roof, and consists of a central square and four 'butterfly' diagonal wings. A pleasure garden near it, unusually small for the period, contained four turrets that echo the wings; it was surrounded by concentric rings of woodland (now mostly farmland), interspersed by avenues, some of which led to a 'great pool', which Kip and Knyff put at 120 acres, though modern estimates suggest half this.

Marriages between aristocrats also went wrong. Lady Elizabeth Hamilton, daughter of the 6th Duke of Hamilton, married Edward Smith Stanley, 12th Earl of Derby, in 1774; it was apparently a love match. She bore him three children but then left him for John Frederick Sackville, 3rd Duke of Dorset, 'the most notorious rake of the day'.[71] Derby refused to divorce her or to give her access to their children, ruining her life. He then courted an actress, who agreed to live with him but not to become his mistress; they married in 1797, eighteen years later, when his wife died, and had three children. Derby seems to have sought solace as an inveterate and extravagant gambler and as the greatest sportsman of the day, devotee of cricket, hunting and cock-fighting and the founder of England's best-known horse race. Knowsley Hall, where he paid Brown £134,000 *(£84)*, is still regarded as one of the best examples of a late eighteenth-century landscape in the north of England, embellished with extensive stables. The estate of 2,500 acres has a stone wall around it that is nine and a half miles long; within it Brown created a string of lakes and water gardens and Robert Adam built a summer house known as the Octagon. Still owned by the family, Knowsley Hall can be hired for weddings.

One of the great scandals of eighteenth-century England overwhelmed Sir Richard Worsley (b. 1751), who spent £84,000 *(£52 10s)* on Brown's work at Appuldurcombe on the Isle of Wight – much of it since destroyed – and held a number of court appointments. In 1775 he married the seventeen-year-old Seymour Dorothy Fleming, daughter of Sir John Fleming, who came with a dowry of £112 million *(£70,000)*. She had numerous affairs – a total of twenty-seven were alleged – and in 1782 Worsley accused George Bisset, a neighbour, of 'criminal conversation' with his wife. This was a civil action, in which Worsley sought damages of £28 million *(£20,000)* for the harm to his 'property' – his wife. The jury found for Worsley, but judged that he had connived at the affair – and probably at others – and awarded him only £70 *(1s)*. Not unnaturally, he left for the continent, while his wife became a professional mistress or high-class courtesan. Worsley used his time in Greece, Rome and Turkey to make the most important collection of Greek marbles then seen in England; they were displayed at Appuldurcombe, where he set up house with his mistress and died in 1805 with massive debts – he could never resist paintings or sculpture.

Business Connections

The wealth underpinning the properties of the early 1700s shown in *Britannia Illustrata* is hardly touched by 'trade' – apart from a few heiresses. By the second half of the eighteenth century, however, the growth of the economy was producing larger numbers of wealthy people who had made their money through banking or as merchants, sometimes as government contractors, and who used it to create gardens.

Sir George Colebrooke (1729–1809), who paid Brown £5.1 million *(£3,055)* for redesigning the park at Gatton (now a school) in Surrey between 1762 and 1768, used his family's banking business (and his wife's inheritance of £330 million *(£200,000)* from investments in slavery in Antigua) to get government contracts and then to attempt to control the East India Company. He overreached himself. Speculation in raw materials proved his downfall: he lost £298 million *(£190,000)* in 1771 gambling on hemp and then failed to corner the world market in alum. His bank closed and his possessions were sold; by 1777 he was bankrupt and living on a small pension in Boulogne. His legacy is the Brown landscape at Gatton, with a large lake and woodland; its kitchen garden – when it was owned in the period between the two world wars by Sir Jeremiah Colman, of Colman's mustard – once employed Arthur Hooper, author of *Life in the Gardeners' Bothy*, published in 2000 and discussed later.

More successful was Robert Drummond (1728–1804), not only Brown's banker but one of his largest clients, paying him £20 million *(£13,231)* for his work in 1775–80 at Cadland in Hampshire, an estate that stretched for nearly eight miles along Southampton Water; most of it has now disappeared beneath an oil refinery. Nephew of the founder of Drummonds Bank, he became a partner; by 1769, when his uncle died, it had 1,500 accounts and annual profits of about £16.4 million *(£10,000)*, some of them from the large number of army agents who banked there. Cadland featured one of Brown's smallest gardens and has recently been restored.

The Gough family benefited from the growth of the economy in another way. Richard Gough (1655–1728) made a fortune as an East India merchant in the early eighteenth century and bought the Edgbaston estate on the edge of Birmingham. His son and grandson employed Brown there in 1776, though only for a small payment, but soon after began to exploit the estate for housing: the first building leases were granted in 1786, ultimately creating a desirable suburb for factory owners and professionals in one of Britain's fastest-growing cities. The Gough family, of course, did not live there but in London and on their estates in Hampshire and, while they became richer, the entire Brown landscape disappeared under new homes.

There is only one manufacturer among the long list of Brown's clients and he was very unusual. Thomas Fitzmaurice (b. 1742) was the second son of the 1st Earl of Shelburne and brother of the 2nd Earl, who became prime minister and later Marquess of Lansdowne. Fitzmaurice had poor estates in Ireland and was in 'very great distress for money' when, in the 1770s, he set up as a linen merchant, bought the Llewenny estate in Wales and established a bleaching factory.[72] In 1785 it was said that he had 'plunged himself into a business which might make even a tradesman tremble. He is a bleacher of linen.'[73] Known as 'the Royal Merchant', he had enough funds to spend £164,000 *(£100)* on Brown at Llewenny, though nothing remains. The business did not prosper and Fitzmaurice spent the last three years of his life until his death in 1793 at Cliveden, one of the greatest of English houses and gardens, which his wife had – luckily for him – inherited in 1790. Sadly, it burned down in 1795, to be rebuilt, then burn down again in 1849, and finally be replaced by Charles Barry's huge Italianate 'villa' in the 1860s;

it became famous in the twentieth century as the home of the fabulously wealthy Astor family, who entertained everyone of importance, and notorious for the 'Profumo affair' of the 1960s, which contributed to the collapse of the Conservative government in 1964. Its gardens, high above the River Thames, are some of the most imposing of those owned by the National Trust while its terrace in front of the house has recently been restored at the cost of £6 million.

WHY WERE THE GREAT GARDENS CREATED?

The great gardens of the seventeenth and eighteenth centuries were the creation of a tiny section of the population, possessors of immense wealth, however it was obtained – from landownership, public office, slavery and the exploitation of India to merchant endeavour and strategic marriage. Their wealth came from different sources, so it is usually impossible to say that, for example, a particular garden was funded by slavery. One notable exception is Painshill, where Charles Hamilton could not have made his fabulous garden without payment from court appointments and sinecures, public expenditure in the form of 'Old Corruption'. The gardens of the East India Company nabobs, too, can be linked very directly to their Indian dealings, legal and illegal; after all, they started off with very little but became fabulously rich. In short, however the money was made, the gardens were the product of an immensely unequal society.

But why were they created? Why did this elite group of aristocrats, courtiers, politicians, merchants, lawyers and nabobs decide to spend their wealth on great gardens such as those shown in *Britannia Illustrata* and later those made by Capability Brown and other designers?

One answer is that they could. Huge though the sums were that paid for the work of George London, Henry Wise, Charles Bridgeman, William Kent, Capability Brown and their colleagues, together with the head gardeners and their large cohorts of subcontractors, foremen, journeymen and labourers, they were – almost always – small amounts of money by comparison with the wealth of the garden owners. Painshill was, again, an exception, as was to be Chatsworth in the nineteenth century, when the 6th Duke of Devonshire got into financial difficulties, partly through plant-hunting; the Chatsworth garden alone – he had several others – had a yearly budget of £1.6 million *(£2,000)* during the 1830s, on top of the wages of the head gardener, Joseph Paxton.[74] Typically, Brown received about £5 million from each client when he was the contractor for the works; this was a small sum in comparison with the annual incomes, let alone the total wealth, of most of them.

Very few of these men were constrained by their incomes. Their biographies abound with tales of their gambling, womanizing and general extravagance, whether it be on entertainment, on the Grand Tour, on purchasing paintings and sculpture or on their houses and gardens. Charles Howard, 3rd Earl of Carlisle (1669–1738) and builder of Castle Howard, featuring one of the largest and most elaborate gardens of the eighteenth century, is said to have obtained 7 per cent of his huge income from gambling. The accounts of Lord Boringdon (1772–1840), owner of Saltram House in Devon, list in minute detail his expenses in the early nineteenth century: he bought pictures, gambled and kept racehorses and managed to spend on these pleasures and the family expenses, such as shoes for the children, all of his income of £9.4 million *(£12,000)* a year. Maintenance of his gardens took up less than £315,000 *(£400)*, a relatively small amount in relation to his annual expenditure, even though he extended them and built his own racecourse by reclaiming land from a nearby river.[75] He died leaving considerable debts. Many members of the aristocracy owned several houses and even the greatest, such as Stowe, were occupied for only a few weeks of the year. Others, such as members of the Grenville family at Wotton House in Buckinghamshire, could create a huge lake and use it to stage mock naval battles. In other words, they could easily afford to create great gardens. Some were temporarily impecunious at times, it is true, but there were always court positions, sinecures or heiresses to rescue them.

That does not explain, of course, why they chose to use their wealth on gardens. Stowe is often described as a 'political garden' and indeed Lord Cobham designed it as a celebration of English liberty. Classical statues, embodying ideals of republican resistance to tyranny, accompany 'the Temple of British Worthies', where King Alfred and the Black

Prince are joined by philosophers and writers, more recent defenders of liberty against the might of the king, and finally by Sir Thomas Gresham, representing the City of London. The intentions of the garden were made crystal clear in the guidebooks – including one in French – that were produced for the eighteenth-century visitors who were welcomed and expected to understand and embrace its message. Others, such as the Lees at Hartwell House, also in Buckinghamshire, did the same, celebrating in particular Frederick, Prince of Wales, for his resistance to his father, George II. There were other messages – Wotton's lake is said to have celebrated English naval power.

But there were, after all, much easier and cheaper – as well as quicker – ways of promoting political ideals than by converting hundreds of acres into a beautiful form of propaganda. The fact that a garden has a particular theme does not prove that it was built solely, or even mainly, to promote that idea. Nor does it seem that most of the great gardens, whether at the end of the seventeenth century or in the age of Brown, had political or similar aims. The most likely, even if more prosaic, motive for building the gardens is that they represented the desire of the elite to 'keep up with the Joneses', to move with the times and with fashion, to demonstrate that they were men of taste and to emulate the kings and queens who, from the time of Charles II, had led the way. In the process, they created jobs for designers, nurseries and thousands of labourers and gardeners. However, this expenditure, welcome as it must have been to those who were employed, did little or nothing to raise their wages – which were set principally by the conditions of English agriculture – or to diminish the extreme inequality within Georgian society.

The gardens were intended to be seen. At both Stowe and Stourhead, inns were provided for the visitors; the one at Stowe has been recreated by the National Trust to serve, once again, as the visitors' entrance to the park. Among its many attractions was the menagerie built in 1781 for the Countess of Buckingham, with its conservatories and aviaries, painted with floral decorations, where parrots perched on classical statues and urns.[76] Engravings by Jacques Rigaud show the family's visitors at Stowe – celebrities, as they would be called today – being viewed from a distance by less privileged onlookers.[77] But even the smaller gardens and houses of the gentry were available to view for the price of a tip to the head gardener or the housekeeper. Noblemen showed off their new plant acquisitions, their greenhouses, fountains and exotic animals: Bowood had, in 1768, an orangutan and a leopard, both living in a menagerie in an orangery designed by Robert Adam.[78] Noblemen also imposed their presence on the landscape, with avenues and rides stretching as far as the eye could see, a real demonstration of power and wealth. Roads were diverted, villages swept away, lakes dug, even ancient churches relocated, all to provide the most tasteful views and approaches to a country house.

There is every sign that kings and queens – and many of their aristocratic subjects – enjoyed making the gardens and then using them: banqueting houses, such as that at Wrest, or the carriage rides that frame many of the gardens, were built to be used; views were carefully devised to be conversation pieces. The gardens had many functions, not least supplying the households with vegetables, fruit, fish and game; they were places to 'take a turn' for amorous or conspiratorial purposes, or simply for exercise.[79] They could be planned and re-planned, generation by generation, sometimes consciously to destroy the designs of a hated father or to stamp the new owner's personality on the property. Not least, their creators wanted to make something beautiful. In short, the great gardens were built for much the same motives that impel every gardener today; the difference is that the elites of the seventeenth, eighteenth and nineteenth centuries had almost unlimited funds with which to pursue their desires.

WIDER IMPACT

A wider question is that of the impact that all this spending on gardens had on the economy of the time. It was, after all, the economy that produced the Industrial Revolution and which saw the transformation of England from an agricultural nation to 'the workshop of the world', with a rapid rise in population and the unprecedented growth of towns and cities. What role did gardens play in these transformations?

Economic growth is – despite centuries of study by economists – still a rather mysterious process, but it is generally thought to stem from technical change and the investment of funds in new products and processes, combined with increases in demand arising from changing tastes and rising incomes. It is facilitated by increased provision of what is called 'infrastructure' in the form of transport systems, markets or financial institutions such as banks and insurance companies, all aided by the rule of law which enables contracts to be made and fulfilled. How do gardens fit in to this?

Although very large sums of money were spent on gardens, similar to the amounts spent on roads and canals, they do not really count as infrastructure. Despite the changes that they wrought on the English countryside, they did not – except possibly through the drainage schemes that sometimes accompanied the provision of lakes – usually assist further economic activity. On the other hand, as Chapter 7 will show, they pioneered changes in technology that were then applied to other parts of economic life, in particular in engineering and construction. In addition, they were an important aspect – though up to now largely neglected – of the rising demand for luxury and new products from home and abroad that is increasingly thought to have fuelled the Industrial Revolution. Just as the wealthier sections of the population filled their houses with furniture, pictures, china and other household objects, stimulating new industries and imports, so their spending on gardens brought new forms of retail and wholesale trade in the form of nurseries and plant importers; garden designers vied with architects, painters and sculptors to equip the houses of the upper and middle classes. Gardens were, in fact, among the most conspicuous forms of the conspicuous consumption that was indulged in by those with money to spend. In the process, they also provided employment which – at least to some extent, given the low wages of most of the jobs that were created – increased demand for other goods.

So much money was spent on gardens by the English upper classes that it also makes sense to consider whether that expenditure diminished economic growth by taking away money that could have been spent more productively elsewhere: as economists would ask, did gardens 'crowd out' other investments? It seems unlikely, partly because even the very large sums that were spent were usually only a small part of the income of the aristocrats whose gardens were shown in *Britannia Illustrata* or which appear in Brown's account book, while even the spending on the royal gardens made up only a small part of public expenditure. It was not large enough to affect interest rates in the economy as a whole. In addition, although a few men such as the Duke of Bridgewater did combine big spending on gardens with investments in canals or mines, there is no sign that there was competition between these uses of funds. Aristocratic funds were not in general directed to industrial activity, partly because of the lack of secure means of investment that could rival the traditional benefits of owning land.

Gardens take their place, therefore, among a range of new forms of demand that characterize the English economy of the late seventeenth century and the whole of the eighteenth. They demonstrated wealth, status and taste. The demand was matched by the supply of goods and expertise displayed by the new garden industry. Let us now investigate the different facets of that new economic activity.

4

Designers

'To gild refined gold, to paint the lily'[1]

On 4 March 1981, the garden designer and journalist Lanning Roper wrote to Mrs Andrew Parker Bowles:

> It is dear of you to have arranged that I should help Prince Charles with the garden of the new house. I had an ecstatic letter from your grandmother with ideas of giving him plants and even propagating things for him.[2]

Those two sentences say so much: about the status of a garden designer; about the networks of clients that help a designer to get commissions; about the enthusiasm of gardeners to share their knowledge and their plants; even about an imminent royal marriage – the letter was written a few months before the marriage of Prince Charles to Lady Diana Spencer that ended in divorce and her tragic death. It is clear that she had little interest in gardening. The 'garden of the new house' is that of Highgrove, the latest in the string of royal gardens. Mrs Andrew Parker Bowles, Camilla, was then the lover and is now the second wife of Prince Charles and holds the title of Duchess of Cornwall. Her 'ecstatic' grandmother, Sonia Cubitt, was the daughter of King Edward VII's last mistress, Alice Keppel. Lanning Roper had worked for Sonia since 1973 at Hall Place, West Meon, Hampshire, but had also, since 1967, been redesigning the garden at The Laines, near Brighton, for Major Bruce Shand and his wife Rosalind, Camilla's parents.

Roper, like many other garden designers, relied on his social connections and networks, as well as his books and newspaper columns, to bring him work. This is entirely natural; people who employ a garden designer – just as they would an interior or kitchen designer or anyone else to work in their house – want someone who will liaise with them and be sympathetic to their tastes. Personal recommendation, from friends and family, is some guarantee of that. More problematic is the exact nature of the relationship: is the designer a friend of the family, to be entertained and treated as a guest, or an employee or something in between? Roper had to cope with clients who had their own staff, and their own gardeners, who needed to interpret their employer's rather vague wishes or demanded a more formal contractual relationship. In June 1981, he wrote to Sir Edward Adeane, Private Secretary to Prince Charles:

> I have no idea of how the Prince of Wales would like me to proceed, and no idea of the money that he is prepared to spend. I have tried to emphasise priorities and to encourage the delay of a number of decisions, until they have had time to live in the house for a reasonable period ... I would be most grateful if you could suggest someone with whom I could have a general conversation, to sort out a multitude of problems, which need to be resolved before I can proceed intelligently.[3]

As it happens, Lanning Roper's 'help' at Highgrove was cut short, after only a few months, by a recurrence of the cancer that was eventually to kill him in the spring of 1983. But these and other letters to a range of clients, preserved with his papers in the Lindley Library of the Royal Horticultural Society, raise intriguing questions about him and his predecessors. What has been the role and status of garden designers over the centuries? Were they employees, professionals, friends, consultants, helpers? What were their social and economic origins and how were they trained? How did they secure their clients, how did they run their businesses and how much money did they make? How did they work with the contractors who carried out their designs and with the nurserymen who supplied the plants?

Conventional biographies of garden designers and landscape architects are largely silent about these matters. Indeed, some of them come close to hero worship instead. Capability Brown has recently been described as 'a great artist', the originator of 'a new art ... its

Michelangelo – an original, self-taught genius – acknowledged in his own time as "the Shakespeare of Gardening" '.[4] This has been an extraordinarily rapid change of sentiment, for Brown was vilified for much of the nineteenth and twentieth centuries and as late as 2000 could be seen 'as the purveyor of a heartless, even philistine, topographical elegance to a sporting aristocracy unconcerned with the social needs of their tenants or the textured charm of their lands'.[5] But in either view, the focus of attention has been almost entirely on his skills – or lack of them – as a designer, rather than on his business methods or his relationships with his clients. Much the same is true of biographies of other great garden designers, including Henry Wise, Charles Bridgeman, Richard Woods, Humphry Repton, Joseph Paxton, Edward Milner, William Robinson, Thomas Mawson, Gertrude Jekyll and Lanning Roper.

It is not usually explained that, as well as being great designers, these men and women were engaged in earning their living. There are several reasons for this apparent oversight, one of which is the paucity of records. Even when plans, plant lists and correspondence with clients have survived, financial and business records have hardly ever been preserved. Few designers seem to have had written contracts with their clients; indeed, at times, one feels, money was not something to be mentioned. There are few invoices. It is natural, therefore, that the focus of biographers should be on what has – at least to some extent – survived, the gardens that were created.

CAPABILITY BROWN AND HIS BUSINESS

It seems amazing, therefore, that among the few surviving business records of the great garden designers are the account book and bank account – covering the majority of his career – of Capability Brown himself; plate 9 shows a typical page from his ledger at Drummonds Bank. They allow us – as I have shown in Chapter 3 – to locate him within the economy and society of eighteenth-century England. They also provide a standard against which we can measure the methods and achievements of his predecessors and successors. The books show him to have been not only an influential and successful designer but a shrewd businessman and a proficient manager of men and of his numerous wealthy and distinguished clients.

Brown lived and worked during the last three-quarters of the eighteenth century, the period that we now describe as the end of the Agricultural Revolution and the early stages of the Industrial Revolution. Both terms are misleading as they seem to imply a period of violent change. In fact, economic growth – as measured by the change in the overall output of the economy or gross domestic product (GDP) – was slow throughout the period, no more than 1 or 2 per cent per year, very much slower than has been the case in some parts of the world in recent years during the rise of the 'tiger' economies of China, Singapore, South Korea and India. So, although the cumulative impact of economic growth was considerable, it was not easily perceptible at the time and it was not measured in the modern way until the mid twentieth century. The growth of the population was more obvious, at least to observers such as the Reverend Thomas Malthus towards the end of the century; it rose by about 70 per cent between 1700 and 1800. This itself helped to fuel economic growth, by providing a larger labour force, but it also put pressure on food supplies and required the construction of larger towns and cities. The average income of the people therefore increased much more slowly, by only about one-third over the whole century.[6] Moreover, as we saw in the last chapter, extremes of wealth and poverty continued to coexist and even worsened.

Even less perceivable than economic growth or population change, but of great importance in understanding Brown's career, was a slow change in the character of the economy. He and the many other designers who emulated or competed with him were part of a growing tertiary or service sector, different in their skills and working methods from the agricultural and manufacturing sectors that still dominated the economy; they were an important segment of what would now be called the 'creative industries', although gardening is rarely categorized as such today. Interior designers, plasterers, furniture makers, architects jostled with garden designers for the custom of the aristocracy and the growing middle class. The boundaries between the different trades were rarely clear – the artist Sir John

Vanbrugh and the gardener Capability Brown both built houses, the architect William Kent designed gardens and furniture as well as houses – but it was a small enough world that the different practitioners knew each other, employed one another, even wed each other's daughters, as the architect Henry Holland Junior did when he married Brown's eldest child, Bridget. Brown's accounts are a window into this world, detailing thousands of payments to different individuals – employees, contractors and subcontractors, foremen, collaborators, friends and family – who were all engaged, in one way or another, in realizing his visions.[7]

A CAPABLE ENTREPRENEUR?

As mentioned in Chapter 3, the Lindley Library of the Royal Horticultural Society in London holds Brown's record – some of it in his own hand – of his commissions, from 1761 to 1783, detailing the payments made by his clients at each site. There are a few references to his expenses for surveys and designs, and travelling costs, but, tantalizingly, no information about what the payments were for. Nor is this revealed by the ledgers of Brown's account at Drummonds Bank from 1755 to 1783, now in the archives of the Royal Bank of Scotland in Edinburgh.[8] This was his personal bank account – in 161 double foolscap pages – covering his household expenses, his gifts to family members, his investments and even his purchases of lottery tickets. But the bulk of the bank account consists of payments to the men who worked with or for Brown, or supplied him with plants or materials – just a list of names with no indication of where they worked or what they did.[9] The two sets of records complement each other, but neither is complete; we know that Brown was sometimes paid in cash and that not all of it was recorded. Similar problems arise in disentangling the accounts of Brown's contemporaries such as Richard Woods or, later, Humphry Repton.[10]

The level of earnings recorded is stupendous. At the height of his career between 1762 and 1779, Brown's annual income was always above £20 million in modern values and in two years surpassed

£50 million.* The receipts recorded in his account at Drummonds total about £840 million.[11]

This is big business indeed, comparable to or surpassing the growing industrial and mining enterprises of the time. Although Brown did, on some occasions, simply supply designs, to be carried out by estate labour forces, the majority of his work was what today would be called 'design and build'. In other words, he took control of the whole process from his initial survey of the 'capabilities' of an estate, through the – often sketchy – design plans, marking out the design on the ground, the earth-moving, lake-making and planting of trees and shrubs.

Brown's clients were scattered across England and Wales – he seems to have declined almost all commissions from Ireland and Scotland – at a time when the road network and postal system were still primitive and travellers might be threatened by highwaymen. Although transport by river was more reliable, Brown was working at the outset of the canal age, so the carriage of heavy or fragile goods such as plants and trees was still slow and cumbersome. He had to establish a network of dependable men all round the country to carry out his commissions; he then needed to control the cost and quality of their work in order to satisfy demanding and sometimes capricious customers. His subcontractors were themselves working with a constantly changing labour force and wages that varied with the season, since labourers were in a strong bargaining position during the harvest.[12] All this was on a greater scale than that of other eighteenth-century entrepreneurs, most of whom operated within a very restricted geographical area even when – as was the case in the textile industry before the coming introduction of the factory – they relied on large numbers of 'outworkers', such as individuals spinning or weaving in their own homes.

On top of this, Brown faced some specific problems. Probably the most significant was that his commissions were 'lumpy': he could not rely on a constant flow of income – such as might come, for example,

* The original money values are not given in the remainder of this section on Brown, since the accounts cover many years with each year's income and expenditure requiring a separate conversion to modern values.

from selling cotton goods or metal manufactures. Although he seems to have been very successful in extracting stage payments from his clients when the work was in progress – rather than waiting months or years for his money – this still produced a very irregular pattern of credits into his accounts. There were months in which no money came in, yet he had to make between ten and fifteen payments each week – though of much smaller sums – to the people doing the work. In modern parlance, managing his cash flow must have been a nightmare.

All this was made more difficult by the primitive – one might even say non-existent – nature of his accounting systems. This was typical of an age before the invention of cost or management accounting. Brown used the method of other great enterprises of the time, particularly the large landed estates, namely the 'master and steward' system. All forms of income were listed on one side of the ledger and all forms of expenditure on the other; the two columns of figures were totalled once or sometimes twice a year and a balance 'struck'. This meant that Brown knew overall whether he was making or losing money, but it was impossible to attribute a single expenditure to any one project or source of income; a profit or loss could not be calculated for a particular job, which must have made it very difficult for him to estimate the costs of future commissions. We know that he did provide estimates, even if some of his clients were essentially willing – and rich enough – to pay whatever he asked. But the estimates must have been very rough and ready.

How did Brown cope with these problems? Four strategies seem to have been particularly important: subcontracting, ensuring high profits, maintaining a good bank balance and liquid investments, and – for the latter part of his career – being able to rely on a regular income from King George III. Brown was not unique in adopting such strategies, but he certainly seems to have been successful in deploying them.

Brown's projects were dominated by earth-moving and water engineering, as were other large projects of the period, such as river improvement, the drainage of the Fens, canal-building and road and bridge construction. He used the same methods. Subcontracting was increasingly common in building and civil engineering in seventeenth- and eighteenth-century Britain. Essentially, any large job was split up into 'bite-sized' pieces. When Sir Christopher Wren rebuilt St Paul's Cathedral after the Great Fire of London, he parcelled the work into small contracts so that he could keep close control of them. In the 1730s, when Westminster Bridge was built, each of the fourteen piers was the subject of a separate contract. Later in the century, the Lancaster Canal had as many as thirty-five subcontractors on a stretch of thirty miles. In each case, a subcontractor – with the skills to carry out the specific task – would undertake it either for a fixed sum or on a cost-plus basis, charging for materials and labour and adding on a profit.[13]

The advantage of subcontracting was that it spread the risk. If one contractor was unsatisfactory, or couldn't manage his workforce, or got his finances wrong and went bankrupt, he could easily be replaced. The risk could be reduced, also, by choosing reliable subcontractors: Brown built up long-term relationships with 'the capability men', as the garden historians David Brown and Tom Williamson have recently dubbed them.[14] Some, such as Nathaniel Richmond, went on to successful careers of their own as garden designers. Others, such as Jonathan Midgeley and Cornelius Dickinson, who may have been particularly expert in water engineering, worked with Brown for several decades, while yet others moved – on Brown's recommendation – to become head gardeners of some of the greatest gardens: Benjamin Read went to Blenheim and Michael Milliken to Kew, where he was commended by George III. Sometimes called 'Brown's foremen', they are probably best described as his subcontractors.[15]

The downside of subcontracting was that it required great attention to quality control, which may account for the incessant travelling to different sites for which Brown was famous. Someone must also have checked the bills that the contractors submitted; the 'master and servant' system in the great estates relied on the submission of vouchers, detailing expenditure, which are still preserved in enormous quantities in the archives of estates such as Chatsworth or in local record offices. Brown must have operated a similar system, though apparently few of his vouchers have survived.

Once the vouchers had been checked, Brown would have had to pay. This meant that he needed to have the right amount of cash in his bank account. This would not usually have been a problem, because Brown made large profits and then kept them available in a liquid form – they weren't all tied up in land or other investments that could not be sold quickly. Careful study of the account book makes it possible, though not easy, to estimate his profits. If we extract all his household expenses, his payments to family members and his investments, and also 'pay' him a salary, the profits that he made on his business come to about £139 million in modern values, which when set against total receipts of £840 million gives a profit rate of about 17 per cent.

This may seem high by modern standards (which, in similar industries today, may be 12 per cent or less), but it was the norm for eighteenth-century businessmen. This was because of a relative scarcity of producers – tradesmen and contractors – who faced, as Brown did, high demand for their services. But it was also a reasonable reward for the high risks involved. There was no limited liability, whereby an investor cannot lose more than the money put in; that was not introduced until the middle of the nineteenth century. So Brown, like other businessmen, was personally liable for any debts and, if he couldn't raise the money to pay, the debtors' prison awaited.

Brown's profits were substantial, but they were also very variable. There were peaks of nearly £25 million in modern values in 1768 and 1774. In the first of these years, his clients included the Earl of Bute at Luton Hoo, the 4th Duke of Marlborough at Blenheim, Lord Palmerston at Broadlands and Francis Herne at Flambards; in 1774, it was Marlborough and Palmerston again, together with the Earl of Donegal at Fisherwick, Lord Clive at Claremont and Lord Craven at Coombe Abbey. But there were also years of losses of a million or two, in 1772 and 1779. So Brown had to make sure that he kept liquid assets, built up in the good times, to meet his obligations in the leaner years.

This was not nearly as easy as it would be today. There were far fewer safe havens for money. Banks, such as Drummonds, where he kept his account, were entirely private and funds deposited in them were not protected from default or fraud on the part of the banker; many private banks around the country failed. If a rumour spread that the banker was in trouble, depositors rushed to withdraw their funds and those who got there too late lost all their money. So it was unwise to keep large deposits in a bank, although actually Brown had to do so because he had to make so many payments. Some nabobs, such as his client Clive of India, were reputed to keep their ill-gotten gains in diamonds, but this was unusual. There was no opportunity for safe and liquid investment in commerce and industry, since enterprises, with the exception of a few chartered companies, could not issue tradable shares; partnership in them, the only way of investing, was risky and inflexible.

There were two kinds of investments that offered both safety and some flexibility, and Brown used both of them. They were investment in land and in government stock. In 1767 he bought Fenstanton Manor, a property of 1,000 acres, from the Earl of Northampton for £20.9 million. This purchase has often been interpreted as social climbing, but Brown was already well advanced in terms of social status; much more likely, particularly as he never lived there and did not make alterations to the estate, is that it was a good investment, at a time of rising land prices. Moreover, if he had needed money, it was reasonably easy to raise a loan secured on such a property. In addition, he could expect to receive a steady income, perhaps of £1 million annually in modern terms, from rents. The estate provided an annuity of £579,000 for his widow Bridget before it passed on her death to their eldest son, Lancelot Junior.

Much more liquid than land as an investment, however, were government bonds or consols. Brown seems to have invested most of his profits in consols, his peak holding being £13.8 million in 1774. His bank account shows that he also invested heavily, for a short period, in the more risky bonds of the East India Company, but cannily avoided the crash of the early 1770s, which ruined many of his fellow investors.

Brown's final defensive strategy was to obtain an appointment as one of the royal gardeners. This would give him access to part of the £11 million or more that the royal gardens were costing annually. He succeeded at the second attempt. The first had been in 1757–8, when William Pitt the Elder, Secretary of State for the Southern Department (foreign secretary, in modern terms) at the time of the Seven Years War, still found time to persuade thirteen other members of the 'great and the good', all 'well-wishers of Mr Browne', to apply to the prime minister, the Duke of Newcastle, for Brown to be appointed 'to the care of Kensington Garden'.[16] They included Pitt's two brothers-in-law, Richard Temple of Stowe and George Grenville of Wotton, where Pitt had apparently worked at garden design with his wife, Hester (Temple), and with Brown. Despite this support, the Duke of Newcastle refused the application.

In 1764 Brown tried again. George Grenville was by now First Lord of the Treasury (prime minister) and acted quickly to appoint Brown at Hampton Court, though not – as he had wanted – at Windsor. The post was well worth lobbying for. Brown was not *the* royal gardener, as there were five, but he still received £2.1 million a year, together with a house – Wilderness House, which survives – that became the base for his business for the rest of his life. By his death, he had received £54 million for his work, with a further £25 million for work at Richmond and Kew. He did have to pay for the upkeep of the garden at Hampton Court and for the works at Kew, but other evidence shows that royal gardeners could expect a profit of at least 30 per cent, so during his tenure he probably reaped £25 million from the public purse. Equally valuable, possibly – since he was not short of lucrative private commissions – is the fact that the post of royal gardener gave him a regular income with which to smooth out the variable income stream from his other jobs.

What did Brown do with his money? To the disappointment of some, he seems to have been a model of rectitude. He did spend £184,000 *(£115 2s 6d)* on lottery tickets, though some of this – it has been suggested – may have been for charity; there is no sign of the gambling debts that appear in the accounts of so many aristocrats at the time. One biographer has suggested that he may have had an illegitimate daughter, the result of a liaison before his marriage, when he is said to have spent time in Lincolnshire learning water engineering, but the evidence is not strong.[17] There is certainly no sign in the accounts of any payments to her – if she existed – although they could have been concealed in the substantial sums that Brown withdrew for himself. Otherwise, about £35 million was given during his lifetime to his family, including his son-in-law Henry Holland Junior and his nephew, with a great deal to his two sons, Lancelot Junior, who after

Brown's death inherited Fenstanton, and John, who received another, smaller, Lincolnshire estate.

BEFORE BROWN

Capability Brown emerges from his account book as a canny and successful businessman, both in making and then in safeguarding his fortune. He deserves a lot of attention, both for his success and because of the unique information that we have about him. But how unusual was he? Has landscape gardening been a route to riches for many others? No one else has left such detailed accounts, but there are other indications of how well they did.

The greatest of Brown's predecessors were George London and Henry Wise. They created many of the formal gardens of the late seventeenth and early eighteenth century that are celebrated in *Britannia Illustrata*. Longleat, with its avenues of trees stretching to the horizon that are shown in plate 10, was just one of them. Like Brown, they worked for the royal family and for many of the leading aristocrats and politicians of their day. By contrast to him, they were not just designers, but also nurserymen. They were based at the famous Brompton Park Nursery, referred to in Chapter 1, reputed to have covered 100 acres in what is now South Kensington in London and to have possessed, at its peak, 10 million plants. Such claims may be exaggerated but there is no doubt that both London and Wise were very successful. Thousands of trees and shrubs were sent to the gardens that they and their partners designed – Longleat, Badminton, Chatsworth, Blenheim, Wanstead – to the gardens of the royal palaces and to other estates across the country. George London toured – he is said to have spent six months of each year in the saddle, sometimes riding up to sixty miles a day – supervising contracts and collecting his fees; he also acted as deputy to the Earl of Portland, who was responsible for the royal gardens under William III. Henry Wise looked after the business in London and, from 1702, as royal gardener, was responsible for Hampton Court, Windsor, Kensington and St James's, latterly in partnership with Joseph Carpenter and then Charles Bridgeman. Plate 11 shows the detailed contract that he signed and to which he frequently referred when asking for extra payments.

Born in c.1640, George London is a rather mysterious character. He worked for John Rose, an early royal gardener to Charles II, and was sent by him to study French gardens, including the work of André Le Nôtre at Versailles, which many of his gardens sought to echo; he later visited the Netherlands. In 1688, as gardener to Henry Compton, Bishop of London, he famously helped his employer to spirit away the future Queen Anne from her father James II's court; his 'healthy, strong constitution' was clearly valued by Compton, who had a garden at Fulham Palace with over a thousand species of 'Exotick Plants in his Stoves and Gardens' and was an active botanist and gardener.[18] But apart from the fact that London married three times and had five children, including a daughter who was a botanical artist, and that he died in 1714, little is known about him.[19] However, there is no reason to think that he was less successful in business than his partner, Henry Wise, and his will shows that he owned a house and garden in Spring Garden, in central London, and farms in Surrey as well as an 'interest in Lead or other Mines in Wales called Sir Car-bury Price's mines'.[20]

The life and work of Wise, who had been London's apprentice and worked with him for thirty years, is much better documented.[21] He was born in Greenwich, then outside London, in 1653 and became a partner in the Brompton Park Nursery in 1687, but the first indication of his financial success is the house that he occupied on the nursery site after his marriage in 1695, which was transformed into a small mansion of twenty rooms. By 1718, when an inventory of the house was made and entered into Patience Wise's household book, it was full of furniture and pictures, including a portrait of Wise by Sir Godfrey Kneller and many works by Antonio Verrio, who had been court painter to the later Stuarts. Nine years earlier, in 1709, Wise had also bought a country estate, Warwick Priory. He paid £21.9 million (£10,601 17s) for the house, estate and other land around Warwick and Leamington and in Surrey. Unlike Brown's Fenstanton, the Priory was to be Wise's retirement home after 1727 and he redesigned the garden to his taste, including a parterre, plantations, avenues and geometrical pools. He died there in 1738 and – aside from debts owed to him by the new owners of Brompton Park Nursery – was said to be worth £359 million *(£200,000)*, although his biographer prefers the more reasonable £179 million *(£100,000)*.[22] Even at that smaller sum, Wise was probably as wealthy as Brown was to become and as George London may well have been too.

Another indication of the potential for making money from landscape and garden design in the early eighteenth century comes from the career of Stephen Switzer, a former apprentice and great admirer of George London. Born in 1682, Switzer was both a designer and nurseryman, but his main claim to fame is as a writer of gardening books and as proprietor – though for only a year, in 1733–4 – of one of the earliest garden magazines, *The Practical Husbandman and Planter*. He possessed nothing like the reputation, nor the client list, of London or Wise, but was none the less capable of designing and managing large projects: a contract of 1718 for work for Lord Cadogan at Caversham Park, Oxfordshire (now Berkshire), envisages a summer workforce of 170 and expenditure of £2.6 million *(£1,392 4s 9d)* with a substantial further contingency of £1.6 million *(£836 7s 11d)*, illustrating how difficult it was to make accurate estimates.[23] Switzer worked at Castle Howard but his most significant projects were at Grimsthorpe in Lincolnshire and Cirencester Park in Gloucestershire, applying a more naturalistic style than that of his mentor, George London. At Cirencester, working for Earl Bathurst and in conjunction with the essayist and garden fanatic Alexander Pope, he developed large sweeps of woodland intersected by broad avenues. Switzer's career seems to have continued successfully, partly because he developed a lucrative career as a seed supplier; his Hoare's bank account for 1743, two years before his death, shows an average balance of £2.2 million *(£1,241 7s 6d)*, although it had fallen to just under £1 million *(£545 6d)* by the time he died.[24] We know nothing about any property or investments that he may have had.*

Switzer's contemporary Charles Bridgeman, born in 1690 and another collaborator of Henry Wise, was a much more distinguished designer than Switzer, responsible for some of the most beautiful parts of Stowe garden and for its overall planning. He also worked at

---

* Switzer's work with water and steam in the garden is discussed in Chapter 7.

Blenheim and many other great estates, including Rousham and Claremont, and is (possibly) responsible for the widespread use in those estates of the ha-ha, a ditch with a vertical wall on one side that prevents animals in the park from entering the pleasure garden, while not obstructing the view; it is often attributed to Capability Brown but Bridgeman has at least an equal claim. Bridgeman and Wise had an annual contract of £5.2 million *(£2,960)* from 1726–8 as the royal gardeners and apparently made a yearly profit of £1.1 million *(£555 10s 5½d)*, shared equally.[25] Then, from 1728 to his death in 1738, Bridgeman was the sole gardener, on £4 million *(£2,220)* a year, no doubt with a similar profit of close to 20 per cent. Yet while Wise became wealthy, Bridgeman left his widow Sarah penniless, at least according to her letters to the Treasury and to Sarah, Duchess of Marlborough, for whom he had worked at Blenheim and was still owed money for the work. However, just before his death, Bridgeman left a will in which £10.9 million *(£6,000)* was bequeathed to his three daughters and £545,000 *(£300)* and the residue of the estate to his wife. His biographer thinks he may have had an overinflated sense of his financial worth, but his widow's pleas to the Treasury – notoriously hard-hearted and bad payers – extracted £9.1 million *(£5,000)*, despite her admission that Bridgeman had not kept regular accounts. Sarah Churchill proved more difficult to move and threatened to bring a lawsuit if Sarah Bridgeman did not stop badgering her, but £960,000 *(£539)* was paid by the Duke of Bedford at Woburn, Bedfordshire, and probably more came from other clients. So Bridgeman may not have been such a failure as a businessman as he is usually portrayed.[26]

A note of caution here: wills can be misleading. People can say anything in them. They can, as Bridgeman's biographer thinks he did, overestimate or overstate their assets. On the other hand, it was normal in the eighteenth century for men to make their wills when they felt death approaching and when they might have a clear view of their likely assets when it came. But a much more common difficulty is that wills give specific bequests – as Bridgeman did to his daughters and Brown to his wife and children – but do not give the value of the residue of the estate, the sum left over after all the debts and funeral expenses have been paid and the specific bequests honoured. The residue can often be a substantial sum. So wills tend to understate the value of an estate, though this appears unlikely in Bridgeman's case.

A further problem are lifetime gifts, such as those which Brown gave his children and which are very common today. These usually don't appear in wills, nor in another source for wealth at death, the probate value.* That seems, otherwise, to be a better source for assessing the wealth of a historical figure, not least because probate values were usually published. But even when we know the probate value – as we sometimes do in the eighteenth century and more usually in the nineteenth – it is misleading as a guide to the total wealth of an individual because, until 1898, it did not include the value of someone's house or other landed property. So in general it is fair to conclude that most valuations before the twentieth century – and probably some since – underestimate the scale of someone's wealth at death. This is worth bearing in mind.

AFTER BROWN

Probably the most-read and most-quoted money value in the entire economic history of gardening appears in in the sixth chapter of Jane Austen's *Mansfield Park*. Mr Rushworth, the fiancé of Miss Maria Bertram and owner of Sotherton Court, 'had been visiting a friend in a neighbouring county, and that friend having recently had his grounds laid out by an improver, Mr Rushworth was returned with his head full of the subject and very eager to be improving his own place in the same way; and though not saying much to the purpose, could talk of nothing else'. When Miss Bertram comments: 'Your best friend upon such an occasion . . . would be Mr Repton, I imagine', Mr Rushworth responds: 'That is what I was thinking of. As he has done so well by Smith, I think I had better have him at once. His terms are five guineas a day.' The odious Mrs Norris, who has an

---

* Today, to be precise, the value of lifetime gifts is included in English probate values, and subject to inheritance tax, on an amount that declines from 100 per cent of the gift immediately after it has been made to zero after seven years. This was not a problem in the eighteenth century, yet lifetime gifts sometimes do appear in wills when the maker wants to explain his bequests.

opinion on every subject, then chimes in: 'Well, and if they were *ten*, I am sure *you* need not regard it. The expense need not be any impediment. If I were you, I should not think of the expense, I would have everything done in the best style, and made as nice as possible. Such a place as Sotherton Court deserves everything that taste and money can do.'[27]

This passage has given the real-life Humphry Repton (1752–1818) literary immortality. But it has also had the unfortunate effect of implying that he came cheaply. Even if modern readers know what a guinea was – £1 and 1 shilling or £1.05 – they are very unlikely to have made the translation to modern values. Repton's quoted daily rate in 1814, when *Mansfield Park* was published, is equivalent to £3,800 today, a sizeable consultancy fee; expenses were extra. But in fact he normally charged more – 10 guineas for a first visit within one day's stagecoach ride of London, 50 guineas for up to 100 miles, 70 guineas for up to 140 miles, and so on.[28] The Duke of Portland paid him about £87,000 *(£100)* a year for twenty years for his work at Bulstrode in Buckinghamshire.[29] In 1814 he was paid £150,000 *(200 guineas)* by the Duke of Bedford for a visit to Endsleigh in Devon and got £22,500 *(30 guineas)* more for a consultation in Bristol on the way. In his early years, Repton amassed more than ten new consultations each year – in 1790 nearly thirty – and he estimated that for twenty years he travelled more than four thousand miles each year.

Repton's main marketing tool was his Red Books, over 120 of which survive.[30] Most contain 'before and after' watercolour pictures of a landscape – as it was, and as it would be if Repton's designs were put into practice – together with an explanatory text and detailed instructions. He charged his clients between £6,000 *(6 guineas)* and £40,000 *(£42)* for them,[31] and often used the designs again as illustrations in his several books on landscape gardening. His commissions tailed off in later years, and many of them were for middle-class bankers, lawyers and manufacturers, rather than aristocrats, but an income from his books continued. He was certainly at the top end of the market – Richard Woods, 'the master of the pleasure garden', was charging at the same period £2,200 *(2 guineas)* a day[32] – and, although Repton's letters show money worries, he was able to maintain the lifestyle of a country gentleman.

This seems to have been true of the other fashionable garden designers who followed Repton over the course of the next two centuries. Let's pick a few of them. One outlier is Joseph Paxton, head gardener at Chatsworth in the middle of the nineteenth century and architect of the Crystal Palace built for the Great Exhibition of 1851. Born in 1803, he was an innovative gardener, working with the 6th Duke of Devonshire to bring plants to Derbyshire from all over the world and designing greenhouses – including the so-called Great Stove – to house them. He used his novel technology to build the Crystal Palace, although his attempt to license his metal and glass greenhouses for the masses was less successful. Particularly after the death of the duke, he became a designer and architect for huge houses such as Mentmore in Buckinghamshire and the Château de Ferrières in France, as well as a number of public parks. He used his growing wealth to engage in railway speculation, formed a corps of navvies to support British troops in the Crimea, became a Liberal Member of Parliament, was given a knighthood and died, in 1865, worth £112 million *(£180,000)*.

By the end of the nineteenth century and into the twentieth, the 'bulk trade' in garden design, for both private owners and public authorities, was dominated by two firms – T. H. Mawson & Son and Milner-White & Son – although individual designers such as Gertrude Jekyll continued to cater for the 'carriage trade'. In addition, and emulating Brompton Park, large nurseries such as Veitch and, later, Hillier began to develop design businesses as adjuncts to their main activities; they will be discussed in the next chapter.

Thomas Hayton Mawson was born in 1861, brought up in a poor family with a small nursery and fruit farm in Lancaster. He worked for a number of nurseries in London and then set himself up, with difficulty, as a landscape architect – as he wanted to be called – in the Lake District, Cumbria. He capitalized on the development of lakeside estates by rich northern manufacturers and then moved successfully into the design of public parks. His greatest gardens were probably for the soap magnate W. H. Lever, later Lord Leverhulme and a major philanthropist, at Thornton Manor in Cheshire, Rivington Terraced Gardens near Bolton and The Hill on the brow of Hampstead Heath in London. Mawson's clients included Queen Alexandra and the

Maharaja of Baroda, but he and his firm also made many smaller gardens. He developed a big international practice, in Canada, the USA and increasingly in Greece; he designed the gardens of the Palace of Peace in The Hague, Holland, worked on royal gardens in Athens and led the reconstruction of Salonika (Thessaloniki) after its disastrous fire in 1917. He championed, in his books and by supporting Liverpool University, the cause of education for landscape architects and designers. He died in 1933, leaving £622,000 *(£3,553).*[33]

Edward Milner was born in 1819 on the Chatsworth estate and became an informal apprentice of Joseph Paxton; he worked with him in Liverpool at Prince's Park, one of Paxton's early public commissions, at Tatton Park and then on the huge Italianate gardens on the new site of the Crystal Palace in Sydenham, London. He was in charge of the geometrical planting, using combinations of yellow with scarlet and blue with purple, which employed about 450 gardeners, as well as of the waterworks, which included an artificial tidal lake, two water towers, a reservoir and ten miles of piping that powered nearly twelve thousand fountain jets. The largest of them produced a jet of water 280 feet high.[34] After the 6th Duke of Devonshire died in 1858, Milner set himself up in private practice as a designer, particularly of public parks; he had apparently learned from Paxton that 'public works could both provide more work and be more financially rewarding' than private commissions.[35] He also established a school of gardening within the Crystal Palace School of Practical Engineering and trained Fanny Wilkinson, who became the first professional female landscape gardener in the 1880s.[36] The firm grew substantially under Edward's son Henry and grandson Edward and, when joined by Edward White in 1902, was employing 'many hundreds' of men as Milner, Son and White.[37] They seem at least at first to have had even greater financial success than Mawson but this gradually tailed off: Edward Milner died in 1884 worth £3.7 million *(£8,191),* Henry in 1906 leaving £1.3 million *(£3,693),* although Edward White was worth only £85,000 *(£1,282)* when he died in 1952.

The two most influential independent garden writers and designers of the late Victorian and Edwardian periods were William Robinson and Gertrude Jekyll. Robinson is an enigmatic character: born in Ireland in 1838, he started out as a garden boy but moved to London in 1861 to work in the garden of the Royal Botanic Society and, surprisingly quickly, became prosperous. How he achieved this is not known. By 1868 he had a house in Kensington and was able to hazard £513,000 (£900) on publishing his first book, *The Parks, Promenades and Gardens of Paris* the following year. But his real wealth seems to have stemmed from the magazines that he founded, *The Garden* and, particularly, *Gardening Illustrated*. Robinson used his publications to advocate what became known as 'wild' gardening, in reaction to the formalism and vivid colour combinations of Victorian carpet-bedding; he recommended the use of hardy perennials rather than annual plants and also popularized alpines. Although the circulation of his magazines is not known, it must have been enormous, since by 1885 Robinson was able to buy a fourteen-bedroom Elizabethan house with 360 acres, later extended to 1,100; this was Gravetye Manor near East Grinstead in Sussex.[38] He never married and his biographer suggests that his death in 1935 was caused by syphilis, although he managed to live to the age of ninety-eight;[39] he left £16 million (£95,954).

Born in 1843, Gertrude Jekyll admired Robinson. By contrast with him, she came from the prosperous upper-middle class – her father was an army officer – was trained as an artist and gained her initial experience in redesigning her parents' estate; she later moved nearby to Munstead Wood, in Surrey. Her background gave her a network of contacts and this, together with her collaboration with the architect Edwin Lutyens – and her very evident skills and plantsmanship as well as her prolific garden journalism – made her the most successful and famous designer of the first quarter of the twentieth century. Her works with Lutyens, many of which survive in at least attenuated form, were one of the greatest achievements of the Arts and Crafts movement of the period. Like Robinson, she laid her emphasis on the plants rather than on formal layouts. One garden writer thinks she was as influential as Capability Brown.[40] She 'advised on the laying and planting out of about 350 gardens, of which about 120 were collaborative projects with 50 architects'.[41] She also – unlike Brown – wrote about her design philosophy and this meant that her influence spread even more widely. She was herself a plant hunter – on Capri in the Bay of Naples, for example – and established her own nursery, supplying plants for her own designs and for general sale; she forfeited her amateur status at plant shows and undercut her rivals, showing herself as a shrewd, if rather old-fashioned, businesswoman at a time when 'well brought up' girls were not expected to enter into trade. She died in 1932, leaving £3.5 million (£20,091), but it is not clear how much of this was inherited from her parents and how much was made by her work in garden design.

Finally, as a further example among many, there is Lanning Roper. Born in 1912, he was the son of a Wall Street banker from a family of American 'aristocrats' – his father's family had arrived in the colonies in 1637 and his mother's before 1647. He studied fine arts at Harvard and then spent a year on an architecture course at Princeton before deciding he wanted to make landscapes rather than buildings. Summers were spent at Newport, Rhode Island, among some of the greatest of American gardens and their wealthy owners. A volunteer in the Second World War, he joined the navy and commanded a landing craft on D-Day before working on shore at a naval headquarters established in the Rothschild house and garden of Exbury, Hampshire. His background gave him an entrée into English society and a close friend in the 4th Marquess of Normanby, lord-in-waiting to King George VI.

Roper decided to stay in England after the war and, later, studied at Kew and then the Royal Botanic Garden, Edinburgh, before beginning in 1950 to write, for *Gardening Illustrated* and then *Country Life*. He worked for the Royal Horticultural Society and, after publishing *Successful Town Gardening* in 1957, became an independent writer and designer, the career he followed until his death in 1983. He wrote, notably for the *Sunday Times*, and became the most fashionable of garden designers, often working with the great nurseryman Harold Hillier. His client list was not quite as stellar as Brown's, but it included, as well as the Prince of Wales, work for the Aga Khan at Aiglemont in France, for Lord Rothermere at Daylesford, Gloucestershire, for the Sainsbury family at their private houses and at the University of East Anglia, and for Michael and Anne Heseltine at Thenford, Northamptonshire.

Roper had family money and does not seem to have been very concerned to make more; he was gentle in asking to be paid, writing to one client when enclosing an invoice: 'Unfortunately, I am at one of those moments when everything is going out and nothing is coming in. We've all experienced it, but I must say I find it slightly alarming.'[42] He lived comfortably, not ostentatiously, particularly after the break-up of his marriage; much of his time was spent with clients as their guests. He had collaborators but not employees, apart from a secretary. Invoices to his clients have not been preserved among his papers, but in 1977, to work at Thenford for the Heseltines, he quoted a fee of £1,035 (£120) per day plus travelling expenses – modest by the standards of Brown or Repton centuries earlier. Similarly, he was paid £1,886 (£460) for a few days' work at Highgrove in 1981 before his final illness, while the fee for the much larger project for the Sainsbury Arts Centre at the University of East Anglia was £33,800 (£3,500). So it is not surprising that his probate value in 1983 was only £318,000 (£77,658), although he also had assets in America.

DECLINING INCOMES

Over the course of 350 years, therefore, landscape and garden design seems to have become less and less lucrative. Great fortunes were made in the seventeenth and eighteenth centuries, and by Paxton in the nineteenth, but not in recent decades. In part, this reflects a decline in what economists call 'rents', payments for scarcity value that stars such as Brown and Repton were able to attract. It is also likely that it stems from a change in demand, as the aristocratic clients of the 1700s were gradually displaced, in the nineteenth century, by manufacturers, bankers and the wider professional classes. Their estates were smaller and their pockets less deep than those of eighteenth-century grandees and they were, partly as a result, more careful with their money, less willing to give carte blanche to their garden designers. They continued to spend, as the Rothschilds did in the late nineteenth century, and indeed a surprising number of country houses have been built and their grounds landscaped in the twentieth century and now in the twenty-first, with many more gardens being renovated. But the new owners, and bodies such as the National Trust, were more likely to question profit levels, keener to hold contractors to their estimates. Mawson found this in the late nineteenth century and it was one reason why he turned, as did his competitor Edward Milner and later designers, to public authorities as they built parks and cemeteries. This continues: by far the largest and most expensive garden design project in recent years was the 2012 Olympic Park in east London.

A further reason for declining personal profits has been the changing structure of garden design. London, Wise and Brown employed many thousands of men, though principally through subcontractors. Repton did not and he was the forerunner of many, through Jekyll to Roper, who produced designs for their clients to execute, sometimes helping with sourcing plants and ornaments, sometimes doing some planting on site, but mainly leaving it to clients and their gardeners. Meanwhile, the larger projects, and particularly those for public authorities, became increasingly the responsibility of the bigger partnerships, like T. H. Mawson and Milner-White, and then of private and public companies. Planning and employment laws, and health and safety regulations, also demanded larger teams to meet their requirements.

LIFE AND WORK

Even if London and Wise, Brown, Repton, Paxton and many others made a good living out of landscape design, they worked hard for their money. They had to be expert – or know how to buy expertise – in a wide range of disciplines: horticulture, surveying, arboriculture, water engineering, botany, architecture. They had to travel endlessly, to seek commissions, to make designs, to choose subcontractors, to supervise works, to placate clients. Their incomes were irregular, dependent on acquiring and satisfying demands that could and did change on a whim. Their reputations were only as good as their last successful project. When they were starting out in business, they were very much on their own; if they became successful, they had the burden of knowing that hundreds, or even thousands, depended on them – Wise is reputed to have had 1,500 men working for him on the Blenheim gardens. It was not an easy life. Richard Woods, Brown's contemporary, describes it well:

... considering how many broken days, in a year I have, for example, take out Sundays, many days ill by getting colds etc. how many days & nights, in Town at expences, mearly to wait on Gentlemen without ever charging anything for it, how many days in a year are spent at home, only in answering letters . . .[43]

This way of life took its toll. Repton's travels – 'I flew from Cornwall to Cumberland – or to Kent – and from Hampshire to Derbyshire – and found no rest for the soles of my feet' – brought alarming symptoms of stress, which his doctor put down to his paying 'too much attention to business'. Since the remedy of 'visiting new scenes to amuse the mind' was hardly appropriate, the prescription was for 'rest and some total change of habits'.[44] But this was exactly what a self-employed consultant could not do, and Repton was forced to continue his exacting travels even after a carriage accident in 1811 – returning with his daughters from a country ball – left him ill and ultimately in a wheelchair.[45] Joseph Paxton's doctor similarly feared a 'brain fever' brought on by his frenetic activity.[46] By the end of the nineteenth century, the railway made it possible for Thomas Mawson to travel 20,000 miles a year, but the strain was much the same: long journeys, writing letters on the train, working late on a folding portable drawing board in a country inn. As Mawson sought, in addition, to establish himself as an author, the self-doubts that seem to have plagued him became overwhelming:

The strenuous life which I was living would not allow of this additional strain, already heightened by the enthusiasm which I felt for my work. The result was a breakdown which laid me aside for many weeks during one of the most critical periods of my career. Success was bringing its penalties as well as its rewards.[47]

Mawson was coping with the expenses of a wife and six young children and had to 'practise rigid economy',[48] but the marriage seems to have been a happy one. Brown's wife, Bridget, and Repton's wife, Mary, had to cope with similarly workaholic husbands; Mary Repton, at least, assisted with Repton's business, such as keeping track of his travels and forwarding letters to where he was next likely to be. Sarah Paxton was even more important, essentially running the garden at Chatsworth when her husband was travelling with the duke or working elsewhere, managing his business interests. The impact on other marriages was more damaging: Lanning Roper's biographer ascribes his separation and divorce from his wife, Primrose, in 1965 to his work schedule. With no warning, she packed and left their flat while he was away working, 'but Primrose's friends only wonder that he did not see it coming'.[49] The strains could show in other ways: Paxton's son George became increasingly unmanageable from the age of nine, was prone even as a teenager to gambling, womanizing and swearing; he finally added drunkenness and, allegedly, fraud to his bad behaviour, before eloping with an apparently unsuitable wife. His childhood behaviour, at least, was attributed to his father's frequent absence.

Clients could be capricious, demanding and unreasonable. As we saw in Chapter 1, Wise had to cope with William III's command to lower the level of the Privy Garden at Hampton Court, after it had been entirely planted and furnished with fountains. Repton arrived expecting to stay with a client for four days and was met with the news that the client had departed for London to consult a doctor and no longer needed Repton's services. With another client he carried on a lengthy correspondence that became increasingly acrimonious, with the result that he wasn't paid for two years' advice. Thomas Mawson described Sir William Cunliffe-Brooks of Glen Tanar, Aberdeenshire, as 'the most generous and yet the most tyrannical client for whom I have ever worked'.[50] Days began at 7 a.m., when Cunliffe-Brooks's piper perambulated the house, followed by breakfast and prayers, until work began promptly at 8.30 with

> the most minute discussion of every detail of plan and construction, making any alteration which suggested itself to his fertile brain. This was one of the difficulties experienced by every adviser, whether architect or engineer. There was no finality. A poor man cannot afford to change a contract; a rich man can do so as often as fancy suggests.[51]

Lanning Roper would have concurred; he had to keep his temper when clients changed their minds, failed to explain what they wanted or, worst, left matters to their staff, who then argued over the plans and the costs.

Thomas Mawson hit another problem: the client who demanded his undivided attention. He first met Lord Leverhulme in 1905; Lever was an extremely successful businessman, founder of the conglomerate that still survives as Unilever. Mawson was fund-raising for his local nonconformist church and wrote to Lever asking for a contribution. Lever obliged and then asked Mawson if he would

> advise me upon the improvement of my garden at Thornton Manor? I have wanted to consult you for the last two years, but all my friends warned me that it would be useless, as you never worked for anyone holding less social rank than a Duke, whereas I am only a poor and indigent soap-maker.

This false modesty was belied, as soon as Mawson met him, by his 'strong personality . . . a veritable Napoleon'. His scheme for Thornton Manor was of 'heroic proportions' and it was followed, over the years, by Rivington Terraced Gardens, Bolton – Lever's native town – and The Hill, Hampstead, which incorporates an enormous pergola, part of it shown in plate 12, over 750 feet long and over 20 feet high, constructed on a brick-encased mound made with earth excavated from the Northern Line; Lever characteristically managed to be paid for taking the soil. Mawson saw Lever not only as his best client but as a friend. But in 1912, just before Mawson was to sail to America to undertake some commissions, leaving his projects in the hands of his sons, Lever wrote abruptly dismissing him, complaining that 'his interests were not promoted with that assiduity which he had come to expect of [Mawson]'. Mawson replied, explaining – quite reasonably – that he wanted to give his sons 'wider opportunities and greater responsibilities by removing for a time [his] personal influence'. In true Napoleonic manner, Lever did not apologize but behaved as if nothing had happened.[52]

In most cases, garden designers seem to have absorbed criticism from their clients, even to the point of writing off their fees; it clearly didn't pay to get a reputation for being difficult. But sometimes they couldn't restrain themselves. Brown was tactful and diplomatic and rarely fell out with clients. But, like all the royal gardeners, he argued with the Board of Works over the scope of his duties in the contract and about the state of the gardens when he had taken them over. His famous patience snapped when he received, in 1770, a letter from the board stating that they had

> found to their great surprise His Majesty's Gardens at Hampton Court by no means in so good a condition as they ought to be according to your contract for keeping the same and finding notwithstanding the time elapsed since the Survey that nothing material has been done towards putting them in proper order but that the complaint against the manner of keeping them is still general, none of the Walks being fit for use, and most other parts of the Gardens most neglected.

To which Brown replied:

> I recd. your Letter and must acknowledge to you that I have lived long enough not to wonder at anything, therefore it did not surprize me, I believe, I am the first King's Gardener that the Board of Works ever interfered with . . . I seldom use epithets otherwise I would translate that Letter as it deserves because I know both the Author's meaning and his Conduct on that subject . . . I have stopped at no Expence in procuring Trees and Plants, nor grudged any number of Hands that were necessary and this day went through the Garden and my Foreman told me that they had more hands than they knew how to employ. But you, Sir, have only done your duty. You will be so good as to inform the Gentlemen of the Board of Works That Pique I pity, That Ideal Power I laugh at, That the Insolence of Office I despise and that Real Power I will ever disarm by doing my duty.[53]

PROFESSION OR JOB?

What was that 'duty'? Deep in the woods that make up much of the Great Garden at Wrest Park is a column with an inscription: 'These Gardens, originally laid out by Henry, Duke of Kent, were altered by Philip, Earl of Hardwick, and Jemima, Marchioness Grey, with the professional assistance of Lancelot Brown, Esq., in the years 1758, 1759, 1760.' What do these words tell us? Was Brown 'assisting' his clients, as a friend might do, or was he exercising a 'profession'; he is described as 'Esquire', but what status did he – and his predecessors and successors – really have? Were they 'professionals' – with the additional sense of exercising a vocation within a creative field, as words such as 'genius', which are regularly applied, might suggest – or merely tradesmen?

Several books have been written about the evolution of the professions in Britain.[54] It is an important aspect of the growth of the service or tertiary sector – professionals are the leaders of the knowledge economy. It is also a very confusing and muddled process, in both terminology and history. Even today, 'professional' can have the very wide meaning of 'someone who is paid', as in 'professional dancer' or 'professional cricketer'; the implication is slightly pejorative. It can mean a member of a profession, self-regulated or regulated by law; or it can refer to a code of behaviour, a set of written or unwritten rules that a 'professional' is expected to follow. It can also signify a place in the class system, part of the upper-middle class, someone who earns a living by selling knowledge rather than manual labour but is not 'of independent means'. As the novelist Anthony Trollope put it: 'The word was understood well enough throughout the known world. It signified a calling by which a gentleman, not born to the inheritance of a gentleman's allowance of good things, might ingeniously obtain the same by some exercise of his abilities.'[55] However it is defined, to attain the status of a 'professional' has been desirable since the eighteenth century or before.

Until the nineteenth century, there were essentially only three recognized 'ancient' or 'learned' professions – the Church, medicine and law – although the term was not confined to them; it could be a synonym for 'occupation', although it was usually applied only to the middle classes. No one was called a professional plumber. During the nineteenth century, an increasing number of occupations acquired the title of 'profession', usually signified by the existence of a regulatory body licensed by royal charter to confer membership. The status of membership was acquired by a mixture of formal educational qualifications and 'on-the-job' training or apprenticeship; the balance varied over time and even within each profession. Most of the institutes or institutions that regulate the professions (other than the original three) began informally, perhaps even as dining clubs, and evolved into the grand bodies – with imposing buildings scattered around London – that they are today. The first to be formally recognized was probably the Institute of Civil Engineers; riding on the back of the canal age, it was formed in 1818 and granted its royal charter in 1828. Architects soon followed, then engineers of other kinds, surveyors, veterinary surgeons and many others.

But not landscape gardening or garden design. It was not until 1929 that the Institute of Landscape Architects was founded, appropriately enough after a meeting at the Chelsea Flower Show, and it wasn't given a royal charter until 1997, under its current name of the Landscape Institute. Thomas Mawson had resisted its formation, arguing that his firm and that of Milner-White could provide for all the demand and no further recruits were needed; he was eventually placated by being made its first president. But this was only a temporary difficulty; it doesn't explain why professionalism, in a formal sense, came so late to landscape design. What does that tell us about the status of its members in earlier times?

London and Wise, and later Bridgeman and Brown, were part of a band of skilled craftsmen employed by the royal household, with titles such as Chief Smith, Sergeant Plumber, Master Bricklayer; Antonio Verrio, whose paintings Wise bought, was a Chief Painter. Sometimes the gardeners reported directly to the Board of Works, at other times to a Surveyor of the Gardens, although they were always at the beck and call of the monarch and the royal family. They were treated as employees. Nevertheless, with their increasing wealth, they acquired the title of 'Esquire', a sign of increased social status that Brown clearly held as early as 1758, at the age of forty-two and only seventeen years after his appointment as gardener at Stowe. By that time, he was on terms of friendship with many of his clients, was entertained in their houses and courted by them. As the statesman William Pitt the Elder – who had worked with Brown at Wotton – put it in a letter to Lady Stanhope: 'The chapter of my friend's dignity must not be omitted. He writes Lancelot Brown, Esquire, *en titre d'office*: please to consider he shares the private ear of the King, dines familiarly with his neighbour of Sion and sits down at the tables of all the House of Lords &c.'[56] It was after dining with the Earl of Coventry, of Croome, one of his longest-standing patrons and friends, that Brown collapsed and died in 1783.

Brown and his contemporaries were valued, as their successors were to be, for their expertise; as Pitt put it: 'you cannot take any other advice so intelligent or more honest'.[57] They were typical, in fact, of the enterprising, entrepreneurial men of the Industrial Revolution, who came overwhelmingly from what would now be called the upper working class. They had little financial capital, at least at the outset of their careers, but acquired – principally through formal or informal apprenticeship – a mastery of their trade, indeed, as William Kent, Capability Brown and others demonstrate, a mastery of several trades. They knew their value; Richard Woods, a contemporary and rival of Brown, refused Lord Arundell's suggestion that he should halve his fees and charge only a guinea a day: 'It is a tender point to me, the same as it wd be to your Lordship to have the half of yr estate curtailed. My science is my estate, & as long as I am able to do business, I will do justice & deserve all I have from every Imployer.'[58] It is a good definition of a professional.

As the economy grew, the social origins of professional men changed. Garden historians have made much of Brown's possibly humble origins and the fact that, despite his background, he was on friendly terms with the king; he is compared with his seventeenth-century predecessor André Le Nôtre, who was similarly honoured by the Sun King, Louis XIV. They've contrasted him with the middle-class Repton, who, allegedly, was an unsuccessful social climber, and have drawn the conclusion that the humble, homespun Brown didn't put on airs and so appealed to his royal and aristocratic clients; one author talks of his 'earthy lower-class competence'.[59] Repton, meanwhile, in a piece of inverted snobbery in which the working classes are elevated above the middle, 'was actually a bit of a toady for whom one of the rewards of his job was the week or so he might spend in the company of the rich and titled, sharing their tables, their conversation and their home entertainments'.[60]

But Repton was part of a growing trend. A 1960 study of 272 leading architects mentioned in the *Oxford Dictionary of National Biography* found a marked decrease in the proportion of working-class recruits between the 1770s and 1780s.[61] The children of the growing middle class and gentry sought respectability, and a sufficient income, in professional occupations – surveying, engineering, architecture. Far better a larger income as a professional than scraping by as a 'gentleman' without the means to justify the title. In most such occupations, training took place 'on the job' and 'book learning' was often looked down upon. In some cases, there was a formal apprenticeship with indenture, but this system was declining. Instead, boys were employed – as both Mawson and Milner were – in garden nurseries or in the great gardens, working their way up to become a foreman in charge of a section and then either becoming a head gardener or branching out into garden design and landscaping. It was a world in which personal recommendation was crucial – Mawson praised the tyrannical Sir William Cunliffe-Brooks for getting him 'the right sort of introductions to his friends'[62] – and the cachet of having worked with a famous 'improver' such as Brown or Paxton was very valuable.

By the end of the nineteenth century, paper qualifications were becoming ever more important, the professional institutions were introducing qualifying examinations and the middle classes were aspiring to send their sons to university; Mawson's desire was that his sons should become architects, as he wished that he had been able to do. At the end of his life, he even admitted that his greatest regret was that he had not done the same for his daughters. He worked enthusiastically with Leverhulme to found a school of landscape architecture at the University of Liverpool, one of a number of such courses – others were started at Reading and Sheffield – established in the twentieth century. These provided the training, validated by the award of degrees, which was the final underpinning of the long progress towards professional status.

It was not so much the social origins of garden designers, and other professionals, that caused difficulty, or at least awkwardness, with some of their clients; it was their current status. Were they friends, advisers, employees or just servants? How should they be treated – should they be invited to lunch or dinner, or to stay the night? Brown and Repton, and later Mawson, seem to have expected to be seen as guests. Mawson describes how, when he was supervising a project, he would stay in the client's house, his son would use a village inn and his workmen would find 'digs'. Gertrude Jekyll, Lanning Roper and other twentieth-century designers such as Sir Geoffrey Jellicoe, had no difficulty; they came from the upper classes and were naturally treated as such. If they had a problem, it was in broaching the delicate topics of their fees and unpaid bills; they rarely negotiated a formal contract with their clients and Roper's project folders contain a number of copies of letters to clients that include the words 'I think you may have mislaid the account that I sent you . . .'. He rarely had to ask more than twice, however. Kitty Lloyd Jones, whose political connections in the 1930s led to a number of commissions, felt that she was expected to give advice in exchange for little more than board and lodging.[63]

The profession of landscape architecture or design thus followed much the same trajectory as other professions, but more slowly than most. Why was this? One reason, certainly not unique to this field, may be a mixture of jealousy, lack of confidence and boundary disputes. Landscape architects, or individuals such as Mawson who aspired to be called that, felt themselves to be looked down upon by 'real' architects dealing with buildings; they tended to be called in after the main contract had been negotiated, employed to prettify the surroundings, rather than having their work seen as integral to a project. At the same time, they craved the recognition of a professional institute, such as the Royal Institute of British Architects. They therefore needed to develop the self-confidence to branch out on their own and this took some time to happen. It also required numbers; while Mawson may have been self-interested in arguing that an institute was not needed, he was right in pointing out that landscape designers were still few in number and the market for their services still limited. That changed during the twentieth century with the new towns, with reclamation projects and, most recently, with the explosion of landscaping in private gardens stimulated by garden makeovers and, less happily, by the popularity of paving.

There are other reasons for the slow development of professional status. From the eighteenth century onwards, garden design was often seen as something that could be done by amateurs. Alexander Pope, Charles Hamilton, Uvedale Price, Lord Cobham, William Pitt and his wife Hester, played a major role in the design and even the planting of their own gardens and were always willing to offer advice to their friends. Philip Yorke and Jemima Grey at Wrest Park were, after all, 'assisted' by Capability Brown and were implicitly claiming the main credit for themselves. Some of the greatest gardens, such as Stowe, Rousham or Stourhead in the eighteenth century, Waddesdon or Old Warden in the nineteenth and Highgrove and Thenford in recent years, have indeed been the result of a happy partnership between a gifted designer and an enthusiastic owner, even if Lanning Roper sometimes had cause to complain of overenthusiasm and a tendency to rush to make changes; he told the Heseltines – as he had told Prince Charles – that they should really live in the house for a year before drawing up any plans: 'I urge you and Mrs Heseltine not to rush decisions. The garden and landscape should evolve gradually.'[64]

The problem was compounded when the amateur in question was a woman. Royal and aristocratic women were certainly involved in gardening and garden design from the seventeenth century, if not before, even if the credit for landscaping was usually given to their husbands or their efforts were ridiculed, as were those of Queen Caroline, wife of George II, at Kew. Queen Charlotte, wife of George III, was a serious botanist, as Queen Mary had been before her. By the end of the nineteenth century, women were beginning to be trained as gardeners or were, like Gertrude Jekyll, developing their own expertise; some of the greatest twentieth-century gardeners and garden writers, such as Rosemary Verey and Vita Sackville-West, were women and there are many working today. But in a misogynistic tactic that has been employed through the ages, the fact that women were excellent gardeners was used to imply that gardening was not a serious occupation but rather a polite pastime. It thus fed into the tendency to belittle gardening as an economic activity and to neglect the need for it to be organized on a proper professional basis.

IMPACT

When it comes to the history of the English garden, designers inevitably take centre stage. Their taste and preferences, sometimes bolstered by their writings – for many have published books, articles or advice columns – are seen as pivotal in the many twists and turns of gardening fashion over the centuries. In some cases, as exemplified by Capability Brown at the height of his career, they seem to have been given a free hand, and unlimited resources with which to achieve the results that they desired. But most were at least to some extent constrained by their customers, by the sites at which they were working, by the garden that they were modifying, by the plants that were available to them. The author and garden designer Rosemary Verey once told Sir Roy Strong that the most important thing she needed to know about a client was how much money he or she had and was willing to spend, though taste and a knowledge of plants were also crucial.[65] A designer's impact is also affected by the skill of others, such as the gardeners who have to implement the designs and those who, over succeeding generations, modify the original vision of the garden as the plants grow and the vistas change.

In economic terms, as well, the designer has to be seen as part of a team. This is reflected technically in the fact that the economic impact of a professional's work is measured by statisticians as equivalent to his or her income from a project, which is frequently only a small fraction of the overall cost. Much the same is true of the work of other professionals, such as architects, whose creative input is thought to be crucial but whose economic input is measured merely by the size of their fees. Even assessing the creative input is problematic, since designers from Repton to Roper have emphasized that their work depends on an interchange between them and their clients, together with the staff actually putting the decisions into practice, and that the original design is rarely implemented in full but rather modified during the course of the project and even more in succeeding years.

The emphasis of this chapter on the financial success of the great designers is thus only a very partial guide to their overall impact. Its main purpose has been to emphasize that it was possible to make money, sometimes very large sums, from the creation of successive generations of great gardens. For every Brown, Repton or Paxton there were, of course, tens or hundreds of designers who were less successful financially and less known to posterity, but who have worked like them to please their customers and to create beauty, however transient their work has been. They, as well as the great designers, have been a crucial part of the garden industry.

5

The Nursery Trade

'Catalogues bright with colour and with hope'[1]

In June 1818, at the height of that year's London Season, Mrs Henry Baring – wife of a banker turned politician – gave a party in her new house in Mayfair. Maria was a phenomenon, one of the first American heiresses to marry into British high society but also, very unusually for this period, a divorcée. She was the second daughter of Senator William Bingham, thought to be the richest man in America in the early 1780s. His fortune had been built on privateering, trading and land speculation. In 1798, at the age of fifteen, Maria had married a French émigré aristocrat, but it didn't last and by 1802 she was married again, to Henry Baring; her sister Ann had already married Henry's brother, Alexander, who with William Bingham negotiated in 1803 the Louisiana Purchase, the largest land sale in history.

Ann made the better match: Alexander was one of the richest men in Europe and was later ennobled as Baron Ashburton. Henry, Maria's husband, was a spendthrift and a gambler. As the normally sober History of Parliament Online puts it: 'His wife, divorcée heiress to £200,000 (£200 million today) ... was divorced from him, on account, it is said, of a liaison she had with a Captain Webster. Baring, who managed to keep most of her fortune, "had no sense of honour or delicacy, threw her into the most dissipated company, was glaringly unfaithful himself and, it is said, laid a plan for her divorce when he had fallen in love with a young lady who became his second wife".'[2] Maria, undeterred, married another French aristocrat.

Maria was still Henry's wife, however, when 600 guests, led by princes of the blood royal, were entertained in rooms bedecked with

1,004 plants, hired for the night from James Cochran, nurseryman and florist of Duke Street and Paddington. The display, installed by five men and costing the Barings £50,000 *(£65 15s)*, included 'supper tables superbly decorated by *or-molu*; the most striking were flower baskets, supported by female *caryatides*'.[3] Splendid as was the Barings' floral display, it made up only a small part of Cochran's sales of about £1 million in modern values during the four months of the Season that year. The Duke of Argyll, HRH the Duke of Gloucester and the slave-owner Sir Bethell Codrington also gave lavish parties and hired huge displays. Meanwhile, two miles to the west of Cochran's shop, Harrison's Brompton Nursery Gardens, covering at least 27 acres around what is now South Kensington Underground station, were providing trees, shrubs, seeds and bulbs to the aristocracy and gentry, as they had been doing for several decades. In 1808, at the height of the Napoleonic Wars, Samuel Harrison and his partners took orders for plants worth at least £8 million *(£9,938)* during the six months from April to October.[4]

The Season – at its height between April and August – was the name of the annual migration of the rich, and their retainers and hangers-on, from their country estates to London for the parliamentary sessions, for royal levees, for balls, the opera and pleasure gardens, for debate, political intrigue, dalliance and other forms of enjoyment, licit or illicit. But it was also a glorious shopping opportunity in the leading commercial and manufacturing centre of the country, a time for ordering clothes, hats, haberdashery, boots and shoes, saddlery and harnesses and all the other paraphernalia of aristocratic life. Among the tailors, seamstresses, bootmakers and saddlers, were at least forty-nine nurseries, visited during the Season by the cream of society. When they were not riding in Hyde Park, lunching, dining or dancing, they went in search of the latest plant novelties from Virginia, China or the Himalayas. They were met by plantsmen arrayed in top hats and tailcoats, conducted round the displays that rivalled or surpassed those of botanical gardens and offered a glass of tea; it was hoped they would respond by ordering hundreds or thousands of plants to be delivered, in autumn and winter, to the head gardeners who had been left behind to tend the parks, wildernesses and pleasure grounds of their employers' stately homes. They had plenty to choose from: the plants alone in the Brompton Nursery were valued in 1812 at £2.7 million *(£3,663 19s 4d)* and Harrison and Cochran also sold tools and even choice fruit such as pineapples.

These nurseries and their competitors, predecessors and successors were among the largest retail businesses – shops – of their times, occupying hundreds of acres of increasingly valuable land near to the city centre.[5] They occupied much more land than any other kind of shop, had large stocks and sales, paid large rents to landowners (as well as substantial taxes during the time of William Pitt the Younger's short-lived shop tax between 1785 and 1789) and sat at the centre of a complex web of farmers, seed growers, toolmakers and, of course, gardeners.[6] Like most retailers of the time, such as tailors or bootmakers, they started off as both producers and retailers. Some nurseries, even today, retain this dual function, though many others have followed the path towards the modern garden centre, growing little or nothing themselves and buying in their stock from home and abroad. Cochran and Harrison's businesses were well on the way to becoming garden centres, nearly two centuries before the term was widely used.

Despite their scale and complexity, nurseries at this time were family businesses, some surviving as such over many decades or even centuries. They had to surmount all the vicissitudes of inheritance that can afflict family-run concerns, such as lack of interest or management skills and diversions of capital to consumption or other purposes. But when they succeeded – and naturally we know most about those that did – they made their owners wealthy. They, and their customers, transformed the face of England, importing tens of thousands of species – gathered by the nurseries' plant collectors from the four corners of the globe – and nurturing and adapting them to English conditions. Today nurseries sell over £7 billion of goods every year, a substantial contribution to the British economy.

It all began in London.

LONDON BEGINNINGS

In 1747, R. Campbell wrote and published *The London Tradesman: Being a Compendious View of All the Trades, Professions, Arts, both Liberal and Mechanic, Now Practised in the Cities of London and Westminster.* In Chapter 68, 'Of the Gardener, &c.', he discusses the gardener – 'a healthful, laborious, ingenious, and profitable Trade' – the fruiterer and the land surveyor, 'employed in measuring Land, and laying it out in Gardens and other kinds of Policy about Gentlemen's Seats'. Completing the picture are the seedsmen and nurserymen:

> The Seed-Shopkeeper sells all manner of Garden and Grass Seeds, Gardener's Tools, Matts, &c. and some of them are Nursery-Men, and furnish Gentlemen with young Trees, both Fruit and Forest, with Flower-Roots, &c. It is a very profitable Branch and in few Hands; requires no more Skill than other Retail Trades, if they are not in the Nursery-Way; but if they are, they must be compleat Gardeners.[7]

One such compleat gardener 'in the Nursery-Way' had flourished in London a century before Campbell was writing. That was Captain (the source of his title is unknown) Leonard Gurle, introduced at the beginning of Chapter 1. He had established his nursery in east London in the 1640s when he was in his twenties; it occupied 12 acres across what is now Brick Lane in Whitechapel. The area is notable in London's history: it was about to become the residence and workplace of many Huguenot silk weavers, expelled from France after the Revocation of the Edict of Nantes in 1685. Later it was occupied by Jewish immigrants from eastern Europe and today it is the centre of Banglatown, a street of Bangladeshi shops and restaurants. Gurle had additional nursery ground to the north-east in Hackney, presumably to supply plants for the main nursery. His speciality was fruit trees and his catalogue, published by his 'Loving Friend' Leonard Meager in 1670, contained 300 varieties; he anticipated modern branding techniques by naming a hardy nectarine, which he had raised, after his own name spelled backwards – the Elruge. It was a staple of kitchen gardens – plate 14 shows a plant label from an eighteenth-century wall – and is still available today.

A precursor of later nurserymen in many ways, Gurle used what we might think of today as inducements to buy, providing not only a catalogue but a guarantee that he would replace any plant that failed or died. His nursery was large – although many were to be much larger; the inventory taken after his death in 1685 lists over 12,000 plants – among them 295 asparagus crowns, 179 young walnut trees and 290 young horse-chestnuts.[8] Like many other nurserymen, he combined his own nursery with running a great garden when, in 1677, he became the king's gardener at St James's; he was paid £655,000 (£320) annually for maintaining the garden there, together with a salary of £491,000 (£240).[9]

Gurle's probate inventory is the first confirmation of Campbell's view that the trade of seedsman or nurseryman was a 'very profitable Branch'. He occupied two houses, one coming with his job as a royal gardener and the other at the nursery in Whitechapel; his combined household goods, including bedsteads, trunks, coffers, hangings and 128 tankards and spoons, were valued at £274,000 (£138 13s 3½d). These are typical of the increasing range of consumer goods that were becoming available at the time, although the quantity marks him out as a successful member of the nascent middle classes. Gurle had somehow amassed sufficient capital to buy a long lease on the nursery grounds, which still had thirty-two years to run when he died, and to pay £115,000 (£58) a year in rent; he had, in return, an income from houses he had built on the site. In total, his assets (not including his landed property) were valued at £1,997,000 (£1,011 10s 8½d).

However, there were two problems. The first was entirely normal for the time, a build-up of debt. Customers wanted their goods immediately, but expected to be billed no more than twice, or preferably only once, a year. Aristocrats, in particular, could then procrastinate for months or years before paying, with shopkeepers reluctant to press for their money in case their reputation was blackened. Debts might have to be written off. A final account accompanying Gurle's inventory shows that his assets included debts due to him 'supposed to be good', amounting to £356,000 (£180 10s 1d) and 'supposed to be desperate', £272,000 (£137 18s 6d). So at least 10 per cent of his assets was at risk.[10]

But this was dwarfed by the second problem, the failure of Charles II to pay his gardener before both of them died. Gurle was not alone: numerous aristocrats holding court positions, as well as tradesmen, found that the king was a bad payer. Gurle's inventory includes, on top of his assets, 'Due to ye deceased from his late Majesty . . . for his salary board wages and otherwise the sum of two thousand five hundred pounds' (just under £5 million in modern terms). The final account shows that the 'Accomptant' administering the will had tried to recover this sum 'but cannot receive the same or any part thereof'.[11] No doubt, like many others in her position, Gurle's widow kept on trying and probably later got some of the money out of the Treasury.

Gurle was, particularly if the debt from the king is taken into account, a rich man. But the stocks of plants in his nursery were very small by comparison with those of some of his successors; perhaps he was concentrating on his royal role. The Brick Lane nursery was certainly soon to be dwarfed – on the other side of London – by Brompton Park, founded in 1681. According to Stephen Switzer:

> This vast Design was begun some Years before the Revolution [of 1688], by four of the Head Gardeners of England, Mr. London, Gardener to the . . . Bishop of London; Mr. Cook, Gardener to the Earl of Essex at Cassiobury; Mr. Lucre, Gardener to Queen-Dowager at Somerset House; and Mr. Field, Gardener to the Earl of Bedford, at the then Bedford House in the Strand.[12]

The design work of George London and Henry Wise, who became London's partner by 1687, was discussed in Chapter 4; we know that Wise became very wealthy indeed. The Brompton Park Nursery, which provided part of this wealth, was said to have covered 100 acres of Kensington, the site today of Imperial College and many of London's great museums. Fifty acres is more plausible, but it was still said to contain 10 million plants. Switzer says that it was valued, some years before 1715, at 'between 30 and 40000 L (perhaps more than all the Nurseries of France put together)'[13] (between £56 million and £75 million in modern terms). This may have been an exaggeration, but the nursery was clearly a valuable asset: we know that Henry Wise leased the nursery, on his retirement, to William Smith and Joseph Carpenter for thirty-eight years at an annual rental of £632,000 *(£332)* and a lump sum of £11.6 million *(£6,100)*. Wise kept quite a lot of the plants and his Brompton Park house on the site.[14] A 'vast design' indeed.

Although Brompton Park predominated at least until the 1720s, the first fifty years of the eighteenth century saw a proliferation of nurseries in London, both in the east around Whitechapel, Hoxton and Hackney and in the west around the villages of Kensington, Chelsea, Hammersmith and Chiswick. Probably much more typical than Brompton Park was, even further west, the nursery of Peter Mason in Isleworth. Mason died intestate in 1730 but his inventory survives.[15] He was growing plants on three 'nursery grounds' and had about 140,000 in all, worth £512,000 (£285 5s 10d), but apparently his wife had 'a considerable stock of Nursery Trees not menconed, which stock the sd. Intestate's Widow claims as her own the same standing on her Jointure Lands'. They had household goods worth £276,000 (£154) in their ten-room house, including £58,000 (£32 9s) worth of plate and two silver watches with a combined worth of £5,000 (£2 12s 6d).

But perhaps most significant is that 114 debts were listed as part of the estate, to a total value of £1.12 million, over twice the value of Mason's own stock.[16] The largest of these, for £307,000 (£171 3s 2d), was owed by the Earl of Burlington, the 'Architect Earl', garden enthusiast, aesthete and patron of William Kent; Lady Waldegrave, whose lover – soon to be husband – fought at Waterloo, owed £50,000 (£28 1s 8d) and the Dowager Duchess of Hamilton £117,000 (£65). The latter was the widow of the 4th Duke, who had been killed in a duel in Hyde Park in 1712. For Mason, carrying debts of this scale must have been a serious problem.

These three nurseries, for Gurle's survived at least until 1719, probably then run by a grandson, were among about fifteen in London by the 1690s and between twenty-five and thirty by 1729 (as compared with twenty in the whole of the rest of the country). After that, although more and more provincial nurseries were established, the London nursery trade continued to grow. The garden historian John Harvey, who made a meticulous study of early nurseries, estimated that by 1760 there were at least 30 nurserymen and 12 seedsmen in London, with 84 by the 1790s, although he could identify only 49 nurseries on Thomas Milne's land-use map of 1795–9. By 1839, the situation had changed; it was then estimated that there were 19 principal nurseries and 11 seedsmen around London, as compared with about 150 in the rest of England and 8 in Scotland.[17]

Campbell in 1747 estimated that the capital required to set up as a master nurseryman was between £898,000 *(£500)* and £1.8 million *(£1,000)* and that a master seedsman required between £180,000 *(£100)* and £898,000 *(£500)*. This implies that substantial amounts of capital were flowing into the garden industry as it expanded and spread throughout the country. If there were 30 nurseries worth at least £1.8 million in London, 20 in the rest of the country and 12 London seedsmen worth at least £0.9 million each, this would come to at least £100 million; but this would be a serious underestimate of total investment, since Campbell was giving a figure for the cost of setting up in the nursery and seed trade, not for future investment and increasing value. After all, we know that Brompton Park Nursery alone was worth between £60 million and £80 million.

Can we believe Campbell? He certainly did not regard nurserymen as exceptional. There were thirty-seven trades, ranging from coach makers to grocers and sugar bakers (refiners of raw sugar), whom – he thought – required at least as much capital as nurserymen and his estimates for them do not seem to have been challenged at the time or since. His figures suggest, also, that it was possible for a journeyman – in the nursery trade or elsewhere – to accumulate substantial capital, or to borrow it, in order to set himself up in business; this is another sign of a healthy trade faced with growing demand.

The London nurseries were only part of what has been called a 'retailing revolution'.[18] Napoleon is famously said to have described the British as a 'nation of shopkeepers', although the phrase was actually coined by the great Scottish economist Adam Smith in 1776. Smith was writing at a time when there was a shop for every 43 people, probably the largest proportion ever; today Britain has a shop for every 225 people. But this understates the change that has taken place since the eighteenth century, when shops existed side by side with many market stalls, pedlars and hawkers. Most retailers were in a small way of business; they worked alone, or perhaps with family members and an apprentice or two. Many were manufacturers as well as retailers, just as the nurserymen grew many of their own plants. But shops increasingly bought in at least part of what they sold; food shops led the way, with commodities such as tea depending on a complex network of growers, shippers and dealers.

The great number of shops and other retailers in the eighteenth century was the natural outcome of an economy in which storage of perishable goods was difficult, transport was slow and cumbersome, and consumers needed – and could afford – the convenience of a local shop that they could use on at least a daily basis. Britain was already far from being a subsistence economy and the famous Industrial Revolution was preceded as well as accompanied by a consumer revolution in which goods – both perishable and long-lasting – multiplied in every household. As the economy became more and more urbanized, markets and fairs were replaced by fixed shops that gradually became more specialized in what they sold.

At the beginning of the eighteenth century, much of the trade of Britain – the nurseries included – was focused on London. Provincial shopkeepers had to travel annually to London, from as far afield as Lancashire, to replenish their stocks and order new goods. Even later in the century, London remained 'the major source of supply for groceries, household wares, and drysaltery'.[19] In other trades, including sugar, provincial centres such as Bristol and Liverpool were gaining ground, but the crucial commodity of tea was still monopolized – so far as the legal trade was concerned – by London merchants, although there was also a flourishing smuggling business. London retailers fought back against their competitors, but it was, in the long term, a losing battle.

This was a matter of demographic change. London, in 1700, was the largest city in Europe, with between 550,000 and 600,000 people, and at that time accounted for about two-thirds of the entire urban population of England. It was the main centre of external and internal trade: London handled 80 per cent of England's imports, 69 per cent of its exports and 86 per cent of its re-exports. By 1800, after another century of growth and urbanization, London accounted for only half the town-dwellers in the country. Ports such as Liverpool, Bristol, Glasgow and Newcastle were taking an increasing proportion of trade, while manufacturing – in which London was still very important – was shifting to the Midlands and the industrial north.

But the early predominance of London in the nursery trade raises one major question: why was this trade, which required large areas of land – unlike most retailing or manufacturing – based around the biggest urban centre in Britain, far away from the estates that bought many of its products? How can it have made sense to produce tens of thousands of forest trees around London and ship them to far-flung parts of the country? Nurseries clearly thrived in London: the area of land they occupied alone makes them the most visible activity, while the value of their stocks and the income they earned suggest that they were among the most successful of London retailers.

London was the centre of the nursery trade for four main reasons: fashionable consumption, transport, networking between nurseries and nurserymen, and manure.

Fashion came to the fore in the annual London Season and London's nurseries seized their opportunity. Here were rich clients, with time on their hands and anxious to make sure that their country houses – to which they would be returning in a few months – would outshine those of their neighbours. Although there were nurseries to the east of the City of London, more began to establish themselves to the west, where the fashionable houses of Belgravia and Kensington were starting to be built. A particular focus was the King's Road, later to be the centre of world fashion in the 'swinging sixties' of the twentieth century. But it was built originally by Charles II as literally the king's own road, his private route from the palaces of Whitehall and St James's out west to Kew and Hampton Court. Access to it was granted originally to the favoured few, who from 1731 were given tickets, and then more generally, but it did not become a public road until 1830.

Long before then, however, nurseries had set themselves up along the road, with their nursery grounds behind them; others clustered a short distance to the north, from Knightsbridge to Gloucester Road – the areas now occupied by Harrods, by the museums, the Albert Hall and Imperial College. There are said to have been at least twenty-five nursery gardens established along the King's Road between the 1750s and 1916. An account of Chelsea in 1869 observes that 'the line of the road was almost exclusively occupied by nurserymen and florists, and thus it became a fashionable resort for the nobility and gentry'.[20] Like modern garden centres, the nurseries were open to visitors and displayed their wares to their wealthy customers; they became, again like today, tourist destinations, complete with shop assistants eager to show off the new arrivals bred from imported seeds and plants.

As with all fashion industries, novelty was enormously important. In the nursery trade, much of it depended on the importation of plants and seeds, which began in earnest in the seventeenth century, expanded with American imports at the start of the eighteenth[21] and, by the end of the nineteenth, encompassed the whole world. Imports were accompanied by the first attempts at plant-breeding and hybridization, associated particularly with the name of Thomas Fairchild of Hoxton, famous as probably 'the first nurseryman to be an outstanding botanist',[22] though he was hampered by his religious doubts at interfering with God's creations. His Mule Pink was a cross between a carnation and a sweet william, but the term 'mule' came to be used generally for hybridized plants. As many of the plants that Fairchild sold were tender, his nursery acted, his biographer says, as a 'boarding house' for those that could not be kept year round in their owners' gardens.[23]

Nurseries spread far beyond the King's Road, however, into the outlying villages to the west, east and south of London. In Camberwell in the south in 1789, for example, the Montpelier Tea-Gardens of Mr John Bendel were said to

> contain about five acres of land, chiefly planted with trees, shrubs, and other choice plants. Large companies resort there in the summer season, in fine weather; and decent people are served with tea at six-pence each. These gardens contain a cold bath; and on the west side is a small labyrinth or maze. The walks are rural, and the whole kept in decent order. Gentlemen are served here with shrubs, flowers, or seeds.[24]

London and its environs thus provided a captive market for nurseries catering for the gentry and aristocracy, who could order or buy the plants that took their eye, to be sent to the care of their head gardener on their country estate. This required, of course, a transport system, but this was exactly the second advantage that London possessed. By the middle of the eighteenth century, if not before, it was linked to the rest of the kingdom by a network of road carriers and, importantly for the transport of bulky and heavy plants and trees, an extensive and well-developed system of river and coastal shipping.

One of the most pervasive myths of British history is the belief that, until the coming of canals and railways, the transport of goods and people over any distance was nearly impossible, with both stuck in mud on roads that had not been mended since the time of the Romans. This was far from the case. In fact, as Dorian Gerhold comments: 'In the 1680s about 205 waggons and 165 gangs of packhorses entered and left London every week, carrying about 460 tons of goods each way and performing a vital service for the country's burgeoning industries and for many other users. Direct services extended as far as Richmond and Kendal in the north, Denbigh, Oswestry and Monmouth in the west and Bideford and Exeter in the south-west.'[25] Long-distance carriers connected throughout the country with regional and local services – every village had at least one carrier – so that goods could be sent anywhere with confidence that they would reach their destination according to a clear timetable. It was not as quick, of course, as the modern 'click and collect', but many goods could be ordered by mail, while coaching inns provided the same safe facilities as modern depots. On 29 June 1816, General Royale ordered six geraniums from James Cochran for £979 (£1 6s 6d), including carriage and packing; among them was one of the 'new Prince Regent' variety at £148 (4s). They were to be sent to the Reverend Barton Watkins, of Lockeridge House, to be left at the Fighting Cocks, Fyfield, near Marlborough.[26]

Waggons with four wheels replaced carts with two and their capacity rose from 4 to 7 tons; they were equipped with variable front axles, springs and brakes, making them safer and easier to use. Relays of horses were available, reducing journey times. The waggons travelled on 3,000 miles of turnpikes – toll roads – by 1750, and 20,000 miles by the 1830s; many roads were paved – Thomas Telford, the famous engineer, used large stone blocks – and road surfaces were gradually improved even before the advent of 'macadam' (tarmac) in the 1820s.[27] Road was complemented by water. First were the river navigations, making usable hundreds of miles of waterways and linking with an expanding system of coastal shipping: coal, grain, groceries and hundreds of other goods – including plants – travelled via a network of coastal and river ports. Then, from the 1760s, came the canals and finally, from the 1830s, the railways. Plants, trees and seeds from London nurseries could travel by whatever means best suited them and their customers, or switch from one to another. It is therefore no surprise that the surviving order books of London nurseries and the records of great estates all over the country record delivery and receipt of large consignments.

But since that transport network could equally well have been used to transport nursery goods from provincial centres to London or each other, transport alone does not explain why early nurseries clustered in London and why the London region continued to be a centre of the industry. One answer seems to be that there are significant advantages for firms from what economic geographers call 'localization and agglomeration': the tendency, observed in many different industries, for firms to group themselves together rather than spreading out across the country.[28] They traded with each other; James Cochran, who rented out floral displays to the elite of Regency London, bought much of his stock from other London nurseries. They exchanged staff, if not always willingly. We know that the great nineteenth-century Veitch nursery, and others, operated something close to a labour exchange and that it was common for ambitious young men to move between the London nurseries; Thomas Mawson, before he returned to the Lake District to establish his business, was one such. Clusters of like-minded firms or individuals also fraternize, as did the founders of the Horticultural Society of London. We are observing a phenomenon that, in the modern world, has given rise to Silicon Valley in California, the Cambridge science complex or the grouping of the great insurance companies in a small area of the City of London close to Lloyd's.

There was, however, a fourth reason why nurseries clustered together alongside the market gardens that provided fruit and vegetables for the growing population of London – a total of about 13,000 acres in the early nineteenth century.[29] The answer lies in the excrement produced by that population and by the horses that provided its transport. There were about 50,000 horses in London in 1800, producing a quarter of a million tons of horse dung annually.[30] Then there was the human waste, probably about another 50,000 tons.* So London was producing, each year around 1800, about 300,000 tons of human waste and horse dung. In addition, waste came from industrial processes such as tanning. What did Londoners do with it?

---

* A human being, on average, produces 124 grams of faeces every day.

London did not have an effective sewage disposal system until the 1850s or later. The horse dung had to be collected, day and night, from the streets and stables, while human waste was gathered at intervals from cesspits under privies by 'nightmen'.[31] Some of it was dumped close by – the ironically named 'Mount Pleasant', later the site of Coldbath Fields Prison and then of the largest postal sorting office in Britain, was a dungheap of 8½ acres by 1780.[32] But much was taken to the River Thames, especially to Dung Wharf in Puddle Dock – later the site of the Mermaid Theatre – and to Water Lane nearby. From there it was loaded on to barges and taken, upriver, to the nurseries and market gardens of west London and probably further afield. This had been going on for centuries. In 1618, Father Busoni, chaplain to the Venetian ambassador to London, visited the city's market gardens and reported that the gravel that had lain below the surface had been dug out and sold, to be replaced with the 'filth of the city', manure 'as rich and black as thick ink'.[33] The manure combined with the high water table, on the low-lying lands close to the Thames, to create ideal conditions for nursery and market gardening. Presumably the citizens of London knew – but did not care – that the fruit and vegetables they bought at Covent Garden market had been nourished by human and equine waste.[34]

SPREADING NURSERIES

London continued to be the site of large nurseries until the First World War, but the 13,000 acres of nursery grounds and market gardens around it at the end of the eighteenth century were an obvious target for house builders. First came the brick makers, seeking the clay. In 1811, the Reverend Henry Hunter described how the built-up area was surrounded by a zone in which 'the face of the land is deformed by the multitude of clay pits, whence are dug the brick-earth used in the kilns which smoke all round London'. At night, a ring of fire and pungent smoke encircled the city.[35] Samuel Harrison, who served the aristocracy of Regency London from his nursery in South Kensington, went bankrupt in 1833. This seems to have been considered to be his fault, but it is much more likely to have been the result of the decision of the Thurloe family, the ground landlords, to develop the area for housing; Thurloe Street, Square and Place are the result. Loddiges in Hackney closed later in the century for similar reasons. Other nurseries faced similar pressures at different periods. At the same time, the smoke of London's coal fires took an increasing toll on plants and spurred nurserymen to move many of their activities away from the centre, out into the home counties, even if they kept a small nursery and shop close to their customers in the city.

The major growth of nurseries was, therefore, outside London. John Harvey dates the start of the expansion to the middle of the eighteenth century, although there had been nurseries in such cities as Oxford and York long before this. By the early nineteenth century, every provincial town would have had its nursery, if not several; a directory of the northern manufacturing town of Leeds, in Yorkshire, lists forty-one 'Gardeners, Nursery & Seedsmen' in 1830. Some provincial nurseries were larger than those in London; Harvey describes the business of William Falla in Gateshead on Tyneside at the beginning of the nineteenth century as 'the largest nursery ever known in Britain',[36] while at the other end of the country, in Devon, the Veitch family expanded in 1830 from their original site in Killerton to a 25-acre site near the centre of Exeter, before their later expansion to London, where a new business became the leading nursery of the late Victorian period.

Provincial nurseries were – as their very growth demonstrates – able to compete with those in London. They benefited, for example, from cheaper land; this enabled the Falla nurseries to pioneer really large-scale production of plants, leading to economies of scale. Despite the centrality of London within transport networks, it was still cheaper as well as less damaging to the stock not to have to carry plants over long distances; in the era before the twentieth-century introduction of container growing, plants were sent bare-rooted and were susceptible to damage and loss, particularly on long sea voyages, until the invention of the 'Wardian' case, a container for plants that kept them in a partially sealed environment. An example of this crucial invention is shown in plate 13. There was also an advantage in growing plants reasonably close to their ultimate destination, in similar soil and climate types, although the great local variation in geological conditions in Britain diminished this benefit. But probably the major advantage of the provincial nurseries was their ability to display and sell their wares to the growing local middle-class populations of the expanding towns and cities, who were unlikely to spend any significant amount of time in London.

In any case, the newly established nurseries were not discrete entities, but rather parts of a complex network of production, supply and retailing, in which businesses bought and sold from each other across the country and abroad. In 1805, for example, the firm of William Caldwell in Knutsford, near Manchester, bought plants worth £105,000 (£120) and seeds and bulbs worth £177,000 (£203) from eleven different London firms, as well as spending £38,000 (£43) with firms in York – across the Pennines – and in the Midlands: the total cost was £320,000 (£366).[37] Caldwell's was still then a relatively small business, although it was to be one of the longest-lasting nurseries, surviving from the eighteenth century until the 1980s.[38] London firms, conversely, bought from the provinces as well as abroad. James Cochran, who had only a small nursery ground in Paddington to service his floral leasing business, spent approximately £15.3 million (£20,544) between 1815 and 1820 on buying plants, bulbs and seeds from other London firms but also from nurserymen and seedsmen as far away as Colchester and Pontefract: £550,000 (£737) was spent on bulbs, largely hyacinths, from one Dutch supplier and £253,000 (£339 12s 8d) from another. Samuel Harrison at the Brompton Nursery did the same. Seedsmen, in particular, bought in large quantities of seeds from growers, especially in Essex and the east of England, as well as in the twentieth century from overseas.

Firms had always been known for a particular expertise, such as Leonard Gurle for fruit, but specialization was taken to new heights in the nineteenth century, when nurseries aspired to supply every possible variety of their chosen species; others boasted of the number of species they could show. Loddiges, at the Hackney Botanic Nursery Garden in east London, can serve as an example. In the 1820s, a quarter of a century before the Palm House at the Royal Botanic Gardens at Kew, George Loddiges built the Grand Palm House, the largest in the world, and equipped it with steam heating, which also provided a sprinkler system to keep the atmosphere and the plants moist.[39] By 1826 it contained 120 species of palm from over thirty countries and by 1829 the collection was said to be worth £155 million *(£200,000)*. In the early 1830s, Loddiges ran out of space there and Joseph Paxton was employed to build another hothouse;[40] by 1845 the two contained 280 species and varieties. But there were also two huge camellia houses built to John Claudius Loudon's designs, orchid houses with over 1,900 species in 1845, eighty different types of exotic ferns and an arboretum that displayed the 3,075 hardy trees and shrubs listed in the Loddiges catalogue of 1830. The arboretum was laid out as a spiral path with the specimen trees arranged alphabetically along one side and nearly a thousand rose species and varieties on the other. No wonder that it was a major tourist attraction: the *Gardener's Magazine* said in 1833 of the arboretum that 'There is no garden scene about London so interesting'.[41]

Later in the century, the Veitch family's Royal Exotic Nursery, Kensington, had shops and hothouses on both sides of the King's Road; there was a long central walk, ending in a *jet d'eau*, a glasshouse full of orange trees, an orchid house, with more houses full of displays of azaleas, rhododendron, conifers, magnolia and many other plants. There was a propagating house, a fernery with rockwork and a waterfall, together with an outside ornamental garden.[42] By 1903 the firm had three growing nurseries on the outskirts of London in Buckinghamshire, Surrey and Middlesex. More typical may have been Cheal's of Crawley, which operated close to the modern Gatwick Airport in Sussex and served mainly a local clientele. It had a number of different nursery grounds – a total of 110 acres by 1912 – so that it could grow plants on different types of soil. Many thousands of plants were moved annually to be sold at the main nursery.[43] Establishments like this could be found all over the country.

RUNNING A NURSERY

The typical nursery of Georgian, Victorian and Edwardian times both grew its own stock and bought in from other nurseries; this entailed, in the case of its own production, forecasting demand for thousands of species up to three or four years ahead of their likely sale. Seeds had to be sown, seedlings transplanted into pots and then into the ground; it was normal to lift and replant regularly, to prepare the trees and shrubs for their ultimate sale. But this greatly simplifies. In many of the leading nurseries, such as Loddiges, Veitch, or Hillier in the twentieth century, complex experimentation took place to nurture plants and seeds brought from all over the world – often by plant collectors sent out by the nurseries – so as to work out the best ways to propagate and grow them in British conditions. A major task, in spring and early summer, was to prepare the plants so that they would be at their best for the shows and exhibitions that were a vital, if time-consuming and expensive, form of advertising and at which many sales were made. Catalogues, 'bright with colour and with hope', as Vita Sackville-West, creator with her husband Harold Nicolson of Sissinghurst Castle garden in Kent, put it, had to be written, illustrations drawn, engraved or later photographed and the thousands of listed plants, often of many different sizes and ages, given their prices. Suttons' catalogues and horticultural guides, as shown in plate 15, set a standard. Nineteenth-century nursery catalogues are several times thicker than those of today, with far more plants and plant varieties included than even on modern websites: the Loddiges catalogue of 1830 lists about 12,000 separate plants, including 1,470 different roses.[44] Orders had to be taken, plants carefully packed and dispatched; most nurseries sold on account, so the debts had to be chased at regular intervals to maintain cash flow. Nursery work, even more than today, was heavily seasonal, with trees, shrubs and bulbs being required for autumn and winter planting, and flowers, herbaceous and bedding plants for the spring. Organizing the workforce – of perhaps one hundred men at a medium-sized nursery – to do all these jobs at the right times was a serious managerial task.

Many nurseries went through a similar pattern of growth. They began by serving a local market, then became known for specializing in a particular species such as orchids or ferns, then developed a mail-order business serving the whole country. Mail order had been available from the London firms from the seventeenth century onwards, but improved transport and postal systems brought its heyday in the nineteenth century. Hundreds of thousands of plants criss-crossed the country by rail. Then, in the twentieth century, higher and higher shipping costs and restrictions on the size of plants that could be shipped, together with the labour costs of packing hundreds of small orders, pushed many nurseries in the direction of a mainly or completely wholesale trade, supplying larger quantities at a time – sometimes under a brand name such as Hillier – to other nurseries and garden centres. Much the same happened with the seed trade, which had been largely based on mail order in the nineteenth and early twentieth centuries but, with improved packaging, moved to supplying garden centres and, increasingly, supermarkets such as Homebase and B&Q.

It was always important for a nursery to maintain good relations with its customers and – perhaps more important – with their gardeners. Catalogues became increasingly elaborate, with illustrations and hints for cultivation. Nurseries such as Loddiges produced beautifully illustrated books of botanical drawings as a further inducement to buy. James Herbert Veitch went further, producing in 1906 *Hortus Veitchii: A History of the Rise and Progress of the Nurseries of Messrs. James Veitch and Sons*, a detailed illustrated account of over 500 pages, giving the histories of the nursery's plant hunters who had scoured the world, of the plants they had brought back and of the hybrids created by the nursery, such as the intrepid search for seeds of the *Davidia involucrata*, or handkerchief tree, by the notable plant hunter Ernest Henry ('Chinese') Wilson on a trip to central China in 1899.[45] Customers who visited nurseries were treated with ceremony and shown round by the owner or a foreman, even if the resulting order was a small one. Perhaps more important – though little known today – were the commissions regularly paid to gardeners or the Christmas presents sent to them; nursery accounts such as those of James Cochran list many such payments. The practice of offering commissions was clearly widespread and may be one reason – discussed in Chapter 6 – why so many head gardeners and designers, who also sometimes received them, retired relatively wealthy. In 1905, Hillier notified their customers that they would only offer commission to a head gardener if the employer explicitly agreed to it. We don't know how many other nurseries were equally scrupulous. Marketing in general remains one of the major tasks of a nursery or garden centre.

However, this brief overview of a nursery's work is only the half of it. The typical nursery did far more than grow and sell plants. From the days of the Brompton Park Nursery onwards, nurseries took on the tasks of contract maintenance of estates and gardens, large and small. Nurseries were often started by ambitious head gardeners: the two leading Exeter nurseries of William Lucombe and John Veitch, both originally head gardeners, continued to maintain the grounds of Powderham Castle and Killerton House, respectively, for many years. Along with garden maintenance went, inevitably for any gardener, redesigns or 'makeovers'; these features of television garden programmes have a history centuries old. Many nurseries advertised their landscaping skills, designing and carrying out hard landscaping of terraces, water features and patios; they sketched planting schemes, supplied or obtained plants and put them in the ground. Joseph Cheal & Son of Crawley were responsible, for example, for the landscaping of 188 gardens between 1890 and 1919, of which 138 were classified as 'main jobs', each costing over £350,000 (£1,000).[46] One job for Joseph Cheal was at Hever Castle, the ancestral home of Anne Boleyn, for W. W. Astor, later the 1st Viscount Astor; between 1904 and 1908, 'thirty acres of classical and natural landscapes were constructed and planted by Cheals . . . Many of the Scots pines now growing near the lake were transported from Ashdown Forest, twelve miles away. Each tree required a team of four horses and ten men to move it – a monumental task.' Numerous pieces of classical and Renaissance sculpture had to be placed in the gardens and 'civil engineering, water engineering, construction work, classical art and architecture all had to be merged with landscape design and horticulture'. It cost £9 million (£25,000).[47]

Many nurseries had shops in London and other large cities, often separate from the nursery grounds, which had to be stocked and staffed. Gardening supplies could be a lucrative business. Most sold numerous tools, from lawnmowers to pruning knives, and, by the middle of the nineteenth century, a widening range of fertilizers and pesticides, often terrifyingly toxic to modern observers. Although the garden centre is thought to be a modern concept – paralleling the supermarkets introduced from the United States at much the same time after the Second World War – many of the nurseries described in the *Gardeners' Chronicle* and other periodicals in the nineteenth century were garden centres in all but name, lacking only the containerized plants which revolutionized the nursery trade in the late twentieth century; not all, however, had the tea rooms or pet shops that nowadays provide much of the profits.

Some nurseries branched out in other directions. James Cochran of Paddington, who supplied Mrs Henry Baring, developed a large business between 1812 and 1820, from his shop in Duke Street, Mayfair, which carried out all the traditional nursery trade but added the supply of plants and floral displays for parties in the London Season. He did this for royal dukes, soldiers and bankers; his client list included the Persian ambassador to London. Another client was Sir Bethell Codrington, MP and slave-owner, whose Gloucestershire estates extended for more than fifteen miles and who nevertheless was rebuffed in several attempts to secure a peerage.[48] The Scottish Whig politician the 6th Duke of Argyll hired 3,721 plants from Cochran between 1816 and 1821.[49] Cochran's annual income from these hires reached a total of £452,000 *(£600 8s 6d)* in 1819. Three months in the summer of 1816 saw orders from the Duke of Grafton (for '104 plants for the night, kept a week – £7-6-0' or £5,400 in modern terms), from the Duke of Leeds ('Potts of Sweet Peas, carnations, honeysuckle, geranium, myrtle, heliotrope and pinks') and from Miss Townsend for Princess Mary at Buckingham House (now Palace) for cut flowers, mignonette and rose trees. Mrs Baring's party required, on top of 1,004 plants, 'the use of 27 green baskets, moss, twine, wire, nails, sticks, etc., to 5 men 2 days *(£2-10-0)*, horse and cart 2 days *(£1-10-0)*, a large quantity of laurel and evergreens, cut flowers for china jars for tables, horse, cart and men clearing away *(£2)*, 2 rhododendrons, 2 hydrangea, 4 baskets, 4 pans'; the whole cost £49,000 *(£65 15s)*.[50]

However, floral displays were only the tip of Cochran's business. He supplied what his customers wanted, large or small, whether it be a peck of Lancashire seed potatoes and half a peck of German 'kidneys' (bean seed) and, later, groceries, oatmeal and shortbread, for the Marquess of Abercorn, or sending a man to do the watering for the Marquess of Salisbury.[51] He sold cut flowers to Mrs Fitzherbert, mistress and illegal wife of the Prince Regent. He even acted as greengrocer to Lord Salisbury, buying fruit from Covent Garden and charging a fee; one order worth £22,300 *(£31 5s)* in 1815 included pineapples costing £7,400 *(£10 8s 6d)* and grapes worth £5,000 *(£7)*, while another was for six and a half dozen Kentish cherries at £713 *(£1)* per dozen.

1. Planting close to the City: the nurseries and market gardens of Southwark, south of London Bridge, in 1658.

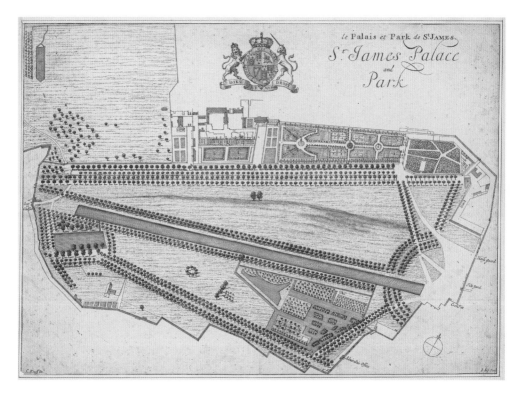

2. The king's works: Charles II's canal in St James's Park, London, 1660.

View of the Great Room &c. at Hall Barn, near Beconsfield in Buckinghamshire.

3. A beautiful contrivance: the canal at Hall Barn, Buckinghamshire, in the eighteenth century.

4. William III's view of the Thames: the Privy Garden at Hampton Court, Middlesex, 1702.

5. The cemetery as landscape garden: a design for Kensal Green Cemetery, London, 1832–3.

6. Reflections in water: the canal and Archer Pavilion (1709–11) at Wrest Park, Bedfordshire.

7. A man of taste: the Honourable Charles Hamilton (1704–86) at the age of twenty-eight, by Antonio David.

CHARLES HAMILTON.
YOUNGEST SON OF
JAMES EARL OF ABERCORN.

8. 'Mr Hamilton's Elysium': Painshill Park, Surrey, 1733–73.

A View from the West Side of the Island in the Garden of the Honble Charles Hamilton Esqr at Painshill near Cobham in Surry.
Vue dans le Jardin de l'honble Monsr Charles Hamilton, à Painshill près de Cobham dans le Comté de Surry.

Lancelot Brown Esq.   Dr

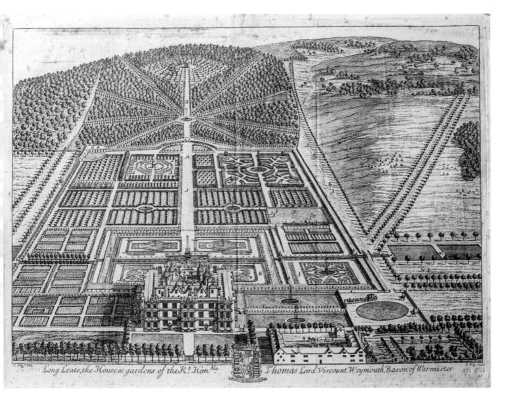

Long Leate, the House & gardens of the R.t Hon.ble Thomas Lord Viscount Weymouth, Baron of Warminster

10. Avenues marching to the horizon: Longleat, Wiltshire, in 1707.

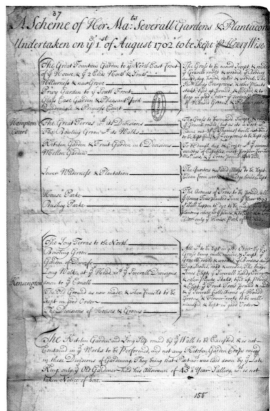

11. Maintaining the royal gardens: the first page of Henry Wise's contract as royal gardener, 1702.

12. An Edwardian extravaganza: the pergola, high on a brick plinth, at the Hill House, Hampstead, 1906.

13. Transforming plant hunting: the 'Wardian' case, c.1829.

14. A Stuart novelty: a plant label for the Elruge nectarine on a wall at Moggerhanger Park, Bedfordshire.

The breadth of his business, and the fact that he had only a small nursery in Paddington that could not have grown much of what he sold, meant that Cochran was part of a complex web of suppliers – buying and selling with each other as well as supplying ultimate customers – around Britain and Europe; he was even prepared to ship vegetable seeds to Bermuda. He seems to have spent an average of about £2.5 million *(£3,400)* each year between 1815 and 1820 on buying in plants, bulbs, seeds, tools and sundries such as twine and nails, together with china and glass for his floral displays.[52] Like Gurle and Brompton Park before him, Loddiges, Falla and Veitch after him and Hillier, Notcutts or Wyevale in recent times, Cochran's nursery was a sizeable enterprise. In 2015, there were 5,350,000 small businesses in the UK, each employing up to forty-nine employees; their average turnover was £227,000.[53] Cochran's was at least ten times this amount. Management on this scale – in the circumstances of the early nineteenth century – was not a simple task.

A FAMILY BUSINESS

The typical nursery, like the typical shop and indeed most other enter-prises before or during the nineteenth century, was a family business. Indeed, many have continued to be until today. If there is anything surprising about the history of individual nurseries, it is that so many of them have survived for several generations, when the typical small firm has a much shorter life. But, of course, most nurseries did not last long enough for their histories to be written, so the histories of, for example, Cochran, Caldwell, Cheal's, Suttons, Veitch, Loddiges, Hill-ier or Aylett are of exceptional firms in many respects and we should be careful about making generalizations based on them.

Family firms have many advantages. The creation of often highly driven individuals, they can be shaped by their founder's enthusiasm and expertise; many nurserymen clearly loved particular species and devoted their lives to studying, propagating and selling them. The fam-ily provides a ready-made labour force, if often underpaid, and – with luck – a clear successor to run the business; as well as wives and chil-dren, brothers and sisters and their children can be drawn in. As a rule, family members are more trustworthy than outsiders and, once the business grows and provides an income for the family as a whole, there is an incentive to support it and keep it going through the generations.

Sometimes it worked well. The Caldwell family of Knutsford managed six generations, for instance. Confusingly, most of the proprietors were called 'William'. The first (although actually he himself came from a family of nurserymen) was born in 1766, the last in 1922; the Knutsford nursery was founded in 1796 and closed, following the retirement of the last William, in 1987. Thirteen different family members, including three women, took roles in running it over nearly two centuries. The much larger and better-known Veitch nurseries started in 1779, when John Veitch – the charismatic founder – was encouraged by his employer, Sir Thomas Acland, to set up a nursery on his family estate at Killerton in Devon; twelve members of the family (including one woman) then ran the different branches of the nurseries until the London branch closed in 1914 and the Exeter branch in 1969. Suttons – one of the best-known and largest seedsmen in the world – was founded by John Sutton in Reading in 1806 and run by twelve family members until it was sold in 1964. Cheal's of Crawley was led by five family members between the 1860s and its closure in the 1960s, while Hillier of Winchester is still in family hands five generations since its foundation in 1864.

The alternatives to relying on the family were often less attractive. Until the mid nineteenth century, business was usually conducted by individuals trading on their own account or by partnerships; the only exceptions, largely confined to international trade, were the great chartered companies such as the East India Company. Individuals were responsible for their own debts; partners were responsible for their own and their partners' debts. The modern limited-liability company – in which a shareholder is liable only for whatever sum he or she has invested – was not introduced until 1855–6 and then only for enterprises with several shareholders. Before then, partners had to be confident of each other as well, of course, as being in agreement about what the business should do. It must often have seemed better for a nurseryman – or other businessman – to keep the business 'in his own hands', to control its management and direction and to keep whatever profits it made.

Things could go wrong in family businesses. Children could rebel and refuse to enter the business or they could have new and different ideas about how it should be run. Illness and death, or, from the twentieth century, divorce, could take their toll. There might be no suitable successor, particularly as girls were hardly ever thought capable of taking a leading role. Biographies of nurseries and nurserymen, often written by family members, are sometimes reticent about such problems. But the Veitch nurseries can provide examples.[54] They were proof of James Veitch Senior's comment, in a letter to the great botanist Sir William Hooker, that he had 'been much harassed of late by family cares and perplexities of which I believe all persons who have families and are connected with business are more or less subjected to'.[55] The youngest son of John Veitch, James had been determined, in the 1830s, that his five sons should follow him into the family business, but all but one of them – the eldest, James Junior – had other ideas. One became a doctor, two took up farming in the Empire (one before ultimately returning to Exeter and the other taking to drink), while the youngest was forced by his father to give up his chosen career of architecture; unhappy at the nursery, he died young.

James Junior inherited his father's autocratic manner and was 'not loathe [sic] to put forward his own views on any matter in an emphatic way'.[56] Robert's return to England from South Africa meant that he could be put in charge of the new nursery that had been opened in Exeter, while James Junior concentrated on the growing London branch.[57] However, he took the opportunity of the death of James Senior in 1863 to remove much of the best nursery stock and the most skilled workmen to Chelsea, where they helped to develop the reputation of the firm for hybridized orchids and for ferns.[58] Robert bought the Exeter nursery; from then on, the two parts of the Veitch business operated separately and James Junior's second son, Harry, ultimately Sir Harry, took an increasingly leading role in London. It was the next generation, Harry Veitch's nephews James Herbert and John Gould Junior, who illustrate some of the other perils of running a family business: John Gould Junior was delicate and retreated into silence; James Herbert became increasingly erratic, alienated staff and customers with his fiery behaviour, had a complete nervous breakdown and died from paralysis, a euphemism for syphilis.[59] Harry Veitch, who had retired, had to return to run the business but, when John Gould Junior died in 1914, no one was left to inherit. The nursery closed and a great sale of its stock and land was held.

Conflicts between father and son, or between siblings, affected other firms. Martin Hope Sutton was forced to enter the family business in 1827 at the age of twelve as his father John was increasingly drunk – the result, it was said, of advice from his doctor to fortify himself with strong drink. John would only allow Martin Hope to study botany or deal in seeds outside working hours, but it was Martin Hope's insistence that led to the ultimate creation of the huge seed business.[60] Similarly, his son Martin John Sutton's repeated illnesses were ascribed to 'his difficult relations with his father, who constantly badgered him and and his brothers over such matters as extravagance in all branches of the firm and the brothers' habit of spending to the limit of their incomes'.[61]

Family firms could get stuck in their ways. A hundred and fifty years after the foundation of the firm of the Caldwell Nursery, David Caldwell joined the family business in the 1950s after horticultural training, but found it 'labour intensive, old fashioned and operated in a paternalistic way' by his father, Bill; David became disillusioned and emigrated to Australia, leaving no one to succeed Bill, and the nursery was sold in the 1980s.[62] There had been worse difficulties. In 1873 Alfred Caldwell found himself in charge of the business at the age of twenty-one, after the early death of his father; he 'married young and found himself in financial difficulties, which he tried to solve by stealing from the business. Eventually he was sent away . . .', though only to Manchester, and his younger brother took charge.[63]

Family members were not always suited to running a business – nurserymen were, naturally, sometimes more interested in plants than in management. The last Cheal, Wilfrid E., 'made no claims to be a businessman, and was always much more interested in the product than the process'.[64] Many gave priority to the condition of their nurseries, the extent of their greenhouses and the range of plants in their catalogues. To be fair, nurserymen faced constant and unusual difficulties; even accountants found it difficult to value stocks consisting of millions of plants – susceptible to drought, pest or disease – which might not be sold for several years, if at all, and this made it complicated to assess the financial viability of a business. The fact that nurseries bought and sold between themselves, at a variety of discounts, made accounting even more difficult. But few of them employed accountants or finance directors.

In the 1970s, the financial condition of Hillier, under Harold Hillier, was precarious. The firm's accountants pointed out that 'the business has grown immensely in recent years and it is now well past the stage where any one man can hope to appreciate all the facets of running the business. The Senior Partner should now be in a position where little, if any, work is undertaken by him' and 'it is . . . essential that a reasonably well-qualified financial man be recruited (or co-opted) as part of the Senior Management Team'. A further problem, common to family businesses, was that the 'family finances [were] intermingled with the business finances'; they must be divorced from them, not least to enable a full set of accounting records to be produced. It was impossible to draw up an organization chart and the 'office accommodation [was] appalling'. There was no adequate system of debt collection.[65]

The bank manager was willing to extend overdraft facilities of £980,000 *(£90,000)* but confided to Harold's son, Robert, that he could not make his father understand the seriousness of the position. Hillier was selling too many varieties – the accountants cited forty varieties of *Hydrangea hortensia* and forty types of hybrid clematis; they were selling fewer than ten of each variety a year of 70 per cent of what they were growing. The costs of wages and the office were not being properly attributed, so prices were too low. The wholesale and council trade had been developed, but profit margins were much smaller than on retail sales. In 1977 it was calculated that sales had to increase by over £2.1 million *(£246,000)*, or nearly one-fifth, just to break even. Harold acknowledged that all was not well: 'Economically, it is unsound and I believe there is no bigger fool in the nursery business . . . but it is, on balance, great fun and well worth while.' It took 'painful and often confrontational sessions' to persuade him to slim down his stock.[66]

Hillier, under Robert Hillier's management, took the necessary steps to return the firm to viability and to lay the foundations of its current

success. But other nurseries, judging by the size of many nursery catalogues, must have suffered from the same problems. All businesses are
prone to rest on their laurels and to resist innovation or the costs of
new investment, but – as with the Caldwells – conflict about change
within a family can be much more difficult to resolve than when there
is pressure from shareholders or independent advisers.

Despite these problems, it was only in the last decades of the twentieth century that family businesses began to be superseded in the
nursery trade; many still exist. A first step was often to bring in an
accountant, which – as with Hillier – might reveal the rocky financial
foundations of the business. Financial problems could lead other
firms to sell themselves to a group or chain such as Wyevale, which
now controls 148 garden centres around the country, or Dobbies,
which has thirty-four. Others realized that their land – irrespective
of their much-loved plants – was a major and underexploited asset.
Several firms, such as Hillier nurseries around Winchester, gradually
sold surplus but increasingly valuable land for housing. Land tenure
was often the crucial determinant of survival. If the nurseryman
owned his land freehold, perhaps on the outskirts of a growing town,
he could benefit from rising land values, perhaps ultimately selling for
housing development and moving his business. The widow of Jacob
Moore, a nurseryman in Birmingham, did this after the death of her
husband in 1827, while her son carried on the business elsewhere.[67]
Leasehold tenants were more vulnerable, as Samuel Harrison and the
Loddiges family had found when ground landlords exercised their
rights, as London grew in the nineteenth century, and turned the land
over to much more profitable housing. In 1847, the firm of Wilson
and Sadler of Derby was forced to move the large stock of its nursery,
adjoining Derby Arboretum, as the land was to be sold for building
'Terrace Houses of a high class'. The fact that the site was close to the
'celebrated' arboretum would increase the value of the houses.[68]
Finally, short-term tenants could find their rents increased as their
businesses prospered, perhaps driving them out of business, or their
tenancies could be terminated, whether for housing development or
simply to find a tenant willing to pay more.

A REWARDING OCCUPATION

Some nurserymen found their reward in heaven. Sir Harold Hillier once said that 'I like to think that, apart from Christianity, there is no other subject which better spreads the gospel of peace than gardening.'[69] Indeed, several of the leading nursery families had strong religious beliefs, which determined how they conducted the business. They had, by the nineteenth century, overcome the doubts of Thomas Fairchild that hybridizing plants was an interference with God's creation. Nor does Charles Darwin's publication of the theory of evolution through natural selection seem to have had much impact on the nursery world, although he visited the Veitch nursery in London and discussed the reproductive system of orchids with James Junior and his expert in the species, John Dominy.[70] Martin Hope Sutton was an evangelical Christian, who had thought of becoming a missionary and much later in life irritated his father by giving 20 per cent of his income from the seed firm to charity.[71] The Cheals of Crawley were a long-established Quaker family and their religion together with their social and business links with other Quakers continued to be important throughout the life of the firm, with later Cheals adopting a missionary outlook that put them at odds with some of their fellow Quakers. Business took second place; the nursery was never open on a Sunday. In the 1920s the Queen of Romania and her entourage insisted on entering while the family were at the meeting house; when Joseph Cheal returned, he 'raged around ... called them back, refused to sell to her and sent them packing'.[72]

Despite the psychological rewards of the nursery business and the esteem of their peers – or the knighthoods awarded to Sir Harry Veitch and Sir Harold Hillier – financial success mattered; after all, in most cases families depended on it. Assessing the profits of nurseries and seedsmen over three centuries is not easy. The archival records are scarce and dispersed; even when they exist, as with James Cochran, they are difficult to interpret. There is also a pervasive view, which can be traced back at least as far as John Claudius Loudon, that nurserymen and gardeners in general are ill-paid.

However, one possible measure of financial success, in view of the numbers of family businesses, is the personal wealth of family members when at death their assets are listed at probate.* On this basis, the nurserymen whom we know about seem to have done well. One of the first, Leonard Gurle, had inventoried assets of £2.05 million (£3,807 6s 4½d) in 1685, not including his unpaid salary as the royal gardener. There is no probate for Henry Wise of Brompton Park following his death in 1738, but a reasonable estimate puts his wealth at £179 million (£100,000). At the end of the eighteenth century, John Russell of the Lewisham Nursery in London, having 'raised himself by his skill and industry to a state of affluence rare among nurserymen and, after keeping his carriage and living many years like a gentleman, died in 1794 aged 73, leaving property to the amount of £20,000' (£24.8 million today).[74] But even Michael Callender, with his relatively small nursery in Newcastle, had a probate value of £340,500 (£300), about today's average among those who leave taxable wealth.[75]

There is much more evidence after the middle of the nineteenth century. Four leading nurserymen died in the 1860s, each leaving over £4 million in modern terms, though none could approach their contemporary, Joseph Paxton, who died in 1865 worth £112 million (£180,000). But the reluctant seedsman and sometime drunkard John Sutton left £9 million (£14,000) in 1863. Conrad Loddiges of Hackney – who by

---

* Some cautions need to be stressed. Most importantly, until 1898 probate values do not include real property – houses or business premises. Second, most people in the past, and many today, did not make wills or, if they did, failed to leave enough for probate to be needed or recorded; even today, government statistics on wealth at death relate to only one-third of the population. Third, assets could be, and often were, given away before death, while those who survived longest after retirement were most likely to have spent at least some of their money during their long lives. Last, we can identify only a very small proportion of nurserymen and many of them did not have their wills proved. However – with the exception of the change to the system in 1898 – all these cautions apply throughout the period since 1660, so that there is no systematic bias in the probate records other than that they relate to the wealthiest people in the population. Most of the evidence relates to older men, so one way of assessing the wealth of nurserymen in the past is to compare their probate values with the average wealth of men aged over sixty-five today. In 2011–13, that was £332,947.[73] The 'identified wealth population' on which this figure is based is approximately one-third of the total population. Men aged sixty-five and over had the highest average net estate values of any segment of the population.

the 1861 census was describing himself as a 'landed gentleman' – left £4.5 million *(£7,000)* in 1865. James Veitch Senior left £16 million *(£25,000)* in 1863 and his son James Junior £39 million *(£70,000)* in 1869. Shortly after, in the summer of 1870, James Junior's eldest son, John Gould Veitch (father of John Gould Junior, mentioned above), died at the age of only thirty-one, leaving £2.8 million *(£5,000)*.[76] The Veitch family were, in modern parlance, seriously wealthy; six family members together left, in modern terms, £35 million between 1880 and 1929.[77] However, other families could rival or even surpass them. The second Conrad Loddiges left £4.1 million *(£10,475)* in 1898 and George Loddiges £5.8 million *(£31,607)* in 1923, although the nursery had actually closed in the 1850s. Most spectacularly successful, however, were the Suttons, supporting the view expressed by Campbell in 1747 that seeds were a very profitable branch of the business. Eight family members who died between 1901 and 1972 left, in modern terms, a total of £128 million.[78] Another, less famous, seed firm, King's Seeds, which also still exists, had produced £19.5 million *(£117,819)* for Ernest William King at his death in 1930.

One of the most successful nurserymen of the early twentieth century is almost unknown today. Alfred W. Smith had 1,000 acres of nursery grounds around Feltham, Middlesex, together with one of the largest ranges of greenhouses in the world; he used no chemical fertilizers but 200 tons of manure a week. Notorious as a stern employer who also worked all hours of the day, sold his own produce at Covent Garden and took no holidays, he left £28 million *(£171,190)* in 1927. On a smaller scale, four Caldwells left a total of £13.6 million in modern values between 1873 and 1953.[79] The Cheals of Crawley were less successful, at least by the 1930s.[80]

In general, nurseries seem to have become less profitable during the twentieth century, although two members of the Notcutt family left £11 million *(£69,096)* and £1.5 million *(£9,324)*, respectively, in 1938 and a third, Maud, £2.7 million *(£46,192)* in 1955. Another example is Hillier, probably one of the best-known nurseries, whose owner in the second half of the century was Sir Harold, only the second nurseryman to be knighted for services to horticulture. He left only £303,000 *(£82,861)* in 1985, not a great deal more than his great-grandfather, Noah, who had left £101,000 *(£200)* in 1873.[81] However, the fact that Sir Harold had established one of the finest arboretums in the country, and generously given it, before his death, to Hampshire County Council, illustrates the perils of making too much of probate figures.

Almost by definition, the nurserymen who have left archives and probate records are likely to have been the successful, or lucky, ones. Others did not do so well. One example – to end this chapter as it began – is James Cochran, the London florist. He has been described as having been an 'astonishing success'[82] whose 'account books have survived as a unique testimony to his success'.[83] But actually those books tell a different story.

In the four summer months of the London Season of 1818, Cochran's order books record sales of over £1 million in modern terms, but sales of different kinds continued throughout the year, so that his annual income must have been between £2 million and £3 million at the height of his success. However, there are telltale signs that his business expanded too rapidly for him to cope. Cochran was born in 1760;[84] nothing is known of his early life, but from 1800 he was in partnership with Thomas Jenkins as nurserymen, seedsmen and land surveyors, renting nursery ground on the New Road – now the Marylebone Road – in north-central London.[85] This partnership was dissolved around 1806 but in 1811 Cochran took the decisive step of opening a shop at 7 Duke Street, off Grosvenor Square in the heart of fashionable Mayfair. By 1816 his business – both plant rentals and other sales – was expanding rapidly, but so were the unpaid bills. In May he listed them – 138 in all, owing a total of £313,000 (£424 5½d) – and noted the efforts made to get payment. His debtors included Lord Dillon, from 'Ditchley, Oxfordshire', noted as 'gone to France', who owed £9,000 (£12 4s); the Dowager Countess Waldegrave owed £720 (19s 6d) and a Dr Symons of Chiswick £15,000 (£20 6s 11d).[86]

As the business expanded, in late 1817 Cochran took over more land in Paddington for a nursery, refurbished it and opened it in 1818 as J. and J. Cochran Nurserymen; he had taken his nephew, also called James, into partnership. It seems likely that, in order to do this, he had to borrow £764,000 (£1,000) from Abraham Chambers (later one of his executors) and Robert Baxter and that he also used – illegitimately – the marriage portion of his second wife, Anna Louisa, whom he had married in 1815 when she was thirty-one, probably wanting her help in looking after his two young daughters.[87] In 1820, after Cochran's death, Anna Louisa sued Chambers and Baxter in the Chancery Court, alleging that on 1 July 1817 they had sold an annuity for £809,000 (£1,073 7s 10d) and 'given it to James Cochran instead of investing it for her as they should have done as Trustees for the Settlement'.[88]

By 1817, Cochran's business was substantial, but precarious. Although he now rented two nurseries – one of them the Argyle Nursery and Arboration, in Paddington Green – he had to buy most of his supplies from other nurseries in Britain and Holland before selling them on, or renting them out, to his clients. Comparison of his order books with his payments suggests that, at a rough estimate, he was spending, in modern terms, £2.5 million on external suppliers while receiving £3 million from his clients. But this doesn't take account of the costs of his nurseries or his shop, nor any bad debts. He probably simply wasn't making enough profit. One possible reason is that his basic accounting methods – one book for orders, another for payments – made it impossible to make accurate estimates or to work out whether he had made or lost money on any job. Another is that he was operating in a very competitive field and couldn't charge enough to cover his costs.

He knew that he was in trouble. On 16 November 1818, he signed his will. His first instruction to his executors (Abraham Chambers and 'John Fielder Gent.') was that they should repay the debt to Chambers and Baxter of £764,000 (£1,000); their second task was to pay £381,000 (£500) to each of his daughters 'if so much shall remain'; and their third to place £305,000 (£400) in the funds to provide an income for his wife, Anna Louisa. So he had sold her marriage portion for £761,000 (£1,000) and was providing less than half of that for her after his death. The will ends, plaintively: 'my property is so out in various ways and I am sorry to say smaller than is expected of me but [I] rest my confidence in my two good and valuable friends Abraham Chambers and John Fielder to act for my daughters to the best of their judgment'.[89] It seems unlikely that his confidence was justified.

Cochran died less than two years later and was buried at St John the Baptist Church, Egham, Surrey, on 26 April 1820. His nephew James carried on the business for a time, but it does not seem to have survived for long and James is next heard of as beadle of the Paddington workhouse. It is a sad end to the story of a nurseryman who, for a time, fulfilled the desires of the elite of Regency London.

6

The Working Gardener

'Better men than we'[1]

When Kipling wrote of 'the gardeners, the men and 'prentice boys / Told off to do as they are bid and do it without noise',[2] he was thinking of men like William Coleman, who was born in 1827 in east Leicestershire, where his father managed the estate of Rolleston Hall, and who began work there at the age of fifteen as a garden boy. Three years later, he was sent away from home to get more experience, working on two estates before returning to spend another year with his father. He was then ready, at the age of twenty-one, to become a foreman managing two or three other men and was sent by Messrs Knight and Perry, nurserymen in the King's Road, London – who were acting as an employment agency – to be in charge of the hothouses at Pontypool Park, Monmouthshire, said to be 'one of the finest forcing places in England'.[3]

It was, Coleman told the *Gardeners' Chronicle* in 1875, 'the hardest place I ever lived in'. The head gardener believed that each greenhouse should be kept at a constant temperature throughout the year, 'independently of climatal changes'. During the day, Coleman had a stoker and an apprentice to help him feed the boilers with fuel and to open and close the ventilators, but at night he was on his own in charge of seventeen boilers, which 'frequently kept me constantly employed until the clock struck two'. Other gardeners describe how a change in the weather could mean that, as soon as they had finished their round of the greenhouses, they had to begin again, aiming to keep the houses to within one degree Fahrenheit of the desired temperature. Not surprisingly, Coleman was 'obliged by illness to give up [his] situation [and was] advised to engage in outdoor employment'. He helped to lay out a new garden for a banker in Carmarthenshire and then spent three years as a general foreman at Crewe Hall, Cheshire. In 1854, he applied to James Veitch Junior of the Royal Exotic Nursery, another nursery acting as an employment agency, and was soon sent – at the age of twenty-seven – to his first post as a head gardener, working for Lord Cloncurry at Lyons Demesne in County Kildare, now in Eire. After a few years, 'having established a good family character', Veitch found him the post of head gardener at Eastnor Castle, Herefordshire, an imposing, indeed forbidding, Gothic Revival pile designed to imitate the medieval castles that had guarded the Welsh border. Plate 31 shows Coleman in that role.

Coleman's first task for the 3rd Earl Somers at Eastnor was to cope with the devastation of the Californian and Mexican trees that had been killed by the hard winter of 1860–61. The earl 'at once set about replacing them with the rich treasures sent home from Japan by Messrs. Veitch and Fortune', while Coleman was also instructed to renovate the fruit garden, rebuilding all the glass structures but contriving to keep the fifty-year-old grapevines in production and satisfying the earl's 'great demand for Figs'. With a high reputation among 'the gardening fraternity' for his fruit, Coleman 'received offers which might have induced some men to change, but so great has been the confidence reposed in me, and so uniform the kindness and liberality of my noble employer and his generous lady, that no monetary consideration' could tempt him away from 'a labour of love'.

That labour, nevertheless, was well rewarded. Coleman died in 1908 and the probate value of his estate was £2.2 million *(£6,147)*. He was significantly richer than many of his fellow gardeners, though by no means the wealthiest.[4] Seventy-six gardeners who died between 1805 and 1972, most of them active during the reign of Victoria, each left over £1 million in modern values; even excluding Joseph Paxton, the average value of their estates was £3.37 million. Yet Coleman's career ran along very similar lines to that of other gardeners – garden boy, apprentice, garden labourer, foreman, head gardener at around the age of thirty. This chapter is about men like Coleman, the working gardeners, how many there were, how they were trained, how they progressed in their chosen career and how well they were rewarded.

THIRTY-FOUR KINDS OF GARDENER

The term 'gardener' has always been ambiguous. In 1822, John Claudius Loudon, who loved making lists, divided gardeners up into thirty-four different kinds, covering between them 'operators or serving gardeners; dealers in gardening or garden tradesmen; counsellors, professors, or artists; and patrons'. The lowest of the grades were garden labourers, the highest were 'proprietors of gardens', though he also remarked that

> Every man who does not limit the vegetable parts of his dinner to bread and potatoes, may be said to be a patron of gardening, by creating a demand for its productions . . . [T]he more valuable varieties [of patron] are such as regularly produce a dessert after dinner, and maintain throughout the year beautiful nosegays and pots of flowers in their lobbies and drawing-rooms.[5]

In other words, Charles II was a gardener, just like the much-maligned jobbing gardener of Victorian suburbs; in fact, every man was a gardener. So too were the 'lady gardeners', even if they were never allowed to touch a spade. Today most people described as a 'gardener' are working, unpaid, in their own gardens.

The ambiguity makes it very difficult to know how many gardeners there were at different periods in English history; in the widest sense used by Loudon, it is most of the population. But this chapter is about paid gardeners, called by Loudon 'operators, or serving gardeners'. The Company of Gardeners, in the City of London, had 500 members by 1622, who by 1649 employed 1,500 men, women and children and 400 apprentices. In 1701, when the company tried in vain to extend its control of the trade throughout England and Wales, it estimated that there were 10 garden designers, 150 noblemen's gardeners, 400 gentlemen's gardeners, 100 nurserymen, 150 florists, 20 botanists and 200 market gardeners, a total of 1,030, but this did not include their apprentices, labourers or foremen. George London and Henry Wise, and later Capability Brown, are known to have employed hundreds of men on their larger projects, although many would have been labourers employed to shift tons of earth. The proliferation of

London and provincial nurseries in the eighteenth century would have added more, so it seems reasonable to guess that there were at least tens of thousands of paid gardeners by 1800, when the population of England and Wales was about 9 million.

Relative certainty about numbers comes only in 1851, with the first British census to record detailed occupations.* Paid gardeners were divided into three types: domestic gardeners, jobbing and other gardeners, and nurserymen. The great estates employed 'domestic gardeners', who were regarded, and at times taxed, as domestic servants; there were about 5,000 of them. The middle classes of Victorian England in their villas and small estates employed nearly 80,000 other gardeners – all male; some would have worked full time, others as 'jobbing gardeners' working on a daily basis for several different employers. Finally, there were 3,000 nurserymen and a few nurserywomen. After 1851, numbers rose rapidly and they had doubled by 1891. By 1911 they had reached over 280,000 and, although the number dropped slightly after the First World War, it then rose again to a peak of just under 300,000 in 1931;[6] it may actually have grown even more, but the intended census of 1941 was cancelled because of the Second World War. There were still nearly 200,000 paid gardeners in 1951 but the number then declined to just over 100,000 in 1981, after which the separate category ceased to be recorded.† In all years, the actual number of paid gardeners may have been higher, since the census enumerators recorded only the main occupation; men working in gardens in their spare time would not have been included, nor would those paid 'cash in hand'.

These are big numbers. Gardening was, in the late nineteenth century, one of the top twenty occupations for men and boys. Successive censuses show that it represented over 2 per cent of occupied males in 1901, 1911 and 1931 and was still over 1 per cent in 1951. Before the First World War, more men were employed in gardening than in each of the census categories of public administration, the armed services,

---

* The census records the 'principal' occupation of the individual.
† Census data are bedevilled by changes to geographical areas (UK, Britain or England) and changing subdivisions of occupations, so these numbers are not as precisely comparable as they may seem.

the chemical industries, print, paper and books, in the docks or at sea or in the provision of gas, water and electricity.

The lives and careers of these men – and very few women – illuminate their place within English society. Biographical material on working gardeners is hard to come by until the publication of interviews with head gardeners in several gardening magazines in the nineteenth and early twentieth centuries; even then, they were the successful ones. So we have to pick examples from records of the lives of paid gardeners, such as the men from the late Victorian period shown in plate 30, where and when such evidence is available. This probably does not matter unduly, because the career patterns, training and lives of working gardeners seem to have changed remarkably little over many centuries, at least until a few horticultural colleges were founded at the end of the nineteenth century.

APPRENTICES AND JOURNEYMEN

When it comes to gardeners working in England, there are two stereotypes: they are the sons of gardeners and they are Scottish. There is a considerable element of truth in this, but neither stereotype is unique to the gardening labour force nor even unusual in the context of the English labour force as a whole.

First, inheritance. It was perfectly normal for sons to follow their fathers into the same occupation. Even as late as 1839–43, nearly half – 48 per cent – of bridegrooms who married in a sample of twenty-nine English counties had identical occupational titles to those of their fathers.[7] The proportion varied considerably between occupations – it was higher for the titled aristocracy and farmers, as it was for miners, who tended to live in isolated communities, and for labourers, who found it difficult to rise up the social ladder – but many skilled and semi-skilled men found jobs for their sons, either with them in a family business or with their own employer.

There were good reasons for this; it was cheaper to train your son yourself than to find him an employer, and most jobs were secured by word of mouth rather than by formal advertising or modern recruiting methods. So it is no surprise that William Coleman, like many other nineteenth-century head gardeners, came from gardening stock; William Cresswell, who kept a fascinating diary of his work at the great house of Audley End, Essex, in the 1870s, was the son of a professional gardener in Grantchester, near Cambridge, and there were many others.[8] Even in the twentieth century, the television gardener Percy Thrower and Arthur Hooper, whose life in the bothy we'll come to later, followed their head-gardener fathers into the trade.

Next, the Scots. Mr McGregor pursuing Beatrix Potter's Peter Rabbit and Mr McAllister making miserable the life of P. G. Wodehouse's Lord Emsworth of Blandings Castle are burned into the English imagination. But from the late sixteenth century there had been complaints about the number of Scottish emigrants who were taking jobs from English gardeners, and the Company of Gardeners argued that their number should be restricted. Some employers agreed and even refused to give them jobs; they occupied the place in English society that the Irish in the nineteenth century, and black and minority ethnic groups more recently, were to fill. It has sometimes been argued that Scots were better educated than the English in the eighteenth and early nineteenth centuries, before the introduction of compulsory schooling, and that it was this that gave them an advantage over their English colleagues. Recent research on literacy rates and the provision of parochial schools tends to discount this, however.[9] The main reasons for Scottish (and Irish) migration to England were that both those countries were relatively poor, that both were ravaged by famine and events such as the Highland Clearances from the mid eighteenth century, and that England was where the jobs were, even if they were often badly paid. In fact, migration from Scotland and Ireland to England even in the late nineteenth century largely replaced – in arithmetic terms at least – English migration to North America and Australasia.[10] They were the economic migrants of their time. But there is no way of knowing whether there really were more Scots among gardeners than can be expected from the proportion of Scots in the English labour force as a whole.

In many other ways, the training and careers of gardeners mirrored those of the rest of the labour force. It was normal for boys to begin full-time work between the ages of thirteen and fifteen – or sometimes earlier, William Coleman being unusual in starting so late – after a period of schooling that left most of them able to read and write.[11] Employment in a wide range of occupations began with a period of formal or informal apprenticeship. Only the learned professions – law, medicine, the Church – relied on formal education. There was no university or college training in other vocational areas until the end of the nineteenth century; the factories, canals, railways and machines of the Industrial Revolution were built by men who had no such training in architecture or civil or mechanical engineering.[12] It was 'learning by doing' or 'learning through looking'. The apprenticeship system had its origins in the regulation of trades by the medieval craft guilds, at least in larger towns. As a means both of regulating numbers in the trade and of assuring quality, apprentices were 'bound' by a contract to serve a master – who would normally provide food and lodging – through a period of training which, under the Statute of Artificers of 1563, lasted for seven years. Successful completion meant that you could then be employed as a journeyman, a skilled worker who had not yet become a master or employer. By the eighteenth century, formal apprenticeship of this kind was breaking down; the length of an apprenticeship had decreased and the system had disappeared in some trades. But training continued, in gardening as in many other occupations, through informal apprenticeship.

Percy Thrower, who began work in 1927, at the age of fourteen, under his father at Horwood House in Buckinghamshire, recalled how

> To begin with, I worked in the greenhouses under one of the local lads who was training to be a gardener, sweeping floors, crocking pots, cleaning out the stokehole, clinkering the fire, making sure there was always plenty of fuel for the boiler, and suchlike basic jobs. Gradually I was given more responsibility – a little pricking out, some potting up, perhaps.[13]

At Aynhoe Park, Northamptonshire, in 1915, Ted Humphris had to lead the pony that, with leather boots over its hooves to protect the grass, hauled the giant mower across the lawns of the pleasure garden.[14] Arthur Hooper, who began work at Bemerton Lodge, Salisbury, in 1922, spent – like many trainees – a year in the glasshouses, a year in the kitchen garden, another in the pleasure garden, before returning, now an 'improver' – not yet a journeyman – rather than a 'boy', to the glasshouses.

A hundred – and probably two hundred – years earlier, careers progressed in the same way. Loudon advised in 1822 that youths intending to be serving or tradesmen gardeners should be placed under master gardeners, because of the opportunities for instruction this provided. It was better for them, he thought, than employment in market or nursery gardens. The aim was to ensure that the apprentice, and then the journeyman, gained experience in all kinds of garden environments. By Loudon's time, an apprenticeship lasted only three years, followed by a succession of jobs as a journeyman; Loudon was adamant that these should be, year by year, in different places. Only at the age of twenty-five, after this experience, could a man aspire to be a foreman, before achieving the 'rank and title of master-gardener' and, probably, a post as head gardener, around the age of thirty.[15]

Until well into the twentieth century, however, employment for anyone – however well trained – was insecure. Even in highly skilled occupations such as mechanical engineering, men could be hired or fired at any meal-break, at the whim of the employer or foreman. Trade unions could, where they existed and where they were tolerated by the bosses, negotiate for better wages or hours of work and, sometimes, over health and safety issues, but they could do nothing to protect individuals from capricious managers or from the vicissitudes of trade; men and women could be 'laid off' at a moment's notice if demand slackened or the weather changed.

Working conditions for younger gardeners were, by modern standards, arduous, even in the most prestigious of gardens, although they were probably not very different from conditions in many other skilled trades. Hours of work were from 6 a.m. to 6 p.m., six days a week, with breaks of half an hour for breakfast and an hour for a midday meal. In some gardens, men were expected to work longer, without extra pay, in the summer to make up for hours lost by lack of daylight in the winter months. Men had to be at their place in the gardens throughout these hours; time walking to and fro, and time taken to clean one's tools, was not paid. Head gardeners enforced a strict dress code of shirt, waistcoat and tie and trousers, with an apron worn over the top, together with stout leather boots. This was despite the fact that under-gardeners were, wherever possible, kept out of sight of the owner and his guests; at some houses a tunnel ensured that they could not be seen on the way to work. Like servants in the house, gardeners were expected to move away from any area of the garden when the owner, or his visitors, appeared.

Hard working conditions and strict discipline could be found in all forms of employment across the country. But there were particular aspects of training and early employment as a gardener in the great estates that were exceptional.

Young gardeners were expected to devote their lives to the job and to live next to their place of work, in the bothy. This was the cottage, or more often the hovel, in which apprentices and journeymen had to live. Bothies can still be seen, often set against the shady north wall of a kitchen garden on the outside; it is sometimes difficult to believe that young men were expected to live in such damp and cramped conditions. Bothies seem to have varied greatly, since their standard was entirely dependent on the attitude of the head gardener or employer. Owen Thomas, later to be a royal gardener, recalled how when he arrived at Little Aston Hall, Sutton Coldfield, in 1863, he found that the bothy consisted of

> one miserable small room, with scarcely any ventilation, and practic-
> ally the only furniture was a dingy bed in one corner, on which two of
> us had to sleep. In this wretched place we had to do everything our-
> selves as best we could – the cooking, bed-making and washing up.[16]

Daniel Judd found that the bothy at Brocket Hall, the seat of Lord Melbourne, was in the 1830s

> a wretched place, situated between and joining two stokeholes . . . the
> roof covered with the old fashioned pantiles, without any ceiling, so that
> when there came drifting snow it found its way to us as we lay in bed.[17]

At Holkham, boys were expected to live next to the boilers that they had to stoke through the night.

Although conditions are said generally to have improved in the late Victorian period, Arthur Hooper, in his first job at Fonthill House, Wiltshire, in 1926, found himself sharing a bedroom, which was lit only by candles; downstairs was a kitchen and a sitting room containing two wooden chairs, a dartboard and a shove-ha'penny board. There was an outside earth toilet and washing was done with hot water from the kitchen range. 'It was', as he said, 'all so very different from what I had been used to at home.'[18] But when he moved, as a journeyman, to Norman Court in Hampshire, things got worse. There were no furnishings at all in the sitting room, there were not enough chairs in the kitchen for everyone to sit down, he had to share a room – accessed by a door that was only 3 feet high – with two others, and the only means of washing was with a bucket, since the head gardener thought it would be extravagant to pipe water into the house. By contrast, the stable for the garden's horse and all the glasshouses each had taps.

Many apprentices and journeymen found living in the bothy quite exacting. The rules of communal living were enforced partly by peer pressure and partly by the foreman in charge, who answered to the head gardener. No women were allowed inside, except for a cook who provided a cooked breakfast and a hot lunch; tea, at the end of a twelve-hour working day, had to be made by the men themselves and was often just a pot of tea, bread and jam. In some places, the cook was allowed to use fruit and vegetables from the kitchen garden and some minor poaching from the estate might be condoned, but in others nothing could be taken. At Chatsworth in 1838, the sixth rule to be observed by the young men as gardeners was (in the original spelling): 'If any one guathers aney kind of vedgatable fruits flowers speciments or cuttings of any plants without permition he will be fined five shillings.* If repeted discharged.' At Mentmore in 1883 a similar rule stated: 'No articles are to be taken from the Gardens or Hothouses without the direct authority of the Head Gardener.'[19]

At least once a month – depending on the number of journeymen – a man was 'on bothy duty', tidying and washing up in the bothy and, simultaneously, in charge of all the gardens and greenhouses throughout the night, making sure that all the required temperatures were maintained. Even in the royal garden at Windsor in the 1930s, this meant reading thermometers by the light of hurricane lanterns, adjusting the massive roof ventilators until the right temperature was reached and then going round to check each greenhouse every time the weather changed.[20] In many places, although not at Windsor, where three

---

* Five shillings is a huge penalty, equal to £190 today.

stokers worked successive eight-hour shifts, it also meant tending the boilers. Although men received some extra pay for this 'duty', they were not allowed to leave the garden at all for the whole week. Holidays were rare or non-existent. It was not until the late 1920s that Arthur Hooper was given a week off with pay: 'We were all delighted at the news. Not one of us in the bothy had ever had a paid holiday before.'[21]

The men did, of course, have time off in the evenings and on Sundays. Some of this was constrained, however: several recall that they were expected to join the estate cricket team and, in some places, this seems have been essentially a condition of employment, as was acting as a beater for shoots on the estate. Otherwise, there was, Arthur Hooper recalls, poaching for rabbits and the occasional pheasant on those estates where a blind eye was turned to this.[22] Then there was the pub and the occasional village dance where the young men could meet girls, although any serious involvement was discouraged.

The pub, and girls, were seen as a distraction from what the bothy boys should be doing, which was improving their gardening knowledge. John Claudius Loudon urged that employers should provide a library of garden books for the apprentices and journeymen, although in many places this simply amounted – Hooper recalls – to tattered copies of the *Gardeners' Chronicle*.[23] Nevertheless, the trainees were expected to study, by candlelight, in the evenings and also to use the resources of the garden – the plants – to improve their botanical knowledge. Andrew Turnbull, who also became a head gardener, remembered going out into the garden every evening after working hours in the early 1820s to study and memorize the names of fifty plants, checking them the next evening before learning more, and Sue Dickinson, later head gardener to Lord Rothschild at Eythrope, recalls doing much the same in the 1970s. It gave her, she feels, an invaluable training as a plantswoman.[24]

The employer and his head gardener controlled every aspect of the young gardener's life. Apprentices in other occupations could not marry, but the prohibition extended on the great estates to journeymen gardeners as well. Although both Percy Thrower and Arthur Hooper found girlfriends, there was no question of marriage for a bothy boy. When Percy Thrower joined the staff of the royal garden at Windsor, he was warned: 'Keep your eyes off the head gardener's daughters'; he didn't, and ultimately married one of them, but the relationship had to be kept secret for years.[25] One of his colleagues was less prudent and 'came – almost on his hands and knees – to the foreman to inform him that he had to get married and would have to leave'.[26] A colleague of Arthur Hooper's, who married one of the housemaids, was allowed to have a house on the estate but demoted from his position and wages as a foreman, while his wife had to give up her own job after marrying.[27] As the average age of first marriage, both for men and for women, fell significantly during the eighteenth and nineteenth centuries – for men from a peak of twenty-eight in 1670–79 to below twenty-five in the 1830s – the complete prohibition on marriage for a trainee or even journeyman gardener must have become increasingly irksome.[28]

The absolute power of the employer, or of his head gardener, is encapsulated in a warning note contained in the rules of the garden at Mentmore, Buckinghamshire, in 1883 set out by the head gardener J. MacGregor (a real-life 'Mr McGregor'):

> Note: The Head Gardener will not forward the interests of those who do not perform their duties to the best of their abilities, or who are not diligent in carrying out his orders, or those of the foreman, and he hopes that all men, except those on duty, will attend Divine Service at least once on Sunday.[29]

The head gardener could, and did, make or break the career of anyone working under him, as could any employer or manager throughout the English labour force as a whole. The power of the head gardener extended even to enforcing a move to the other end of England. Working for the Duke of Devonshire at Chiswick House, London, in 1848, Robert Aughtie was 'much surprised, on returning home, to hear from Mr Edmunds that I would be going to Chatsworth'.[30] Although that was at least a transfer from one house to the other of the same employer, the Duke of Devonshire, the transition from London to the wilds of Derbyshire was still clearly traumatic. Other trainee gardeners found themselves suddenly instructed to go to another employer, possibly because one of the head gardeners had asked a friend to help or perhaps, ostensibly at least, for the good of the young man. Arthur Hooper, working in Wiltshire in the 1920s, was 'called into the Head's office and told that I was to be moved to Norman Court, Hampshire ... Mr

Parsons had been in bothy with Mr Mills some years earlier . . . and Mr Mills thought I would benefit by going there.' Hooper realized that 'there would be no use protesting, as the Head obviously had someone in mind to take my place. That was the system; I suppose it was to some extent feudal, yet we were really the lucky ones, having a home and a job when so many people had neither.'[31]

There were advantages to apprenticeship, to the rotation of journeymen from job to job, and even to the bothy. The system seems to have been effective in training young men in plantsmanship, covering the whole variety of tasks in the open air as well as in the greenhouses. As they progressed from one department to the other and from boy to apprentice to improver to journeymen, to under-foreman and then departmental foreman, they were also gaining experience in the management of others, which they would need to manage a large labour force as a head gardener. The bothy system, however unpleasant at times, enabled men to build up a network of friends within the wider labour force who could help in finding new jobs and in learning new methods.

The system, if a man persisted in it, could also bring reasonable rewards in the short term and substantial benefits in the long run. It is often alleged that gardeners were badly paid. Indeed, it is sometimes said that it was John Claudius Loudon, in the first edition of his *Gardener's Magazine* in 1826, who drew attention to the poor pay of gardeners. But in fact Loudon was careful to make a distinction between gardeners who were no better qualified or educated than a common labourer and those who had been educated, or had educated themselves, to a higher standard of knowledge:

> It is a common complaint among gardeners, that they are not sufficiently paid, and that a man who knows little more of gardening than a common labourer, is frequently as amply remunerated as a man who has served a regular apprenticeship to his business. This is perfectly true where the gardener is nearly or equally devoid of elementary instruction with the labourer. But the remark *does not apply* [emphasis added] to gardeners who have either received a tolerable scholastic education, or have made up for the defect of it afterwards by self-improvement; or if it apply to them, the blame is their own.

He went on to argue that although 'some noblemen . . . do not allow their head gardener more than the wages of a servant in livery', there were 'a number of proprietors . . . who cannot get gardeners so well qualified as they wish, and who would gladly increase the emolument for a superior class of men'.[32]

It seems that ordinary garden labourers were commonly paid about the same as agricultural labourers and, indeed, were interchangeable with them on large estates.[33] Trainee or trained gardeners seem to have received more, around the average annual wage at the time (about £25,000 in today's values), with more being paid to those working in botanic gardens and, of course, in the royal gardens. The issue is complicated by payments in kind. In 1926, for example, Arthur Hooper was paid £195 *(£1 4s)* a week as first journeyman at Fonthill House, but he lived rent-free in the bothy and had to contribute only £57 *(7s)* a week to the cost of his food, leaving a reasonable surplus for a young man who was, perforce, without dependants. He was able to build up some savings, which he later used to buy furniture after his marriage. Few other young men from the working classes would have been able to do the same.

One example of a gardener with several dependants who has been intensively studied is Joseph Allen, who worked in the village of Bolton Percy in Yorkshire, although apparently not on a large estate. He and his wife Jane were paid to keep careful accounts of his income and expenditure over an entire year in the early 1840s, with details of what they bought. These have been used to study their diet and nutrition.[34] They had five living children at the time. He was, perhaps unusually, in full-time work, taking off only Sundays and a day each at Easter and Christmas, and earned just above the average annual wage, £26,470 *(£36 4s 4d)*. However, this was not the total of the family's income as Jane and their ten-year-old son, William, were both earning too; the total was £37,170 *(£50 16s 11d)*, which was comfortably above their annual expenditure. Even so, calculations of their consumption of calories – flour and bread were the main items of their diet – suggest that they were still only at the margin of having enough to eat to maintain normal activity and work. Another, earlier example is the gardener from Ealing, in west London, described by Sir Frederick Eden in 1797: he combined several jobs to earn a labourer's wage, which just covered his outgoings, although it has also been argued that he had an income 'far above bare bones subsistence', particularly by comparison with his counterparts in other European countries.[35] Neither example supports the view that male gardeners were particularly badly paid.

However, women gardeners – throughout the centuries – fared much worse. At Hampton Court in the 1690s, they were paid less than half as much as men for a day's work and this continued to be the case.[36] At Chatsworth in 1918, the maximum daily wage for a woman was £19 (2s) while for men it was £47 (5s).[37] Women were employed on a casual or seasonal basis and were principally used for weeding because of their smaller fingers. There seems to have been no possibility of their serving an apprenticeship or climbing the career ladder; they were not allowed into the bothy and there are no recorded examples of women as head gardeners or even foremen until the twentieth century. Horticultural colleges admitting women, or private gardening colleges, were not established until very late in the nineteenth century. Gertrude Jekyll, one of the most famous garden designers of that period, was largely self-taught in her family garden, apart from some botanical drawing at a school of art. Kew did not admit women trainees until just before 1901.[38] There was clearly a great deal of pent-up demand and Swanley College, Kent, which had been formed in the 1880s to give horticultural training to men, had by 1903 become all-female.

By this time, gardening was seen as a respectable occupation for middle-class girls, akin to nursing or teaching, and it was indeed only the middle classes who could afford the fees at the colleges; no grants were available for women.[39] Very little is known about the students or their future careers, but Sue Dickinson remembers that the regime at Waterperry, Oxfordshire, where she trained just before the college closed in the 1970s, was 'spartan', probably like that at the girls' boarding schools from which many of the students had come. Training there was modelled – by the formidable Miss Havergal, who founded and was in charge of the college throughout its life – on the apprenticeship of a male gardener. The day began at 7 a.m. with chores – everyone had their own glasshouse and was expected to water the plants individually in their clay pots – then work a week at a time in each of the four departments: fruit, vegetables, flowers and glass. The training was very practical, with lectures held only after tea.[40] For Sue and others, it was the prelude to similar careers to those followed for centuries by the men, working in different gardens and gaining knowledge and responsibility.

Although life in the bothy and on the staff of one of the large estates could be taxing, the life of a gardener could be really unpleasant for anyone who tried to go it alone or was obliged to do so. 'Servitude' was the word that Archibald M'Naughton used when he wrote from Hackney in London to the *Gardener's Magazine* in its first edition of 1826. He had left Edinburgh in 1777 and had a series of jobs in London, including one in Putney which he was obliged to leave when the owner's daughter married, 'her husband having an aversion to Scotch servants'. He then tried, with a friend, to establish a nursery in Epsom but that failed and they lost all their money.

> Not liking to go into servitude again, I began jobbing on my own account, and a poor business I have found it ever since. When I first began, the highest wages I could get were 3s. a day,* and obliged to find my own tools. I had a good deal of employment at first, partly from the circumstance of being a Scotchman . . .

But M'Naughton suffered a series of misfortunes and, when he wrote, was contemplating being forced into the workhouse. He hoped that his tale would serve as

> a warning to gardeners when they are in good situations to keep in them . . . And, especially, let them never give up any place whatever for the condition of a jobbing gardener, for that is greater slavery than being a common labourer.[41]

Jobbing gardeners worked for one or a few employers, mostly in the suburbs, small towns and rural parts of the country. Many garden writers were dismissive of their skills or lack of them. Some gardeners, such as Archibald M'Naughton, were trained in the larger gardens but then decided, or were forced, to drop off that career ladder. Others found

---

* Three shillings in about 1790 equates to £197 today or about £16 an hour for a twelve-hour working day, significantly above the current minimum wage.

jobs initially in nurseries or market gardens before moving either to estate gardens or into jobbing. Then, from the middle of the nineteenth century, came new opportunities in the public parks. Although Arthur Hooper continued to find work in private gardens, Percy Thrower recalls how, in the 1930s, he and other young gardeners at Windsor concluded that 'the future lay in the public parks' as private estates, hit by high taxation and the economic depression of the inter-war years, reduced their gardening workforces. Thrower asked for the help of a representative of a seed firm and soon found work with the City of Leeds Parks Department.[42] After the Second World War, which disrupted many careers – although both Thrower and Hooper were exempted from military service because their gardening skills were employed in growing food – he became parks superintendent in Shrewsbury, where he was spotted by the BBC, commissioned to do radio talks and soon became one of the first of the television gardeners.

Apprenticeship and life as a journeyman gardener could certainly be hard. But what stands out from biographies of working gardeners is the enjoyment that so many of them found in learning to nurture plants and the skill that they gained in the process. The ability to identify hundreds of different plants and to know how to grow them, adapting their methods to soil and weather conditions, was developed without formal education but men still acquired great skill and expertise, as they did working in so many other occupations in eighteenth- and nineteenth-century England.

THE HEAD

All the hard work was rewarded if, usually in his late twenties or early thirties, a man became a head gardener; Joseph Paxton, who as an under-gardener at Chiswick charmed the Duke of Devonshire by speaking loudly enough for the deaf aristocrat to hear, was entirely exceptional in his appointment to Chatsworth at the age of twenty-three. In another way, he satisfied tradition; one of his first actions was to marry. As he recalled in a much-quoted passage, he arrived at Chatsworth at 4.30 in the morning and climbed over a gate into the garden; he surveyed the place and set the men to work before he

> got Thomas Weldon to play me the water works, and afterwards went to breakfast with poor dear Mrs Gregory [the housekeeper] and her niece. The latter fell in love with me, and I with her, and thus completed my first morning's work at Chatsworth before nine o'clock.[43]

Unlike the men who worked under them, head gardeners had to be married. In the early 1930s, Arthur Hooper applied to be head gardener at Heathfield Park, Worcester, where he had been working for some time. He spoke to his employers,

> who were taken by surprise and did at first not know just how to answer, then they pointed out that I was not married, and they must have a married man as Head. I assured them that I could be married in a matter of a month or two, as I was already engaged, and I was quite sure Dorothy would marry me when I could offer her a home.[44]

He didn't get the job, however, and the marriage was put off; it finally took place in 1934.[45]

The position of head gardener was an exacting one. On a big estate, he was in charge of a labour force at least as large, if not larger, than that employed in the house – domestic servants under the control of the butler – and on the estate, where workers were managed by the bailiff or agent. There could be as many as a hundred men working in the gardens. The domestic hierarchy was strict: the butler at the top, then the gardener and the housekeeper – Paxton's choice of bride respected this status – followed by layer upon layer of their respective staff, each knowing his or her place. The cook stood slightly outside in terms of status, below the butler and housekeeper, but was sometimes paid more, particularly if he was male and French, as at Shardeloes in the eighteenth century, when French chefs were rare. The head gardener had to be on good terms with the butler and particularly with the cook, responding to her – it was usually a woman – demands for not only particular produce, but also of particular size and quantity, to meet the whims of the master and mistress.

Within the pleasure and vegetable gardens, the control of the head gardener was absolute. He not only managed a large staff but also a large budget, easily the equivalent of half a million pounds a year today, often much more. He appointed the staff, bought plants and seeds, organized the work, all in the service of providing whatever his employer might desire or what would impress visitors. He was essentially in charge of a nursery, since plants had to be propagated, nurtured and pampered, often with the aim of extending – with the aid of hothouses, cold frames and all the other technology of the garden – the growing, flowering and fruiting seasons.

The head gardener and his men were responsible for providing and arranging flowers for the dinner table and to decorate the house in general. This meant working closely with the butler and footmen, often early in the morning before there was any chance of being seen by the family members or guests and then returning to the house after lunch to dress the table for dinner. Victorian flower displays were extremely elaborate and, although it became acceptable in the villas of the middle class for ladies to arrange vases and bouquets – a complicated language of flowers evolved that had to be learned by young ladies – in the big houses it remained a man's job to grow, arrange and display flowers throughout the house and throughout the year. Part of the reason was the weight and size of the vases and the amount of water that they held; at Waddesdon Manor, for instance, two Meissen vases each held three buckets of water and had to be emptied and refilled each week when new flowers were arranged.

The head gardener was provided, on top of his salary, with a free house and fuel and usually with free fruit and vegetables from the garden. Joseph Paxton at Chatsworth was given a horse. Jonathan Denby has identified and photographed many surviving houses built to accommodate a head gardener and shown that they became larger and more imposing as the nineteenth century progressed.[46] Houses in the royal kitchen garden at Frogmore and at the Rothschild house at Mentmore were, for example, particularly splendid, but the standard was in general that of a servant-keeping gentleman. This was perhaps no more than could be expected for someone who was managing more than a hundred gardeners and a substantial budget.

Percy Thrower recalls that, even in the 1920s, 'a head gardener of a large estate in those days was almost like the Lord of the Manor. He was looked up to in the locality and feared by most of his staff.'[47] His friends were the other head gardeners in the neighbourhood and those with whom they had trained and worked in the bothy. Thrower remembers C. H. Cook, at Windsor, as 'a gentleman in his own right. Dressed in black jacket, pinstripe trousers and bowler hat, he would pace through the gardens with the dignity of any lord . . . monarch of all he surveyed.'[48] One of his tasks was to escort the king (George V) and Queen Mary, and their guests, when they visited the gardens, while the other staff were expected to stay out of sight.

Cook, who would become Thrower's father-in-law, seems to have been reasonably benevolent; others were not. Arthur Hooper had great respect for his first head, Mr Mills at Fonthill, and was unpleasantly surprised when told to move to another garden. This feeling intensified when he met his new head, Mr Parsons, at Norman Court:

> He looked at me for a while, then said sharply 'Can you work?' and, without giving me time to reply, said 'What use were you to Mr Mills?' immediately followed by 'Are you ever ill?' Without pausing for an answer he went on 'You will always do as you are told, and quickly, make sure you are never late, and remember to wear a collar and tie at all times, I will not have my men coming to work half dressed.'[49]

Hooper remembers him as morose and reserved, except when the men were working in the garden after hours and Mr Parsons would chat with them about gardening, revealing his depth of knowledge. But his ruthlessness was shown when one of the old-established gardeners died, at the age of seventy-six: Parsons refused to grant his colleagues permission to use any flowers from the garden as a wreath for the funeral and only one gardener was given time off to attend.[50]

The head gardeners, together with the owners of leading nurseries and seedsmen, largely controlled the market for gardeners, at least in the larger establishments. It was the nurseries and seedsmen, with their travelling salesmen or, as they called themselves, 'representatives', who acted as employment agencies: they got to know promising and aspiring gardeners, notified them of job opportunities and sometimes employed them for short periods between permanent jobs. Nurseries seem to have favoured the employees of estates from whom they received good orders. Andrew Turnbull reported that his master at The Haining in Selkirk had recommended him for employment at an Edinburgh nursery and that he was promised the work by a traveller:

on the faith of his promise I walked about forty miles, and at the end of my journey was told that a great many more young men than were expected had come for employment from places in which they had a much greater interest than Haining, and in consequence they could not take me in.[51]

With power and status came substantial rewards. Many large estates, including the royal gardens, employed their head gardeners essentially as contractors, paying them a yearly sum to maintain the gardens, which would include their own profit as well as the wage costs of the under-gardeners, tools and equipment. This is not always obvious in the archival record, but it can often be deduced by comparing the sums paid to gardeners with those to other members of staff. For example, at Cannons, Middlesex, in 1722, the Duke of Chandos paid his gardener £200,000 (£100), while his chaplain received £150,000 (£75) and his chief steward £100,000 (£50); the duke also had a private orchestra of twenty-four players and had recently employed George Frideric Handel as his resident composer, so there was no shortage of money. The gardener was probably on a contract and the total sum included other garden wages and expenses.[52] He was responsible for a garden of 83 acres (some of which survives in altered form as Canons Park) with grand avenues more than 1,000 yards long, a canal and lake or 'great Bason' and an aviary that contained tortoises as well as birds: storks, flamingoes, ostriches and a range of wildfowl. More difficult to deal with must have been Virginian deer, a Mexican muskrat and a 'tiger' from Ghana.[53] At the much smaller establishment of Shardeloes, Buckinghamshire, in 1748, the gardener received £37,720 (£20), the same as the butler and the 'gentleman' (presumably the valet) but less than half what was paid to the French cook.[54] Sometimes the distinction between salary and expenses is explicit: at Chicheley, Buckinghamshire, in 1752, the 'gardiner' was given £35,500 (£20) together with £106,500 (£60) for expenses in the garden, in this case apart from the sum paid for the wages of other gardeners.[55]

If contracts are excluded, head gardeners received on average an annual wage – in modern values – of £32,000 for the period from 1660 to 1700, £47,000 for the eighteenth century and £53,000 for the nineteenth; all are above the average wage, though they are certainly not munificent.* However, all these sums exclude the many benefits in kind that head gardeners enjoyed. At Powderham, Devon, in 1730, William Lucombe received £90,400 *(£50)*, a horse and 'his and another's dyatt [i.e. diet or food supply] when ye family is at Powderham'.[56] The head gardener at Stow Bardolph in Norfolk in 1712 could dispose of surplus vegetables.[57] At Saltram, Devon, in 1814, David Smith received £56,000 *(£70)*, which included 'Wages, Board Wages and Keeping a Horse'.[58] Moreover, a major perquisite through the centuries was the head gardener's house, provided with free candles and fuel, and often with daily milk. The houses were usually built close to the walled kitchen garden so the head gardener or his servant needed only to step outside to help himself to fruit and vegetables, only occasionally extending this privilege to the men in the bothy nearby. Then, if the head gardener was on good terms with the gamekeeper and cook, a haunch of venison or a brace of pheasants might arrive. Thus housing, rates† and a lot of food came free to the head gardener and his family.

In some cases, a head gardener could combine his duties with running a commercial nursery. This was the case with William Lucombe at Powderham, who founded one of the largest nurseries in Exeter. The Veitch dynasty, probably the most successful nurserymen of the nineteenth century, owed their start to Sir Thomas Acland, who employed John Veitch at Killerton in Devon and who suggested in 1779 that he should set up a tree nursery.[59] Thomas Ayres, gardener to Lord Scarsdale at Kedleston, did the same in Duffield, Derbyshire, in the early years of the nineteenth century and his son advertised his skills as a landscape designer.[60] In the twentieth century, between the wars, a number of landowners saw establishing a nursery in the old kitchen garden as a solution to financial problems and employed their head gardeners to do the work and share the profits. But the greatest contribution to the head gardener's income and the probable source of the large sums that many of them left at their deaths was less well publicized.

* But as Ambra Edwards suggests, in *The Garden* of March 2018, a modern head gardener would be lucky to earn more than £48,000 and might well not be supplied with a house, so wages may have fallen in the long run.

† The rates (local taxes) on the house were normally paid by the employer.

During March and April 1871, the editorial and correspondence columns of the *Gardeners' Chronicle* discussed 'the gardener's discount', sometimes disguised as a Christmas present. Nurserymen and gardeners each blamed the other for the system by which nurserymen and seedsmen paid head gardeners a commission on their orders, sometimes as much as 25 per cent. A builder pointed out that the same was true of what were likely to be even more valuable orders for constructing greenhouses. The words 'bribery' and 'corruption' were used. It was pointed out that, if an employer was so foolish as to order plants or seeds from a new source, depriving the head gardener of his commission, it was easy for him to ensure that the plants did not thrive and to blame poor stock. Editorials argued that the solution lay in a more professional garden labour force and that part of the problem stemmed from the low pay of gardeners. But letter writers countered that it was just a normal way of doing business.[61]

Clearly nothing was done to stop the practice, as in August 1895 Hillier wrote to all their customers to say that in future they would only give a Christmas box to a gardener with the approval of his employer: 'This practice . . . lends cover to those Nurserymen and Seedsmen who offer gardeners as much as 10 per cent to 15 per cent commission as an inducement to secure their orders. Such transactions are a disgrace to our business.' The letter goes on to refer to legislation before Parliament that would outlaw such commissions; however, 'as a temporary arrangement . . . we now say that we can only give the usual 15 per cent commission to gardeners <u>with the knowledge and consent of their employer</u>.' This was despite the fact that, as members of the newly formed British Gardeners' Association, the firm wished 'it to be clearly understood that we have no desire to limit or curtail the incomes of the gardeners, who, in some instances are already underpaid, but our object is to have all transactions carried on in a clear and straightforward manner'.[62]

Commissions of this kind were not in fact made illegal until 1906 and it seems that throughout the nineteenth century, and probably earlier, they had been paid routinely to head gardeners. In some cases they were probably larger than Hillier was used to paying; the 1871 correspondence seems to imply that 25 per cent was quite normal. If that is the case, whoever was the head gardener to Alfred de Rothschild at

Halton, Buckinghamshire, in the 1880s could have received as much as £8 million commission on the £33,230,000 (£72,216 1d) that the firm of Veitch was paid for work done in that garden, now sadly much degraded.[63]

There were other ways to bolster a head gardener's income. Several became successful authors, writing books or columns for local and national newspapers. In the twentieth century, men such as Percy Thrower and Roy Lancaster were the face, and the mainstay, of television programmes viewed by millions. But commissions – together with free housing and other perks – seem likely to have been the major reason for the substantial wealth at death recorded for head gardeners.

CONCLUSION

To people today, used to a system in which men and women are prepared for the world of work by vocational training in college or at university, one of the most extraordinary aspects of the world of gardening is that the working gardeners of the past were not trained in that way – at least not until a few attended horticultural colleges at the end of the nineteenth century. Yet they achieved an extremely high level of skill in a wide range of activities, from botany to landscaping, water engineering to the management of a large workforce. Men who had left school at the age of fourteen with no formal qualifications, just able to read, write and do simple mathematics, learned how to identify thousands of plants – using both their common and Latin names – and how to nurture them. They were expected to be able to undertake basic surveying in order to lay out lakes and ponds, to become carpenters and glaziers to maintain the greenhouses, and to display their artistic talents to satisfy the exacting standards of Victorian flower-arranging.

Many did much more. The biographies of head gardeners mention time and again how their subjects became nationally renowned for their skill with particular species of plants, be it orchids, ferns or geraniums. They showed off their skills at horticultural and flower shows around the country, or at the pinnacle of the system in the annual competitions of the Royal Horticultural Society, even if it was their employers who took the credit for the prizes that they won. Working gardeners had to take solace in the esteem of their peers or, sometimes, to delight in naming a new plant variety or even having one named after them. Meanwhile, head gardeners had to satisfy the exacting demands of their master and mistress, or their cook, for a steady supply of fruit, flowers and vegetables and for manicured lawns and coppiced trees fit to be shown off to critical visitors. To do this, they had to oversee the daily tasks of perhaps a hundred men and boys and manage a large budget, negotiating with nurseries and other suppliers – of manure, plant pots or gardening tools. Some, such as Joseph Paxton, designed and built greenhouses as well as bought and grew new imports in a way that rivalled the great botanical gardens.

Even if all this was paralleled, in other professions such as engineering, shipbuilding or architecture, by men who had been trained in a similar way by apprenticeship and emulation, it was still a tremendous achievement, replicated as it was in thousands of great and small estates across the country and from generation to generation. Equally skilled men – and later some women – worked in nurseries, in botanical gardens and in public parks. Others scoured the world, risking their lives while developing a high level of botanical knowledge, to feed the insatiable appetite of English gardeners for new species and varieties.

What shines through the biographies and autobiographies of English working gardeners is their enthusiasm, their love for their work. Gardening is physical, back-breaking, wet, and immensely frustrating when plants fail to thrive or cold takes them unawares. The long years of training, twelve-hour days and life in the bothy had to be endured, while the financial rewards – at least before the status and perquisites of the head gardener were established – were adequate at best, and probably meagre by the standards of equally skilled professionals in other areas. But gardeners – together with the employers who encouraged or sometimes infuriated them – were always conscious of their creative role, making and maintaining beauty.

7
Technology

'Here from dry rocks, like Moses at a blow,
Command the cool translucent streams to flow'[1]

Thomas Savery took out a patent for his new 'Engine to raise Water by Fire' in 1698 and described it to the Royal Society in 1699; shown in plate 16, it was the first working steam engine, precursor to those that powered the factories, ships and railways of the Industrial Revolution, to the widespread use of fossil fuels and, less happily, to global climate change. Savery thought it would be useful for 'draining mines, serving towns with water, and for the working of all sorts of mills where they have not the benefit of water nor constant winds'.[2] But it was not long before it became part of English gardening. By 1729, as Stephen Switzer wrote, Savery's engines had been installed and were working at 'Cambden House' and

> in the Garden of that Right Noble Peer the present Duke of Chandois, at his late House at Sion-Hill; where the Engine was plac'd under a delightful Banquetting-House, and the Water being forc'd up into a Cistern on the top thereof, us'd to play a Fountain contiguous thereto in a very delightful Manner.[3]

Savery's engine, which had limited capacity and uses, had already been improved by Thomas Newcomen; Switzer also describes the Newcomen engine, which was, with modifications, used more and more widely for most of the rest of the century. It was probably a Newcomen 'fire engine' that was being installed at Stowe – to pump water from a well – when Capability Brown arrived there in 1741.[4]

Gardening was quick to borrow the new technology. But it was also no slouch at devising new methods of its own that would then be adopted in other fields: waterworks installed in the great gardens were the precursors of canals; builders of greenhouses and conservatories pioneered the use of central heating; garden buildings were among the very first to use metal and glass in construction. In the process, gardeners transformed the English landscape and created some of the iconic buildings of the eighteenth and nineteenth centuries.

'Technology' means 'how we do things'. For an economist, it refers to the whole complex of ways in which the economy operates. It is not only about machines such as Savery's engine; it also describes the working methods used in every form of economic activity, from architecture to zoos. It encompasses bookkeeping, ploughing the soil, digging canals. Technology changes because someone, somewhere, has a bright idea about doing things better and manages to convince others to adopt the newfangled notion. We usually call the idea an 'invention' and the process of putting it into practice 'innovation'.

Invention is often very difficult to explain; it's a creative act and as difficult to understand as the genius of Mozart or Jane Austen. It's sometimes just a tweak, such as altering the shape of a spade, sometimes as epoch-making as the steam engine. Quite often, it borrows something that works in another context in order to solve a different problem; this was the case with the lawnmower, as we shall see later in this chapter. Innovation, on the other hand, usually occurs because it pays; it makes it possible to do something better than before and to do it more cheaply. It can let the innovator steal a march on competitors, perhaps by using fewer labourers or economizing on fuel or raw materials. On occasion, an innovation is disruptive and sweeps all before it; much more often, it makes a bit of a difference but needs to be rethought and altered before being widely adopted. People need to be trained in the new ways; suppliers have to be found for the components of a new machine; money has to be borrowed to cover the cost of experimentation to get it right. All this has applied in the development of the technology for garden waterworks and greenhouses.[5]

Invention and innovation, however, are the basis of the economic growth that has occurred throughout human history but which has speeded up spectacularly in the past three centuries. They are the reason why each person in Britain today, on average, produces and consumes eighteen times more in terms of goods and services than in 1700. Understanding how people constructed larger lakes, or devised heating for their greenhouses, or mowed their lawns – topics to which we will turn later in this chapter – helps us to understand the wellsprings of that economic growth and the many ways in which knowledge and practice are transmitted throughout the economy.

That is why a garden writer, seedsman and designer such as Stephen Switzer was interested in a steam engine. He wanted to do things better and to make more impressive effects in gardens. Savery's invention had the potential to solve one of the greatest problems encountered by garden designers in the seventeenth and eighteenth centuries, how to raise water – particularly in the relatively flat countryside of most of England – to power the fountains and cascades and to fill the lakes that were such important features of garden design. But raising water, however important, was itself only one of a number of water engineering and architectural challenges faced by London and Wise, Switzer, Bridgeman, Woods, Brown and their contemporaries. In coping with those challenges, and working out how to grow the many new imported plants, garden designers and their subcontractors anticipated many of the achievements of the canal age and the age of construction in iron and glass. They were at the forefront of technological change in an era of invention and innovation. Here it was that theoretical and applied knowledge came together as the Enlightenment merged with the Industrial Revolution.

WATER

Water engineering, in the service of gardens, made a huge difference to the English landscape. England has many rivers, but hardly any natural lakes south of the Lake District in Cumbria. The expanses of water that we enjoy today are largely man-made, created by landscape designers and their patrons in the eighteenth century, or by water companies building reservoirs in the nineteenth and twentieth, or, most recently, by the landscaping of old sand and gravel pits. There is some evidence of Roman water engineering in Britain, and moats and fishponds survive from medieval times; the royal accounts of 1166 record a payment of £26 9s 4d 'in operatione fontis' (diverting a spring) for installing or maintaining what became Rosamond's Pool on the present site of Blenheim Palace.[6] Monasteries and country houses alike often had strings of fishponds to supply an important part of the diet. Moats and meres surrounded fortified houses and castles. But it was the seventeenth century that saw the beginnings of major waterworks, in the shape of three developments: river navigations, the draining of the Fens and the building of 'canals' in English gardens. The canals that now criss-cross the countryside took their name from these large rectangular ponds dug in royal or aristocratic gardens. Such waterworks were precursors of the lakes that we nowadays assume are natural features of the English landscape.

Good transport is vital for the economy, and so improving the rivers was an obvious first step as the economy of England expanded. By 1600, most of the populated parts of Britain were already within fifteen miles of a navigable river or area of coastline. Although a short stretch of a deadwater* canal was built at Exeter in 1564–6, the major inland investment before 1750 was in dredging, straightening and removing obstacles from rivers; it was carried out by river navigation companies authorized by parliamentary Improvement Acts and applied to at least forty rivers between 1660 and 1750. This gave the companies the right to charge tolls to cover their costs and any damages or flooding caused by their operations. It was a successful model and the number of miles of navigable river increased from about 850 in 1690 to 1,600 by 1750.[7] Problems were encountered, however, even on great rivers such as the Thames – local landowners had to be placated, watermills had to be bypassed and weirs and fish-traps removed – but the engineering tasks were reasonably straightforward.

More complex, although similar in nature, was the major investment during the seventeenth century in draining the Fens around the Wash in eastern England. This was a huge area of peaty marsh, the haunt of wildfowlers, and the motivation – as with the river improvements – was unabashedly commercial, to bring the land into cultivation. It did do that, but in retrospect it was a hugely expensive

---

* To distinguish it from the water flowing in a river; the canal was only one and three-quarter miles long and not very successful.

mistake. Essentially, the problem was that as the land was drained –
by dredging the rivers and pumping water into them from the
marshland – the peat dried out and shrank, taking the level of the
land ever further below the level of the sea into which the water was
supposed to flow. However, it took some decades, if not centuries,
before anyone would admit that it had been a mistake. In the mean-
time, ever more powerful and ingenious 'engines' – for a long time
entirely wind- or water-powered – were devised to raise the water
into the rivers and thence to the sea.

The 'canals' built in the great gardens of the Stuart and early
Hanoverian periods were – by contrast to navigations and fen
drainage – principally decorative, although they also served as fishponds or reservoirs. They were challenging as engineering projects,
however. The canals were typically rectangular, often set at right-
angles to the mansion to provide a reflection of the house from the
garden and of the garden from the house. Beautiful examples have
survived – those at Hampton Court Palace, at Wrest Park in Bedfordshire, at Hall Barn in Buckinghamshire (shown as it was in the
eighteenth century in plate 3), at Shotover Park in Oxfordshire and at
Chatsworth in Derbyshire, among many others – but most were
destroyed by Capability Brown and his peers less than a century after
they were first built. Many had been the creation of London and Wise
and about twenty of them can be made out, as drawn by Kip and
Knyff, in *Britannia Illustrata*.

Ponds, even small ones, are difficult to create and to manage even
today. For a start, there has to be a reliable source of water nearby.
Plastic pool liners have replaced beaten earth linings and mechanical
diggers have supplanted men with spades and barrows, but digging
and then lining a pond to make it watertight is still a major project.
Moreover, as every pond owner knows, that is only the beginning.
Water has to be circulated or allowed to flow through the pond to keep
it aerated. The edges have to be protected; soil must not be allowed to
creep over them. Duckweed, blanketweed and algae can disfigure the
appearance of the water and clog up any circulating pumps. Water
plants proliferate faster than weeds. Fish have to be nurtured and
protected from herons and other predators; in an early example, at
Hartwell House in Buckinghamshire in the mid eighteenth century,

the canal had to be drained to get rid of an otter that was stealing the fish, after shooting and hunting with dogs had failed to kill the animal. Ponds have to be emptied regularly for silt to be removed. The pristine surfaces and beautiful reflections created – at great expense – for the Chelsea and other flower shows, which have recently inspired so many to build their own pools, rarely last long.

The canals created in the great gardens of the seventeenth and eighteenth centuries were not small – indeed, some of them were vast. The Long Water at Hampton Court Palace was originally a rectangle measuring 105 feet by 3,800 feet. At Wrest Park the Long Water is about 60 feet by 600 feet and at Hall Barn the Great Canal is 148 feet by 591 feet, while at Westbury Court, Gloucestershire, one of the best survivals, the canal is 450 feet long. Many others, at such houses as Chatsworth, Longleat or Grimsthorpe, or in the Royal Park of St James's in London, were as large or larger. They were not usually very deep, but even a 3-foot depth of water implies a huge excavation – at that depth, making the Long Water at Hampton Court would have meant filling and carting away nearly half a million barrowloads. But that was only the beginning. The water mustn't be allowed to leak, through the intrusion of tree roots or the burrowing of rodents. Without plastic liners, the pond had to be made impermeable – in much of England outside the areas of clay soil – by 'puddling'. Up to as many as eight layers of clay mixed with lime and flints were spread on the pond bed, each rammed into place by men with large wooden hammers, though at Ashburnham Brown used oxen to do the job. The same technique had to be employed elsewhere and will be covered in more detail shortly.

Then there was the problem of the water supply, not only for the canal itself but for the fountains that were often set within it. In some cases, as at Hall Barn in Buckinghamshire, the canal was created by damming a stream, but this implied constructing a dam – in that case an earth bank – strong enough to hold back the water to the desired level and incorporating a weir through which the surplus water could flow. In other places, existing springs were used, but sometimes the water had to be brought from a distance through pipes or gullies. Stephen Switzer was employed by Lord Coningsby at Hampton Court, Herefordshire, early in the eighteenth century to provide the water supply for the Great Fountain there; he constructed a new river, just under two miles long, to feed the water through a series of pools, and achieved a fall of 4 inches in one mile along it.[8] Amazing though this degree of accuracy in surveying and construction may seem to be, Switzer considered it to be normal. The problem, there and elsewhere, was to build up sufficient pressure to make a fountain work. Sometimes, as at Chatsworth, the surrounding hills did the job, but in the Thames Valley at Hampton Court Palace, even after a river was diverted to feed the waterworks, there was never really enough pressure to put on the show that the royal family demanded.

Part of the problem for the garden designers and water engineers – usually the same people – was the ambitious model that their patrons had in mind. There were two emblematic waterworks that were well known in Stuart and Hanoverian England, illustrated or described in a variety of gardening books and visited by the fortunate few on the Grand Tour. The first were the gardens of the Villa d'Este at Tivoli, built in the hills to the east of Rome in the 1560s, and the second was Louis XIV's superlative seventeenth-century Palace of Versailles, outside Paris. With over a hundred fountains and waterfalls, all supplied by gravity from a diverted river, the gardens of the Villa d'Este were difficult to emulate in much of England, although the cascade at Chatsworth represents one successful attempt that has survived.[9] On an even smaller scale, dams, pools and cascades formed one of the main features of the landscaping of the poet and essayist William Shenstone at The Leasowes, near Birmingham, a celebrated *ferme ornée*, inspired by classical imagery, which was a great tourist attraction in the middle of the eighteenth century.[10] Versailles, where the number of fountains ultimately reached 2,456 and there was an enormous Grand Canal embellished with naval vessels and Venetian gondolas, was equally out of reach even for the English nobility as far as size was concerned, but it was possible to aspire to an imitation on a smaller scale – if sometimes not much smaller – with fountains and basins as well as canals.

Versailles demonstrates the sophistication and power of seventeenth-century water engineering in the service of gardening.[11] Louis XIV used more water there than was supplied to the whole of Paris. The pipework throughout the gardens was made of English lead, supplied by the Duke of Bolton and which, according to Switzer, 'brought him in at least 250000 *l* Sterling, clear of all Expences' – an amazing profit of half a billion pounds in modern terms.[12] The initial supply, in the 1660s and 1670s, was gravity-fed, supplemented by pumps and windmills that raised water to a storage tank high up in a tower. Even so, it could power only a fraction of the fountains at one time and an elaborate system had to be devised to supply only the fountains near to where the king was walking, with gardeners turning some off and others on as he and his guests processed through the gardens. In 1678, however, Louis held a competition for a way to get water from the River Seine, four miles away and lying 330 feet lower than the palace, with a hill in between. The result was the Machine de Marly, shown in plate 20, by far the largest and most expensive machine of any kind to be invented before the advent of steam power. The Seine was diverted to allow the installation of fourteen wooden water-wheels, each 39 feet in diameter, powering 221 pumps; they could raise water only 164 feet at a time, so the intervening hill, more than double that in height, had to be conquered in three stages, with two reservoirs partway up the slope. The wooden machine, which required fifty craftsmen to keep it running, was described as horrendously noisy and was still able to supply only a quarter of the Versailles fountains, even if Louis had not diverted much of the water to his new mansion at Marly. He decided, therefore, to build a 16-foot-wide canal and a series of aqueducts to Versailles from the River Eure, fifty-one miles away; at the height of the construction process, a tenth of the French army and 8,000 civilians were working on the project, which was aborted in 1688 only by the financial and personnel demands of the Nine Years War.

There was nothing like it in England, but, although not enough is known about the techniques used in constructing the English canals and fountains of the late seventeenth and early eighteenth centuries, the French example shows what could be done. The knowledge was transferable: Grillet, the builder of the Chatsworth cascade and the celebrated – but now destroyed – waterworks at Bretby Hall, Derby-shire, was apparently a pupil of André Le Nôtre, the designer and gardener of Versailles. Two Frenchmen, Salomon and Isaac de Caus, had, early in the seventeenth century, designed elaborate grottoes, such as those at Woburn, Bedfordshire, and Wilton House in Wiltshire; the latter contained 'spouting sea monsters, a table with hidden jets to wet the unwary, and hydraulically simulated birdsong'.[13] They also published books on techniques for raising water. Grottoes continued to be fashionable in the eighteenth and nineteenth centuries; they were often elaborately encrusted with tropical shells, and shellwork was a socially approved hobby for ladies.[14] Tunnels, sometimes under waterfalls, linked different areas of the gardens, as at Bowood in Wiltshire, Waddesdon in Buckinghamshire and Old Warden in Bedfordshire. In the late nineteenth century, such tunnels, together with rivulets, waterfalls and cascades, were usually made of Pulhamite, an artificial stone (referred to in Chapter 2) that still deceives visitors to estates all over England into thinking it is real stonework.[15]

A pumped water supply was used at Hampton Court Palace; a pump, supplying water to tanks in the roof of Blenheim Palace, was installed in the Vanbrugh bridge there – probably by Switzer – but submerged later by Capability Brown.[16] Water was needed in houses for normal domestic purposes – washing, cooking – and to guard against the ever-present danger of fire, so that the utility of having it 'on tap' was obvious. This was realized by the inventor of the home-grown British technology of the atmospheric engine, Thomas Savery. He had developed his machine to drain mines but, in his short book *The Miner's Friend*, showed how it could be used for 'palaces, or the nobility's or gentlemen's houses' by pumping water up 20 feet to a cistern. Then, 'in every staircase a pipe may go down the corner, or behind the wainscot, so as to be no blemish even to the finest of staircases. At every floor there may be a turn-cock with a screw' into which a leather pipe could be fixed and used to extinguish a fire.[17] It was this technology that was soon also applied, as Switzer described, to powering fountains. One of Savery's engines cost about £95,000 *(£50)*, which together with the continuing cost of coal and labour may have limited its appeal.[18] Another problem was that it could raise water by only 20 feet, because 'the pressure needed to force water to any greater height was liable to cause the boilers to burst and the soldered joints to melt'.[19]

Technologies based on water and, sometimes, wind power continued to be changed and improved during the eighteenth century. Switzer particularly mentioned and illustrated the 'multiplying Wheel

Bucket Engine' invented by George Gerves in 1725 and installed by Sir John Chester at Chicheley, Buckinghamshire.[20] Nothing survives except its picture, shown in plate 17, but the grounds of Chicheley Hall, now owned – appropriately – by the Royal Society, contain many aspects of its early eighteenth-century garden, including a U-shaped canal. The technology was then developed further by John Grundy at Gopsall Park in 1761, and housed in an engine house between 70 and 80 feet high 'built of Brick with Rustick Coines [i.e. quoins, masonry blocks at the corners of walls]'. Gopsall, in Leicestershire, was 'owned by Humphrey Jennens, an extremely wealthy Birmingham business-man, who had amassed an incredible fortune with his iron foundry business'.[21] In 1761 Richard Woods received a quotation of £83,000 *(£49 15s)*, plus the cost of a building to contain it, for a wind-powered engine to raise water into the lake at Cusworth, South Yorkshire.[22] In 1749 an anonymous visitor to Stourhead was so amazed by the pump-ing machinery that carried water from the lake to the house that he, or she, burst into verse:

> See yonder engine! Mark each curious part
> Where nature's power is overpowered by art!
> Water that downwards tend its boisterous tide
> Woundrous, Ascends a lofty mountain's side.[23]

One of the most ambitious schemes, still to be seen today, was that of Charles Hamilton at Painshill. His small estate ran alongside the River Mole and he decided to create a large lake whose surface would be 15 feet higher than the river, the water being raised by a waterwheel – or rather a series of waterwheels, as Hamilton and his successors experimented with the technology. The first effort was based, appro-priately for a man steeped in classical learning, on a design in Vitruvius' *De Architectura* (On Architecture), written in the first century BC and first printed in 1486, and had a wheel 36 feet in diameter that chan-nelled the water into a leather pipe running from its central core. It was claimed in 1752 that it could raise more water in twenty-four hours than the Machine de Marly.* After a dispute with a neighbour over use of the river water, it was replaced in 1770 by a horse-engine,

---

* This seems unlikely, since the French machine had fourteen waterwheels.

in which the poor animal was forced to walk in a circle for two and a half miles every hour; three further designs followed, the last being the steam-powered cast-iron Bramah waterwheel of the 1830s that still operates today, though now driven by electricity.[24]

However they were developed, the techniques of water engineering were soon applied in a new way that was pioneered in England though later imitated throughout Europe and beyond; this was – as in the example of Painshill – in the creation of lakes. Switzer had provided lakes for his clients, for example at Exton Park, Rutland, in the early eighteenth century, where two lakes of 70 acres were separated by a cascade over a dam, with grotesque rockwork and a two-storey folly.[25] Perhaps the greatest water garden that survives, Studley Royal in Yorkshire, was created by John Aislabie, Chancellor of the Exchequer between 1718 and 1721, following his disgrace after the collapse of the South Sea Bubble, one of the most notorious of all financial scandals.[26] He created a sequence of reflecting pools, lakes, canals and cascades, 'the wonder of the North', which was later brought to perfection by the purchase in 1768 of the nearby ruins of Fountains Abbey by his son, William.[27] Together, it is now justly a UNESCO World Heritage Site. The design of the original garden was basically geometric but, at the same time, made use of the potential of the wooded valley of the River Skell in which it was built and of the Yorkshire countryside around it. Naturalism in the use of water took a further step forward with William Kent's work at Rousham, Oxfordshire, creating ponds and waterfalls in the 1730s, before its full flowering in the second half of the century.

Whatever the intention of the designs, the techniques employed were clearly derived from those used to create the geometric garden canals. They were brought to perfection by Capability Brown and his men and by contemporaries such as Richard Woods. The objective was different in a subtle way, for Brown wanted to create the illusion of a river while actually making a lake or a series of lakes, but the technology and the challenges that it created for designer and contractor were common to both the new fashion and the old. It probably owed little, incidentally, to the draining of the Fens; it is sometimes suggested that Brown, perhaps in the 'missing years' before he moved to Stowe, during which his whereabouts are not definitely known, was in Lincolnshire learning water engineering. But the methods used to raise water into the fen rivers, by dredging, pumps and windmills, had little in common with those Brown was to use in his projects throughout England. His techniques had, however, been common currency in garden design for decades and were clearly well known to the contractors whom he employed to carry out the work.[28]

The novelty of Brown lay not so much, therefore, in the technology that he employed but in the scale on which he used it and in the technical challenges that he therefore had to meet. It is in scaling up a technology that problems often arise. Hall Barn's Great Canal is about 2 acres in extent; Brown's lakes could be 50 acres or more. Wotton's is 47 acres, spread across two miles of lakes, canals and artificial rivers – though Thomas Jefferson when he visited the garden in 1786 thought they covered 72 acres; Luton Hoo has two, of 14 and 50 acres respectively; Trentham in Staffordshire has one that is a mile long and 80 acres in extent; Coombe Abbey in Warwickshire has one of 90 acres; at the peak, as in so many other ways, is Blenheim, where the lake – much bigger than in the original design by London, Wise and Vanbrugh – is 150 acres.

Brown's lakes were usually quite shallow: he recommended a depth of between 3½ and 4 feet (although Repton later preferred 4½) even when the lake was to be used for boating.[29] But the weight of water was still enormous. This implied very big dams. At Wakefield Lodge, Northamptonshire, where Brown worked for the Duke of Grafton at his hunting box – the family mansion is Euston Hall in Suffolk – the great pond was held back by an earth dam measuring 700 feet long, 25–30 feet wide at the top and 80 feet wide at the bottom, that raised the water by 25 feet.[30] At Bowood, Wiltshire, the dam built for Lord Shelburne was 450 feet long and 30 feet high and, once the resulting lake was filled, it submerged the main Bath Road and a village – one notable result being Oliver Goldsmith's poem, *The Deserted Village*, published in 1770, though that distinction has been claimed by other places.[31] At Coombe Abbey, the dam is half a mile long and up to 20 feet high.[32] Brown constructed about forty such dams during his career.[33]

A look at one of the dams at Petworth, Sussex, a diagram of which is shown in plate 18, reveals how these huge barriers were made. At its core is a vertical clay wall. The inward sloping side towards the lake is constructed of rough clay faced with puddled clay and then faced with stone; the outward side is packed with soil.[34] At Blenheim, later, a similar structure had within it a cascade flowing over one end of the dam – still one of the most attractive features of the park – with a system of pumps and sluices that allowed the water level to be controlled. The surplus water was channelled away along a newly built side canal.[35]

Puddling was a crucial technique wherever the underlying soil was permeable. It involved lining the ground under the lake with successive layers of rammed clay. The first layers were mixed with lime to deter rodents. Then a layer of clay 4 feet thick was laid and rammed into place by teams of men with hammers; more, thinner, layers followed – laid at right-angles to each other – followed by a thin layer of chalk and a final layer of rammed clay.[36] The amount of manpower required to do this by hand over an area of 150 acres at Blenheim is almost unimaginable today (especially when it is added to the effort involved in digging up the clay, lime and gravel that was used) and it is not surprising that preparing the lake bed and dam took six years during the 1760s, with a further two years to fill the lake with water. Brown's works cost, apparently to the horror of the Duke of Marlborough, although he had commissioned them, over £35 million *(£21,000).*[37] Puddling, on much this scale, was employed, wherever it was needed, on the network of canals that was soon to be built throughout Britain; as we have seen, the technology was used earlier in England's gardens, though it may in fact have been used in Roman times.[38] Similar techniques have been employed the world over and for centuries: in the fourteenth century, for instance, elephants were used by the Raja of Udaipur in Rajasthan, India, to trample down the bed of the huge artificial fresh-water lake Pichola. At Cusworth in 1764, Richard Woods gave detailed instructions for constructing the dam:

> Let the clay be put in thin courses not more than 6 or 7 inches at each course, and well ramed . . . and as you advance in height with the clay keep filling up on both sides with earth wch. must also be as well ramed.[39]

The dam was an essential component of the three lakes, joined together in the form of a serpentine river, with a cascade and ornamental boathouse, that Woods created within the 250 acres of parkland at the heart of a landholding of 20,000 acres.

Blenheim is, in most ways, exceptional, but waterworks and lake-building were never cheap: Switzer referred to 'the vast Expence which attends the raising up and conveying Water for the Embellishment and Watering of Noblemens and Gentlemens Seats'.[40] There are few reliable costings of lakes, but Brown's estimate for the first that he built at Petworth, including enlarging an existing pond, building a dam and grassing the edges, was £1.2 million *(£700)*. This seems plausible; a recent example of lake-building, at the Center Parcs holiday resort at Woburn, cost £500,000 for a substantially smaller area.[41] There were some compensations: Switzer advocated lake-building for its drainage benefits and argued that the cost was outweighed by the improvement in the value of land.[42]

None of the work was easy or straightforward. Indeed, Brown's first attempt to create a lake, in 1746 within the Grecian valley at Stowe, failed entirely. Lord Cobham wanted a neo-classical temple to be reflected in 'new waters' and 23,500 cubic yards of earth were excavated to create a pond on two levels with a clay dam; but the springs that had been intended to fill the lake failed to do so and the whole scheme was abandoned.[43] At Petworth in 1755–6, it seems that the lake was initially not watertight and continued to give trouble, with the water level falling.[44] In 1722 at Croome, the lake began to leak, apparently because tree roots had caused fissures in the banks, and Lord Coventry wrote to Brown to ask for a 'man of practice and sound direction' to solve the problem.[45] Water broke through the dams at Burghley House in Lincolnshire and at Tottenham Park in Wiltshire.[46] The cascade at Grimsthorpe had been made too steep and, after thirty years, was in need of major repairs.[47] There were particular problems at Harewood, West Yorkshire, in 1777 when, after two years' construction work, the lake began to be filled and 'the water ran out half as fast as it came in', having 'made its way through the clay wall'.[48] At Sherborne Castle in Dorset two men drowned in the 1750s in the course of making the lake.[49]

Lakes were built to be drained, if necessary, to remove silt. At Petworth, the dam has within it a vertical, well-like structure made of brick, which connects at the bottom with a horizontal tunnel leading from the lake bed through the dam; in this is a sliding oak panel, which could be raised or lowered from the top of the 'well' by a rack and pinion, if the lake needed to be drained. Fish added to the complexity: at Rycote in Oxfordshire, digging out the silt from the Brownian lake recently revealed a wall running through the middle;[50] this probably made it possible for half the lake to be drained at a time, while the fish took shelter in the other half. Brown's sinuous lake replaced the three rectangular lakes, with some smaller ponds close to the house, near the 'East India Deere Park' drawn by Kip and Knyff in *Britannia Illustrata*; several of the bodies of water are likely to have been fishponds. Long-term maintenance and dredging of the kind that has been carried out at Rycote has been so expensive that some of Brown's lakes have since been neglected and allowed to silt up. One that has not been neglected is the lake at Blenheim, where removing the silt – 330,000 cubic yards of it – in 2020 from only a part of it, the Queen's Pond, is expected to cost about £6 million.

Almost all these waterworks took place before the classic canal age, signalled by the opening of the Bridgewater Canal in 1761. As we saw in Chapter 3, it was built by one of Brown's clients, the 3rd Duke of Bridgewater, to carry the coal from his mines in Worsley to Manchester. It worked; despite including a section underground and an aqueduct and costing £281 million *(£168,000)*, it greatly cut the price of coal and thus enabled Bridgewater to sell much more of it, making him very wealthy indeed. There were soon many imitators, sparking the so-called canal mania throughout the country as investors lined up to build new routes, and canal-building became one of the staples of engineers and contractors for several decades. The success of the Bridgewater Canal is often ascribed to the genius of James Brindley, its architect – he is even credited, wrongly,[51] with the invention of 'puddling' to make the canal watertight – but, as we have seen, the techniques of canal-building, including puddling, had been developed long before, in the parks and gardens of seventeenth- and eighteenth-century Britain and before that in France and Italy.

Water engineering for gardens and that for commercial canals were not exactly equivalent. Gardens did not need locks to raise and lower barges; commercial canals did not need fountains and cascades. But both used pumps where needed to maintain water levels. Both, too, were used by boats: at Wotton and West Wycombe in Buckingham-shire, as well as at the royal lake at Virginia Water, the lakes were used for 'naumachia' – simulated naval battles; cannonballs still turn up around the Wotton lakes,[52] which also had bastions on the banks from which guns were fired. Others were used for more peaceful pursuits, such as fishing, a fashionable pastime for young ladies in Hanoverian England; in the nineteenth century, there was a craze for breeding goldfish.[53] But the purposes do not need to be identical for engineering technologies to be transferred; indeed, a common and crucial aspect of technological change is that an idea developed in one context is applied and modified in another.

Canals were not, in any case, the only successor to the garden waterworks of the eighteenth century; the next two centuries were to see the building, throughout England, of the reservoirs that today provide so much of our water supply. It was the canals of London and Wise and the lakes of Brown that showed how this could be done and how the landscape of England could be transformed by the peaceful beauty of these large expanses of water. Because of the pleasure they bring, lakes and ponds are still being built, such as those created recently in the garden of Michael and Anne Heseltine at Thenford.[54]

FIRE

Central heating was another form of technology in which gardeners led the way. The plants that they prized, such as the orange trees that adorned Italian gardens, the pineapples that captivated Stuart and Hanoverian aristocrats and the tropical plants brought from all over the world by plant hunters in succeeding centuries, all demanded heat as well as protection from cold. The result was decades of experimen-tation with heating systems in garden buildings, long before they were provided for shivering humans in their chilly homes.

The Romans, and possibly other cultures in different parts of the world, had hypocausts; many public buildings and some domestic houses were built with floors raised on brick pillars, between which circulated hot air and smoke generated by a wood-burning boiler to heat the floor or walls of the rooms above. In some cases, brick or ceramic pipes conveyed the warmed air to upper storeys. It's still possible to see these systems in the ruins of many Roman buildings, in Italy itself but also in France and Spain, in Trier, former capital of the Western Roman Empire, and in numerous villas of the rich throughout the Roman world. But, in an interesting example of a forgotten technology, the system does not seem to have been used anywhere in Europe during the medieval and early modern periods. Its revival came only from the seventeenth century onwards, in the service of plants.

Citrus trees – both orange and lemon – were greatly admired in Renaissance gardens, not only for their fruit but also because of their sculptural qualities and the fact that they were evergreen; as we saw in Chapter 3, they were known as 'greens' and it was for them and for the equally admired myrtles that the first 'greenhouses' were devised, the term being used interchangeably with 'glasshouse' by the nineteenth century, if not earlier. They might be able to weather Italian winters, although even there they were often covered or moved indoors every autumn and out again every spring – a major task for gardeners; but north of the Alps there was no chance that they would survive and in fact they needed not only to be indoors but to be warmed. Other tropical and semi-tropical fruits such as pineapples and melons needed heat to germinate and grow. But, for a much wider range of plants, artificial heat was needed if they were to flower and fruit over as much of the year as aristocratic salons and dinner tables demanded. The technological challenge was taken up, particularly in England, and hot air, hot water and steam were pressed into service, along with direct heat from manure and tan.

Any horse-drawn society has plenty of manure and it must have been used for centuries to warm plants in hotbeds (beds of earth heated by fermenting manure), as well as to refresh the soil. A newer technology used tan or oak bark – a by-product from the tanning industry that had first been used in tanning leather – which was placed in large pits within specially constructed buildings and provided, as it fermented, a constant and economical heat source.[55] Philip Miller, at the Apothecaries' Garden in Chelsea (now known as the Chelsea Physic Garden) in the middle of the eighteenth century, believed that by stirring and adding more bark, the heat could last for two to three months.[56] The records of English country estates, such as Shardeloes in

Buckinghamshire, describe the construction of buildings to house tan pits and regular purchases of tan, almost certainly to make it possible to grow pineapples. Both to serve as the centrepiece of displays of fruit and flowers and – if you could afford it – to eat, these were one of the most prized and expensive products of Stuart and Hanoverian gardens. Other plants could be nurtured at least for part of their growth under cloches or 'bell glasses', or in the cold frames and hotbeds that were an important part of kitchen gardens. Richard Bradley gave instructions, in 1710, about how to make hotbeds for cucumbers, kidney beans and asparagus.[57] Eighteenth-century walled kitchen gardens, such as those to be seen today at Croome in Worcestershire or Holkham in Norfolk, often contained a further internal 'hot wall' – a boiler heated air that was then circulated through channels in the brickwork, encouraging growth and giving some protection from cold.

Many tender plants required, however, a warm atmosphere that could be provided only in buildings with wood- or, later, coal-fired heating systems. It was possible, of course, to heat plants – like people – with open fires, but there were two problems. First, it is very difficult to regulate the heat from an open fire, and plants, unlike people, cannot move to and fro – or change their clothes – to get warmer or to avoid overheating. Second, some plants were damaged by smoke and fumes. The initial solution, described by Isaac de Caus in the first half of the seventeenth century, to the problem of preserving 'greens' that could not be moved indoors was to erect temporary buildings, heated by charcoal stoves. Once permanent greenhouses and orangeries began to be erected, they too could be heated by charcoal; Richard Bradley, in 1710, recommended that 'a small Fire of Charcoal . . . when it burns clear' should be hung up at night near the windows once frosts began to set in. However, 'those Fires are not without their inconveniencies; where there are many of them, they are so suffocating that several Men have been choaked by them' and there was the ever-present danger of setting the house on fire. A better solution, he thought, was to use 'one of those Chimnies improved by the Reverend and Learned Mr Disaguliers' in the form of a hot-air system with a fire in a separate room.[58]

The danger of killing off your gardeners certainly recommended the use of systems of indirect heating, akin to the Roman hypocausts, although there was still a danger, which Bradley noted, of deadly smoke seeping through cracks in the bricks. The diarist and gardener John Evelyn gave an illustration, shown in plate 19, of a greenhouse heated by a hot-air system with pipes running from a charcoal furnace, in the eighth edition of *Kalendarium Hortense* (Gardener's Almanac), published in 1691. Also in the late seventeenth century, John Watts, the Keeper of the Apothecaries' Garden, had a greenhouse warmed by a 'great fire-plate', and the Archbishop of Canterbury had a three-room greenhouse at Lambeth Palace, the central room being the hottest and designed for tropical plants, with stoves and flues underneath.[59] Later greenhouses, such as one at Holkham Hall, Norfolk, contained underground chambers with stoves that were tended by boys who were expected to sleep next to the boilers. This system continued to be used in garden and other buildings well into the nineteenth century.[60]

Heating by steam, provided by coal-fired boilers, developed after the middle of the eighteenth century, although it does not seem to have been widely used in garden buildings. Its most famous early use was in the Salford Twist Mill, designed and built by Matthew Boulton and James Watt in 1799, where steam was circulated through the iron columns that provided the structure for the seven-storey building. Matthew Boulton is said to have installed a steam-heating system in his own house in 1810.[61] From then on, steam was the main means of heating large buildings – and a few houses, such as the novelist Sir Walter Scott's at Abbotsford – until it was supplanted in the last quarter of the nineteenth century by hot-water circulation.[62] The alternative hot-water system seems to have been pioneered, once again, in gardens, for example in the orangery designed by Joseph Bramah for Windsor Castle in the early nineteenth century. The early low-pressure systems – which required large and cumbersome pipes that can often still be seen – were acceptable in greenhouses but not, probably, in domestic or other buildings. Even when such heating systems were installed in private houses, as at Stratfield Saye for the Duke of Wellington in 1833, they rarely extended beyond the entrance hall, corridors and possibly the main downstairs rooms.[63] By contrast, Paxton's huge greenhouse, the Great Stove at Chatsworth, completed in 1840, used hot-water heating conducted from eight underground boilers through a seven-mile maze of 6-inch pipes.[64]

In the first issue of the *Gardeners' Chronicle*, of 2 January 1841, the firm of Daniel and Edward Bailey, of Holborn, London, who became major greenhouse builders, advertised 'Hot-Water Apparatus for Heating Horticultural Buildings, Dwelling-Houses, Churches, and Manufactories, upon improved principles, and at very moderate charges'. J. Weeks and Co., Architects, and W. and I. Walker of Manchester – who 'have lately fitted up the immense Conservatory at Chatsworth' – competed with them in the same issue of the *Chronicle*. In 1860 Cottam and Company advertised 'Horticultural Buildings and Heating by the Circulation of Hot Water' among their enormous range of 'Conservatories, Summer Temples, Arbours, Greenhouses', while, by 1870, it is likely that high-pressure hot-water systems, which used narrow-bore, thinner, pipes, were in use. 'Weeks's One-Boiler System' was advertised in January as providing for '14,471 feet of Piping, equal to nearly 3 miles', in fernery, coach house, stables, orchid house, a variety of plant houses, forcing houses and stoves.[65] The firm of Foster and Pearson, of Beeston in Nottinghamshire, itself a merger of a horticultural and an iron-founding business worth £3.8 million in modern values as early as 1883, diversified into central heating and, under the name of the Beeston Foundry Company, supplied boilers and radiators to tens of thousands of suburban houses in the twentieth century.[66]

In other words, whatever the exact system of central heating that was used, garden buildings frequently led the way, with England, Loudon observed in the 1820s, at the forefront of the technology. Industrial and commercial buildings followed and houses lagged behind. The reasons are not entirely clear. It can hardly have been the fear of fire, since domestic houses routinely had open fires for burning wood, and from the seventeenth century onwards coal, for both heating and cooking. One factor may have been the danger of explosions, a constant threat with early steam boilers; they could not do so much damage in a separate garden building as in the main house. Another problem may have been the cost of retrofitting central heating into old buildings and the cumbersome size of early boilers and pipes, together with the coal-holes needed for their fuel. It may have been problems in obtaining insurance or it may just have been conservatism. As late as 1870, J. J. Stevenson wrote in *House Architecture* that

for heating English houses the best system on the whole is the old one
of open fires. No doubt it is unscientific . . . as well as wasteful. But it
has the advantage that everyone understands it and it is so pleasant
that we are not likely to abandon it as long as the coal lasts.[67]

But the reason may simply have been that garden owners were prepared to spend more on protecting their precious and tender plants
than they were on keeping themselves, and their families and servants,
warm. The Great Stove at Chatsworth consumed 300 tons of coal and
coke each year; as Deborah, Duchess of Devonshire, put it: 'considerably more than that used in the house, but then the welfare of the
exotics in the conservatory was more important than that of mere
humans'.[68] It was the cost of coal that, after the First World War, persuaded the then duke to order the demolition of the Great Stove; the
plants inside it had already died during the war of cold and neglect.[69]

LIGHT AND AIR

Central heating was, normally, invisible or intended to be so. It had
a big impact on the design of buildings but in ways that were and
are largely unappreciated. The boiler rooms of greenhouses, like the
gardeners who tended them, were kept out of sight. So were the steam-
heating systems of factories and other industrial buildings, partly to
minimize the danger of fire. When, in 1818, the Marquis de Chabannes
installed a mechanical system to ventilate the Covent Garden Theatre in
London and used the heat from its gas chandeliers to warm the patrons,
few of them would have been aware of the new technology. Nor, probably, were or indeed are most Members of Parliament aware of the
heating and ventilating systems installed in Charles Barry's new Houses
of Parliament in the 1840s, although it required the building of the central tower to be designed initially as an air exhaust stack.[70]

But the other impact of greenhouses on architecture was highly
visible. Their builders pioneered the use of structural iron and glass.
They were, therefore, the forerunners of the architects whose buildings of all types have changed the shape and appearance of our towns
and cities as well as our parks and gardens.

At first, greenhouses were built of wood. Although they existed at Pompeii, probably glazed with translucent stone, the Roman technology was not revived until the sixteenth or seventeenth century, initially to protect the 'greens'. The earliest recognizable fixed greenhouses or glasshouses date from the end of the seventeenth century, when the invention of sheet glass made it possible to produce larger panes. The greenhouses were built, like the modern 'lean-to' versions, against a south-facing brick wall. The sides were built of brick or masonry, so the only glazed area was sloping to the south; it could be covered with shutters or matting to control the heat from the sun. By the 1690s, glass panes 4 feet by 7 feet could be produced by a casting and rolling process, invented by the Frenchman Louis Lucas de Néhou, bringing about the 'classical long and narrow hothouse with roof and south wall glazed and north wall of masonry'.[71]

During the eighteenth century, garden buildings designed to house plants developed in two different ways. Orangeries, ancestors of the nineteenth-century and modern-day conservatories, were designed for display, even for promenades. They were prominent features of gardens, designed on the best architectural principles to complement the main house and outbuildings and often built in stone; Richard Bradley described in 1710 a greenhouse with Corinthian pillars bearing sliding glass panels, 'agreeable to the rules of Architecture and at the same time . . . rightly adapted to the Welfare of Foreign Plants'.[72] Meanwhile, particularly but not entirely in the kitchen gardens, a plethora of constructions in different shapes and sizes – from greenhouses, hothouses, hotbeds and cold frames to melon houses, pineapple pits* and cucumber frames – catered for all the fruit, flowers and vegetables that rich households required, together with the tropical plants that flooded into England from around the world. Buildings and constructions of all kinds were themselves built to be admired as visitors passed through the gardens.

The constraints on both types of buildings were both technical and financial. It was gradually realized that plants needed light as well as warmth and that it was important to maximize the glazed

---

* Also known as 'pine pits', these were trenches heated by decomposing manure or fermenting tanner's bark.

area – aligning it as far as possible towards the sun – while minimizing the size of supporting structures, whether of stone, brick, wood or, from the second half of the eighteenth century, wrought or cast iron. However, it was still difficult and expensive to produce large panes of glass and much of the older, bubble, glass continued to be used. In England the excise tax on glass, introduced and then soon repealed in the 1690s but reintroduced in 1746, was levied by weight, so that glass-makers had an incentive to produce thinner and thinner panes, suitable for windows but less so for garden buildings.[73] The solution came with the repeal of the glass tax in 1845, which made all glass much cheaper, but also with the simultaneous development of an efficient process, patented by James Hartley in the late 1840s, of producing rolled plate glass that was translucent, though not transparent; this did not matter for its main uses in greenhouses, skylights, railway-station roofs and market halls, so glass production in England boomed, competing with cheap imports from Belgium.

Perhaps even more important for the greenhouse, and then much more widely for other buildings, was the development of prefabrication using both wood and cast iron in the shape of standardized interchangeable parts. Most people today do not realize that, until well into the Industrial Revolution, each part of every manufactured product – each bolt or screw, each table leg, each garden spade, each window frame or roof beam – was made individually on demand by a carpenter or blacksmith. There was some mass production: Adam Smith's famous example of the division of labour was that of pin-making, which used a common design, but even so such a process could not guarantee that one pin would be interchangeable with another. Spare parts had to be individually made and adapted to replace the original effectively. This was partly because the tools that were used, from saws, planes or lathes to primitive metal-working machinery, were themselves not standardized; nor were measuring devices accurate enough to aid the production of exact copies.

The start of the mass production of interchangeable parts was the block-making machinery installed in the Portsmouth dockyard in 1802–5 by Marc Isambard Brunel – father of the great railway engineer – following the designs of Samuel Bentham, brother of the great utilitarian philosopher Jeremy Bentham. During the Napoleonic

Wars naval ships needed huge numbers of wooden pulley blocks in their rigging and the machines made them to a standard pattern. Wood was used first, with iron not far behind. It was only a decade later, in 1816, that the garden writer and designer John Claudius Loudon invented a machine that could produce standardized curved wrought-iron sash bars to hold curvilinear glass; linked together and supported on cast-iron columns, they could span large areas. With Loudon's other idea, the 'ridge and furrow' system of glazing – overlapping panes of glass, set within vertical glazing bars – it formed the basis of the mass production of greenhouses, initially by W. & D. Bailey, to whom Loudon licensed his inventions. He built, at his house in Bayswater, London, a prototype for what he thought – correctly – would be a multiplicity of iron and glass buildings of all types; that is why Loudon – essentially a gardener – has been described as the 'true father of curvilinear iron and glass architecture'.[74] It was not until after the 1890s that steel became the preferred material.

One of the first uses of Loudon's designs was for the huge greenhouses built in 1819–22 in Hackney, London, by W. & D. Bailey for the nurseryman Conrad Loddiges. The camellia house is shown in plate 21. These included the largest palm house in the world, with other houses forming a 1,000-foot walk along which Loddiges displayed his stock. But hundreds of other buildings followed, as the Baileys experimented with different designs and as other designers and constructors entered the market. Innovations in design and manufacture also proliferated, leading directly to the greatest iron and glass (and wood) building of them all, Joseph Paxton's Crystal Palace of 1851. There were also railway stations, winter gardens, conservatories, exhibition halls, shops – including the Pantheon Bazaar Conservatory and Aviary in Oxford Street in London – and of course an enormous range of greenhouses, hothouses, palm houses, pineries (for growing pineapples), camellia houses and greenhouses (from temperate to tropical). After early experimentation, many of them were prefabricated, essentially put together from kits comprising thousands of cast-iron components and sold in their millions throughout Britain and the rest of the world.[75] By the end of the nineteenth century, no middle-class home in Britain was complete without its conservatory: 'a real bower for a maiden of romance, with its rich green fragrance in the midst of winter . . . a picture in a dream . . . a fairy land, where no care or grief or weariness could come'.[76]

Greenhouses gave a glimpse into worlds that most English people – before the modern age of mass tourism – could never dream of seeing. By 1840, the Royal Botanic Gardens at Kew had ten, each acclimatized to different parts of the world, from the Cape of Good Hope in South Africa to Botany Bay in Australia.[77] It was even proposed that greenhouses should be populated by native people as well as plants, imported and displayed like the hermits who were employed to ornament eighteenth-century Gothic garden ruins. Artificial climates, which it was thought could recreate in England the atmosphere of foreign places, also attracted medical interest. A prevalent theory attributed epidemics to a miasma from rotting organic matter.* Miasmas could be combated by providing fresh air and light, it was thought – as important for people as for plants. As a result, 'Madeira houses' – a form of conservatory attempting to replicate the atmosphere and high temperatures of the island – were set up in the 1840s in London to treat pulmonary disorders. After the success of the Crystal Palace, Joseph Paxton was invited to design a large conservatory – nicknamed the 'Crystal Sanatorium' – for the City of London Hospital for Diseases of the Chest in Victoria Park in east London, to provide a 'healthy, temperate climate, simulated through the use of an elaborate heating, cooling and ventilation system'.[78]

The scale of building, in cost, in numbers and in the size of individual greenhouses, conservatories and winter gardens, is difficult to comprehend. Philip Miller's greenhouse at the Apothecaries' Garden cost £3 million *(£1,625)* in 1732.[79] Brown's wood and masonry greenhouse at Ashburnham, built in the 1770s, cost £914,000 *(£600)*. The most spectacular Gothic greenhouse of the late eighteenth century, built at Warwick Castle to hold the ancient Roman Warwick Vase – and now a tea room – cost at least £2.2 million *(£1,500)*.[80] W. & D. Bailey's 1827 wrought- and cast-iron greenhouse at Bretton Hall, West Yorkshire, cost £11 million *(£14,000)*. It was suggested at the same period, by the curator of the Royal Botanic Gardens in Regent's

---

* The miasmatic theory was the prevailing medical orthodoxy until the 1880s, when it was replaced by the germ theory of disease.

Park, London,* that an acre of ground could be covered with green-houses for £7.7 million *(£10,000)* and this was clearly not thought to be an outrageous expense.

It was not only the buildings that were innovative. In his *Encyclopaedia of Gardening* of 1822, Loudon describes the Automaton Gardener, or Regulating Thermometer, comprising a device that gave 'the power requisite for opening the sashes or windows of hot-houses';[81] this technology is still in use to cope with tall greenhouses and variable climates within greenhouses in botanical gardens. The London nurseryman Conrad Loddiges pioneered the use of devices to create a misty atmosphere, while Paxton's fernhouse at Tatton Park, Cheshire, employed a cascade for the same purpose, to replicate a New Zealand forest.

Conservatories could be huge: one designed by Paxton for Capesthorne Hall in Cheshire connected the library with the chapel and measured 150 feet by 50 feet. His 'Crystal Sanatorium' was 200 feet long and 72 feet wide. The conservatory built at the Rothschild house at Halton, Buckinghamshire, in the 1880s had two large and nine small domes.[82] Winter gardens for the public could be on an even larger scale. One early enormous project was the Anthaeum at Brighton, an 'Eden Project' of the 1830s; it incorporated the largest dome in the world at the time, 165 feet in diameter and 65 feet tall and covered 1.5 acres. Inside would have been tropical trees and shrubs, artificial lakes and hills, flowers, birds and fish. Unfortunately, the builders decided not to construct the central iron pillar that had been designed to support the roof; when the scaffolding was removed, the entire structure collapsed. Its cost is not known, but a proposal to rebuild it for £7.7 million *(£10,000)* was rejected. The Great Stove, Paxton's enormous construction at Chatsworth (plate 22), was 123 feet wide and 277 feet long, with an arched roof measuring 67 feet high, and cost £25.6 million *(£33,000)*.[83] It became an instant tourist attraction, with a visit by Queen Victoria in 1843 during which she drove through the huge building in a coach. At about the same time, the Palm House at

---

* Established in 1841 within Regent's Park, the gardens were used for flower shows and for a permanent exhibition of plants; they were closed in 1932 and the site used for what are now Queen Mary's Gardens.

Kew – 25 per cent larger than the Great Stove – designed by Decimus Burton and built with public money by the Irish iron founder Richard Turner, a pioneer in the structural use of wrought iron, cost £22 million *(£30,000)*. Burton also designed the later and even larger Temperate House, estimated to cost £15.8 million *(£25,000)*, although it eventually cost much more.[84] Winter gardens on a similar scale were built in many other European countries.

Smaller garden buildings, but for a client list that was almost as stellar as Capability Brown's, were the speciality of Jones & Clark, later transmuted into Henry Hope & Sons Ltd, of Birmingham, specialists in 'metallic hothouses and horticultural buildings' from 1818. Their order books include sketches of many of their products, although only a few give prices.[85] A very early order was for the Duke of Newcastle, a pine house (for growing pineapples) measuring 49 feet 6 inches by 14 feet on a cast-iron framework, soon followed by orders for hothouses, vineries, pine pits, forcing houses and conservatories from such aristocrats as the Barings, Earl Nelson and Earl Spencer, as well as a substantial number of country clergymen. Hot-water heating systems were common. Jones & Clark collaborated with architects such as C. R. Cockerell, and were not cheap: a metallic lean-to greenhouse with heating for the Countess of Clancarty cost £424,000 *(£535)* in 1834 and a conservatory and attached greenhouses, 100 feet long in total, for the Earl of Bradford were £1.2 million *(£1,525)* in 1840. The firm's greatest coup was an order in 1842 to provide two ranges of greenhouses, each 350 feet long, for the new royal kitchen garden at Frogmore, Windsor, while the fashion-setting role of the queen soon produced an order in 1846 for three metallic hothouses in one range, 150 feet long in total and 16 feet 8 inches wide: 'The whole to be completed in the same style as the houses in the New Royal Gardens including a hotwater apparatus.'[86] It cost £985,000 *(£1,350)*; the client was Sir Robert Peel, the prime minister, who within two months, on 18 May 1846, would be out of office after he split the Tory Party over the repeal of the Corn Laws.

But greenhouses and other garden buildings were, from the middle of the nineteenth century onwards, also products for a mass market. Paxton was the first to realize the potential, writing of hothouses for the millions. In 1860, J. Lewis's Horticultural Works advertised

SUTTON'S
Bulb Catalogue
for 1889.

SUTTON & SONS, Reading.
SEEDSMEN BY ROYAL WARRANTS TO
HER MAJESTY THE QUEEN
AND H.R.H. THE PRINCE OF WALES.

15. A catalogue 'bright with colour and with hope': *Sutton's Bulb Catalogue*, 1889.

16. 'To raise water by fire': Savery's pumping engine, 1698, the first working steam engine.

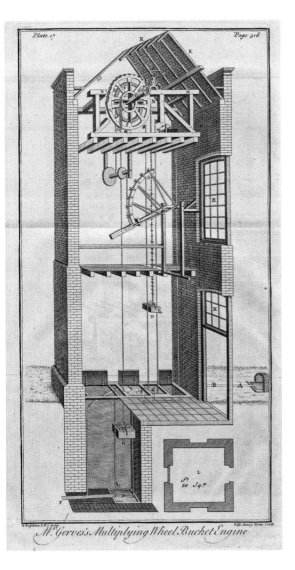

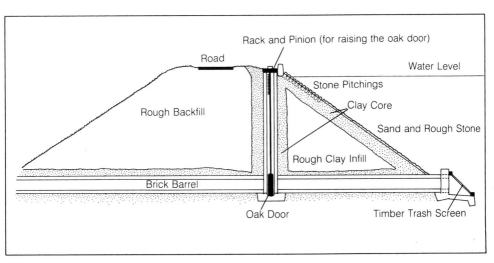

17. A domestic waterworks: George Gerves's 'Wheel Bucket Engine' at Chicheley Hall, Buckinghamshire, in the early eighteenth century.

18. The secret of a Brownian landscape: the construction of a dam at Petworth, Sussex, 1752–7.

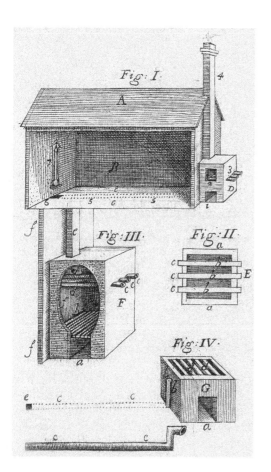

19. To avoid poisoning the plants and the gardeners: heating John Evelyn's greenhouse, 1691.

20. The greatest machine built before the Industrial Revolution: Louis XIV's Machine de Marly on the River Seine, 1684.

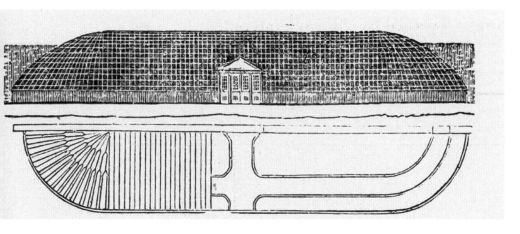

21. A Regency garden centre: Loddiges' camellia house in London, 1819–22.

22. 'A mountain of glass': Joseph Paxton's Great Stove at Chatsworth, Derbyshire, in 1840–41.

23. *Overleaf*: A seventeenth-century tool shed: John Evelyn's garden tools.

Elysium Britannicum.

The Instruments presented to
the Eye by the Sculptor, and
referring to the numbers as
they are described.

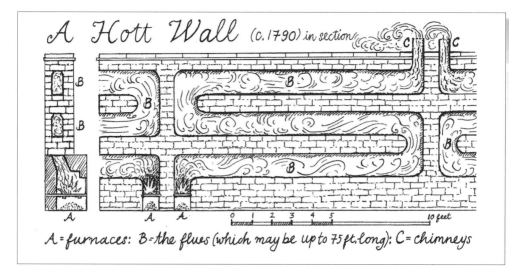

A Hott Wall (c. 1790) in section

A = furnaces: B = the flues (which may be up to 75 ft. long): C = chimneys

24. Extending the seasons: the hot wall in a kitchen garden, 1790.

25. Fire in the garden: the furnaces of a hot wall, probably from the late eighteenth century, at Weston Park, Staffordshire.

greenhouses 'on a much better and stronger principle than those manufactured on Sir Joseph Paxton's plan' at £16,000 *(£30)* for one measuring 30 feet by 16 feet, or £49,000 *(£90)* for 60 feet by 24 feet.[87] It is interesting that what were clearly seen as products for middle-glass gardens could cost so much. A competitor, Samuel Hereman, offered a 100-foot lean-to greenhouse for £40,000 *(£73 12s)*. By the end of the century, Boulton and Paul of Norwich were selling

> a bewildering variety of garden-houses, decorative footbridges, keep-ers' huts, portable bungalows ... aviaries and pheasantries, and, above all, greenhouses ... peacheries, vineries and palm courts, some so elaborate that they took two years to build ... it was the age of the conservatory and the wrought-iron pleasaunce, of the gabled fishing-temple and boat house, of the galvanised iron shooting box, of the panoplied and striped marquee.[88]

They and other firms used their construction of garden buildings as a springboard for a much wider range of products, sold both at home and throughout the world.

The Victorian period represented the high point of greenhouse and conservatory construction in Britain, but the records of Henry Hope suggest that demand was declining by the third quarter of the century. The firm continued to receive hundreds of orders each year – it remained profitable – but more and more were for house heating systems and for metal windows, considered a luxury at the time, and there were fewer and fewer greenhouses; by 1904 horticultural build-ings made up only 5 per cent of their business. There has been a recrudescence recently and greenhouse and conservatory building continues to be a major economic activity, often carried out by firms that – like Henry Hope, later Crittall-Hope – are the descendants of Victorian entrepreneurs. Large greenhouses for public institutions continue to be built: the Princess of Wales conservatory at Kew cost £17 million *(£4.75 million)* in 1985. In the wider market, the higher incomes of consumers in the twenty-first century have sparked an interest – as it has for many other goods – in bespoke or purpose-built garden structures, tailored to the spaces of individual customers; thus mass production, characteristic of the twentieth century, is being replaced by a return to craft modes of production, although often still using prefabricated and interchangeable parts. Many greenhouse makers stress that they are family-run businesses. There has been a return to the wooden greenhouse, on Victorian or Georgian patterns, and even – as purchased by the former prime minister David Cameron – a shepherd's hut as a garden study, an alternative to the ubiquitous garden shed.

EARTH

Most gardeners did not make lakes or cascades; nor did they construct orangeries, conservatories or greenhouses with acres of glass and elaborate heating systems. Their ambitions for garden buildings were confined to nothing more grandiose than a garden shed or, possibly, a small greenhouse. The technology that they used was, at least until the nineteenth century, one of hand tools, probably crafted by the village blacksmith and handed down from generation to generation. The spades, knives, scythes or trowels were very similar to those used by the Romans and very much the same as are used by gardeners today. When, in 1659, John Evelyn listed, in *Elysium Britannicum*, the tools he considered to be necessary for a gardener to possess, there were few that could not be bought today in a garden centre – as plate 23 demonstrates, the shapes might be slightly different but the essential features and purposes remain the same. To modern eyes, the only really unfamiliar item is a kind of portable four-poster bed 'furnished with tester and Curtaines of Greene . . . to draw over and preserve the Choysest flowers, being in their beauty, from the parching beames of the Sunn'.[89] It was much the same in 1702 when Henry Wise signed a contract listing 'Sev. Tools and Materialls that are to be Provided & found by ye Undertaker' as royal gardener; there is nothing surprising in an eighteenth-century context except, perhaps, 'A Sort of Engine to Sprinkle ye Trees and Plants'.[90]

Loudon's *Encyclopaedia of Gardening* illustrates much the same range of tools, but then pulls the modern reader up short. After surveying a range of rat, mole and other traps, Loudon turns to human predators. He recommends that garden owners invest in a mantrap, which he describes as a 'rat-trap on a large scale . . . a barbarous contrivance'. Plate 29 shows just how fearsome it was. If the gardener is squeamish, then a 'humane' mantrap might be preferred, since 'instead of breaking the leg by crushing, and consequently by the worst of all descriptions of compound fractures, [it] simply breaks the leg, and therefore is comparatively entitled to the appellation of humane'. Traps ought to be supplemented by guns. A 'Spring Gun', activated by tripwires, was 'found extremely useful in gardens and nurseries in the neighbourhood of London' but the 'Common Gun, or Musket, is essentially necessary for the gardener, in order to kill birds, or deter many of the enemies of which the above traps and devices may still permit to escape'.[91] These deterrents may have been effective, as the Old Bailey records don't include many thefts of plants, valuable though some of them were. Anyone daring to brave the mantrap would, if caught, have risked transportation to Australia.

Large gardens, with plants that required regular watering, posed a challenge. Lakes and dipping ponds, common in walled kitchen gardens, provided sources of water, but the labour involved in collecting water from them and then carrying it to the depths of gardens to water plants and the earth around them could consume hundreds of man and woman hours. No wonder that, in 1724, John Dickens applied for a patent for a 'Machine and floats for raising water to supply cities, drain mines, water gardens, turn mills, move ships, and for other purposes'. He was followed in 1729 by Thomas Bewley, whose invention consisted of a 'Machine for raising water by alternate exhaustion and pressure of the air, without the help of fire; for supply of towns, gardens, &c., also for draining mines'. More patents followed in 1734, 1774 and 1777.[92] In the nineteenth century, hand-operated pumps served a dual purpose in many large gardens, for watering and to put out fires.

Grass required the first major innovation in garden tools. We tend to think of a well-kept lawn as synonymous with an English garden, but in fact grass was not the predominant feature of the landscapes of *Britannia Illustrata* in the early eighteenth century. There were meadows on the estates, of course, and there was grass in orchards and wildernesses, but these lay outside the areas of the formal pleasure gardens, where there were small areas of grass and grassed paths but the most obvious material covering the earth was gravel, around yew and box hedges. Wise's contract at Hampton Court treats the two as equivalent: 'The Grass to be mow'd Swept and Rowled, ye Gravell rowled and weeded, ye borders dung'd, dug, howed, raked and weeded.'[93] The royal accounts contain payments for vast amounts of gravel, dug up all over the Thames valley, and for the labour required to keep it in good order. The only reasonably large areas of grass were bowling greens, which had been provided in many larger gardens from early in the seventeenth century, if not before.[94]

It was not until the 1730s that larger areas of open space, covered with grass, replaced the planned space of box or yew surrounded by gravel. Even then, for example at Rousham in Oxfordshire or Painshill in Surrey, the design was 'mostly a matter of decorated lawns connected by walks'.[95] The sweeping expanse of greensward that we now consider normal in great gardens came even later, from the 1740s and 1750s, and it came at considerable cost. It was, like so much else in gardening, expensive both to create and to maintain. Lawns need to be smooth, the grass needs to be mown and the grass-cuttings removed, weeds need to be dug out, leaves swept away, edges carefully trimmed. Building a lawn involved a major effort in flattening the earth, by digging and rolling, and then in seeding, but the real, continuing problem is the constant effort required, in the English climate, to keep it mown. For about a century, this new design feature of English gardens had to be kept looking good by men with scythes. They are at work at Hartwell House in Buckinghamshire in plate 26. Scythes cannot cope with uneven ground, so bumps or molehills had to be removed. Grass is best scythed while the dew is still on it, so armies of gardeners had to be up with the dawn, cutting and raking away the grass before their employers awoke.

The landscapes of Brown and his contemporaries and successors relied for their effects on large areas of park meadow and grassland. These were usually separated from the lawns in pleasure gardens and near the house by the ha-ha, a ditch with one sloping and one perpendicular side that animals cannot climb; it gives the illusion that sheep or cattle are grazing on a stretch of lawn sweeping up to the house. Animals could therefore keep the park meadows cropped, but the garden lawns were a different matter and gardeners must have welcomed the invention, by Edwin Budding in 1830, of the first lawnmower, although like many new technologies, it still took many years to come into general use.

Budding is said to have been inspired by a bench-mounted device used in a mill in the cloth-making district of Stroud, in Gloucestershire, in which a cylinder with cutting blades fixed on it was used to trim the cloth to a smooth finish after weaving. This may indeed have been the genesis of Budding's invention, but the engineering challenge that he faced and overcame was to translate a static machine into a mobile one, in which the action of pushing a large roller was transmitted, through gearwheels, to a smaller cutting roller that abutted a knife bar. At almost the same time, in the United States, Cyrus McCormick was surmounting a similar challenge in devising his reaping machine. Both depended on the use of cast-iron parts and on the recent invention of metal-working machine tools that could mass-produce gearwheels. Gears, although an ancient technology, were less used in machinery in the early Industrial Revolution than pulleys, partly because it was difficult to manufacture them sufficiently accurately to minimize wear. All these metal parts – and the machines as a whole – look, to modern eyes, 'over-engineered', clunky and very heavy; it was soon realized that Budding's lawnmower needed two people to power it, one to pull and one to push. The *Gardener's Magazine* was probably right to suggest shortly after its invention that country gentlemen could use it as 'an amusing, useful and healthy exercise', but they would have needed to be both strong and determined.[96]

Budding's machines are well known but it is not often appreciated that they were very expensive; one of the first was sold to the Zoological Society of London for £8,000 *(10 guineas)*.[97] Plate 27 shows that they also needed two men and a horse to operate them, itself an expensive business. Scaling the machines up from the original 19-inch width – a greater width being necessary for the larger lawns – was also a problem; it made them even heavier and noisier, so noisy that they frightened the horses or donkeys that had to be used to pull them. 'Noiseless' machines, driven through chains rather than gears, were available by 1860 – one type was Green's 'Patent Silens Messor or Noiseless Lawn Mowing and Rolling Machine', as advertised in the *Gardeners' Chronicle*, but it remained expensive; the 30-inch machine cost £10,800 *(£16 14s)* plus £665 *(£1)* for leather boots for the pony – essential to protect the lawn. Shanks's 'New Patent Lawn Mowing, Rolling, Collecting and Delivering Machine', with a 4-foot cutter bar, cost £18,100 *(£28)*, although there were smaller, people-powered machines, down to a 13-inch version 'easily worked by a boy'. A similar small machine by Green's cost £4,210 *(£6 10s)*. These high up-front costs of lawnmowers probably delayed their general adoption, although the large number of advertisements for them suggests that many gardeners were willing to pay the price.[98] Indeed, by 1870 the *Gardeners' Chronicle* stated that machine mowers were all but universal.[99]

The lawnmower deserves to be celebrated as the first of what is now a wide range of garden machines. By the end of the nineteenth century, lawnmowers could be powered. Initially small steam engines were used, though they weighed over a ton, but by 1900 there were petrol versions, pioneered by Ransomes, Sims & Jefferies but imitated by many others; particularly successful was the Atco mower, introduced in 1921 and – despite its price of £10,500 *(£75)* – selling in tens of thousands by the mid 1920s. Prices continued to fall: the cheapest Qualcast motor mower cost about £2,600 *(£15)* by the mid 1930s.[100] Millions of smaller and cheaper unpowered machines were being bought at the time. Later came rotary cutting machines, electric mowers and the hovering Flymos; now there are robotic mowers. But alongside them in the garden centres are hedge cutters, chainsaws, rotavators, strimmers, leaf blowers, shredders – an apparently endless list of toys for gardeners to buy.

Garden centres are full, also, of another technological innovation of the nineteenth century, chemicals. Gardeners have, of course, always been concerned about insects and diseases attacking their precious plants and about how to promote growth. There were probably as many folk remedies for plant diseases as for human illness. In 1710, for example, Richard Bradley devoted an entire chapter to 'Blights' – 'the most common and dangerous Distemper that Plants are subject to' – ascribing them to caterpillars carried on an east wind. As befitted a member of the Royal Society, he reported his scientific investigations, using his microscope to investigate 'Beings which are not commonly visible' before recommending a 'Receipt' to prevent wheat from becoming smutty by soaking the seed in brine.

He noticed, astutely, that ants did not cause peach-leaf curl but rather farmed the aphids that he thought did so.* Unfortunately, the remedy – proposed by 'a Person of Honour' – seems more likely to poison the gardener than cure the tree: it was 'to bore a Hole with a small Gimlet sloaping downwards thro' the *Bark*, so as to reach the Wood of the Tree, and pouring into it some *Quick-silver*, about half an Ounce, or more ... and stopping it up'. Bradley, having 'been Witness of the Effect that *Crude Mercury* has had upon *Worms* in humane Bodies' was confident that it would work.[101] He also recommended various types of manure, regarding pigeon droppings, mixed with water, as particularly effective; dovecots or pigeon-houses adorned many large gardens, providing both meat and fertilizer.[102]

Pigeon droppings were still recommended in the early nineteenth century, when Loudon published his *Encyclopaedia of Gardening*. He was keen on different kinds of manure – 'Every thing in culture may be said to depend on the use of manures' – and lists a wide range, from seaweed to fish, bones, blood and urine, although the 'dung of sea-birds has never been much used as a manure in this country'; the age of the large-scale use of guano was still to come.[103] Loudon was also sceptical of fertilizers of mineral origin, said to be 'doubtless of more uncertain use'.[104] More exotic technologies were discussed, including the possible benefits of applying the newfangled electricity to plants, which continued to be investigated throughout the century; the great electrical engineer Sir William Siemens conducted a range of experiments in 'electro-horticulture' in the 1880s.[105] Particular benefits for estate kitchen gardeners were found in the use of electricity to help grow plants 'to a timetable dictated by the social season rather than a natural one'. A more radical approach was to change the English climate; Loudon reported the proposal of a Mr J. Williams to 'erect large electrical machines, to be driven by wind, over the general face of the country, for the purpose of improving the climate' by preventing too much evaporation.[106]

When, in 1850, Loudon's widow Jane published a new edition of the *Encyclopaedia*, much had changed. Part 2 of the book, 'The

---

* Some species of ants farm aphids by protecting them on the plants on which they feed and then consuming the honeydew that they produce.

Science of Gardening', had been entirely rewritten to incorporate new methods. Bone meal was still being widely used, as it is today, but guano – bird droppings dug up from Peruvian islands – was in common use, together with a range of inorganic fertilizers such as gypsum (sulphate of lime) and nitric acid.[107] More changed in the next quarter-century; the supply of guano was running out and it was increasingly being adulterated, but a major breakthrough, in gardens as well as farms, came with the discovery by John Bennet Lawes that calcium phosphate – found in both bones and minerals such as coprolites – could be treated by sulphuric acid to produce a very effective fertilizer that he called superphosphate. By the 1860s, his annual income from this invention was £31 million *(£50,000)*, some of which he used to found the Rothamsted Experimental Station in Hertfordshire for agriculture research. Although it was said in 1901 that excessive profits were being made from fertilizers,[108] English gardeners continued to buy them as well as using their own compost.

The control of insects and plant diseases was also revolutionized. Loudon had been sceptical, arguing that plants could not be easily protected from insects, except by picking them off when they appeared. More broadly, he believed that the 'only way to protect from diseases is by using every means to promote health and vigour . . . Regimen and cleanliness, therefore, in plants as in animals, are the grand protectors from disease.'[109] In 1850 Jane Loudon agreed, but then came the chemical revolution of the late nineteenth century and the creation of the chemical industry in Britain but particularly in Germany. In 1860 insecticides appeared in advertisements in the *Gardeners' Chronicle*: there was 'Tobacco Paper' for fumigating and killing all insects except red spider, while Gilshurst's 'Compound' claimed to be able to do the latter. Keating's 'Persian Insect Destroying Powder' and Page's 'Composition for the Destruction of Blight' appeared alongside other gardener's necessities such as Epps's 'Homeopathic Cocoa' and Heal's beds. Bordeaux mixture, a copper fungicide, was used against potato blight. By 1901 William Williamson was rejecting the Loudons' belief that good cultivation was all that was required and recommending an array of fungicides and insecticides, most of which have since been banned. Liver of sulphur and Bordeaux mixture should be sprayed on plants as fungicides, while insects required even stronger methods: if paraffin emulsion did not do the trick, then arsenate of lead, cyanide of potassium and Paris green (a highly toxic combination of copper acetate and arsenic trioxide) might be employed.[110] Strychnine-baited worms were used by green-keepers to poison moles.

The pendulum swung back at the end of the twentieth century towards what is now called 'organic' gardening, eschewing 'artificial' chemicals. In 1952, Brigadier C. E. Lucas Phillips, in *The Small Garden*, was still recommending a range of preventatives and remedies such as nicotine, derris dust, lime sulphur, malathion, copper solutions, BHC and even DDT, except when flowers are in blossom, all spelled out in twelve pages covering 'The Enemy in Detail' and 'The Medicine Cupboard'.[111] DDT was banned in the USA in 1972, after the publication of Rachel Carson's *The Silent Spring* in 1962 exposed its dangers, but not in the UK until 1984. The 2006 edition of Lucas Phillips's popular book retreats to a range of copper- and sulphur-based fungicides and to derris and pyrethrum as insecticides; the even more popular D. G. Hessayon returns full circle, in 2009, to Loudon's advice in 1822 – 'Prevent Trouble before it Starts'[112] – by good cultivation, rotation and feeding. He advises that there are now no permitted soil pest insecticides and almost no systemic insecticides, together with only a small range of permitted fungicides, but a general return to age-old organic remedies.[113] Innovations have not always been good for us.

INNOVATION IN AND OUTSIDE THE GARDEN

Gardeners have been inveterate innovators. They are always on the lookout for new plants and equally keen to protect and nurture them by every means possible and legal, and at whatever expense. But they are also conservative, attached to the tools, plants and methods that they know and love and which they have tested, year after year, in the type of soil and climate they have to cope with. It is a good combination, preserving what works and discarding what doesn't.

Gardening is a microcosm of the impact of invention and innovation on our society and economy, a transformative force but one which often works stealthily and slowly. Over the years, it has produced innovations – in water management, in greenhouses and conservatories, in central heating and in garden tools – which have borrowed from other areas of life but also, to an extent much less recognized, pioneered changes that have transformed England's landscape, transport, architecture and housing.

8

The People's Gardens

'Some sort of an apology for a garden'[1]

'Dull, conservative, commonplace; provincial, parochial, insular; blinkered, bourgeois, boring' – just nine of twenty-three synonyms in a modern dictionary for 'suburban'. They would all have appealed to the popular but combative journalist Thomas Crosland, who, in 1905, wrote that 'We have it on the best of authority that a love of flowers is a true sign of gentle birth and reasonable breeding. It is possibly for this reason that suburbans of all degrees do their best to secure some sort of an apology for a garden at the back or front of their houses.'[2] He then concluded: 'Horticulture in the suburbs, like most other things in the suburbs, is a fiction and a pretence.'[3]

Attacks on the suburbs and their gardens began almost as soon as the word 'suburban' was first commonly used in England, around 1800. In 1818, John Keats, now one of the most admired of English Romantic poets, was attacked by the critic John Gibson Lockhart as irredeemably suburban, a 'Cockney Poet' writing 'laborious affected descriptions of flowers seen in window-pots, or cascades heard at Vauxhall'.[4] Lord Byron piled on the insults against Keats and his mentor, Leigh Hunt: 'The grand distinction of the Under forms of the New School of poets – is their *Vulgarity*. – By this I do not mean that they are *Coarse* – but "shabby-genteel" – as it is termed.'[5] Byron was offended by the fact that Keats did not himself possess landed acres about which he might write but could only visit the countryside and then return in the evening to a suburban villa.

Keats lived in Hampstead, where the suburban villa and its garden in which he heard the nightingale is still to be seen in what is now called Keats Grove after him. It was one of the earliest of the London suburbs. By the end of the nineteenth century, the houses of Camden Town and Kentish Town had filled up the green fields between Hampstead and the centre of London, but the taint of suburbanism remained. My grandmother Phyllis Ford lived there on the edge of upper-class society – her grandfather was a baronet but her father was 'in the City'. She was asked at a girls' lunch party in Chelsea in central London around 1900: 'Are you up for the season?' My grandmother 'longed to say [she] was but felt forced to admit that, "No, I live in Hampstead." Suburban! Labelled at once.'[6] She felt keenly the fact that her parents' garden was not large enough even for a tennis court.

Attacks on the suburb and its gardens have continued for two hundred years. Writers from both the political left and right have joined in: H. G. Wells, George Orwell, J. B. Priestley, John Betjeman. Betjeman, later poet laureate, celebrated but also regretted, patronized and disparaged Metroland – the new settlements around the Metropolitan Underground line through the north-western suburbs of London that he thought were destroying the countryside. He even went so far as to call in 1937 for their destruction in a well-known poem, 'Slough'.[7] Three years later, his request was granted as German bombs did indeed fall on the Berkshire town and others like it.

The suburbs have been attacked on many grounds other than the snobbery of Byron and Betjeman. They have been seen as trapping women within a stultifying existence, marooned in a suburban semi,* or alternatively as empowering them within their own domain into which the hapless male was then ensnared, doomed to supply the voracious needs of a vampiric female.[8] They have been accused of destroying the communal street life of working-class districts in the cities, replacing them with families cowering within their individual fortresses, ever frightened of offending the neighbours by neglecting their front gardens or – even worse – mowing the lawn on a Sunday afternoon. Suburbs have destroyed, it is said, both the countryside

---

* The English shorthand for a semi-detached house, i.e. one of a pair of houses with a common interior wall. They became the ubiquitous housing form of the suburbs, supplanting the terraces of earlier generations.

and at the same time the vibrant civic pride of the towns and cities around which they spread. While striving for *rus in urbe* they have replaced the rural idyll with serried ranks of soulless, half-timbered Tudorbethan boxes, aping in miniature the long-gone mansions of the estates over which they have spread. There are few sins that the suburbs have not been accused of committing, but what critics saw as their corruption of the ideals of English gardening has been regarded as among the worst.

Yet it is in the suburbs where the greatest increase in gardening in English history occurred. In the process, over two centuries, they have given the English people as a whole one of the things in life that – surveys consistently show – they most desire. They have given them their own gardens.

URBAN GARDENS BEFORE THE SUBURBS

There have been gardens in towns since Roman times, or probably before. Fifteen per cent of the land area of the provincial town of Pompeii in southern Italy was made up of gardens before it was buried under the volcanic ash of Mount Vesuvius, and there is no reason to believe that the provincial towns of Roman England were any different.[9] Nor should we believe – though there is very little evidence – that gardens somehow disappeared from English towns as the Roman troops departed, only miraculously to reappear in the fifteenth or sixteenth century. Garden historians have concentrated on the monastic and herbal gardens of the medieval period, but there were certainly secular gardens in London – and not only royal ones – early in the sixteenth century at the time of the dissolution of the monasteries. Thomas Cromwell, Henry VIII's chief minister and main architect of the dissolution, owned one, as did his religious adversary Sir Thomas More. By the end of the century, Sir Thomas Gresham, Queen Elizabeth's financial 'fixer' and royal agent in Antwerp, had a large garden enclosed within the grounds of his house in the City of London, later to serve as the first site of Gresham College and the first home of the Royal Society. In the middle of the seventeenth century, 70 per cent of a group of houses in High Holborn, just outside the City, had gardens, while further out, in Piccadilly, almost all of them did; most seem to have been for pleasure and recreation, not for growing crops.[10] Maps show gardens large and small within the urban area, as well as market gardens and nursery grounds outside it. This remained true during the whole of the seventeenth century, while at its end Kip and Knyff document in *Britannia Illustrata* some very large urban gardens, both in London and in other towns, such as Pierrepoint House in Nottingham and Burford House in Windsor.

Some gardens in the capital were also described when, in 1722, Thomas Fairchild published *The City Gardner*, based on his thirty years as a nurseryman in Hoxton, north-east of the City. His fame – he is known as 'the ingenious Mr Fairchild' – rests mainly on two accomplishments: his identification of plants which could thrive, or at least survive, in the pollution from 'sea-coal'[11] that already plagued the capital and, even more important, his status as the first person to hybridize flowering plants. He was in business to sell plants, so he must have known what he was writing about when, in the preface to his book, he states that 'almost every Body, whose Business requires them to be constantly in Town, will have something of a Garden at any rate'. 'One may guess at the general Love my Fellow-Citizens have for Gardening', he continues, by observing how they 'furnish their Rooms or Chambers with Basons of Flowers and Bough-pots, rather than not have something of a Garden before them'. This was a desire common to everyone; the more money that someone made, the easier it was to satisfy that desire: 'in Proportion to the Money Men get, so may their Gardens be larger and better garnish'd'. Success in business could 'lead them in their latter Days into Quiet and Ease' in enjoyment of their large estates and many acres, in contrast to their current life 'for the sake of Trade pinn'd down to a narrow Compass of Gardening'.[12]

A historian of London's town gardens in the Georgian era, Todd Longstaffe-Gowan, has documented an expansion of London and its small gardens that was particularly marked after 1750.[13] Referred to as 'little town gardens', 'little walled gardens in streets', 'street-gardens' or 'fourth-rate gardens', they were placed usually at the rear of the rows of terraced houses – which are now very highly valued – in areas of London such as Bloomsbury and Belgravia; these were developed on estates owned by aristocratic landowners. Longstaffe-Gowan writes of the 'ubiquitous development of modest city houses, each with its own garden' from the third quarter of the eighteenth century.[14] One much-studied garden was created on the Bedford estate in 1791 by Francis Douce, later Keeper of Manuscripts at the British Museum, with the help of his friend Richard Twiss.[15] Its exact size is unclear, but a similar garden close by measured 30 feet by 110 feet (3,300 square feet),* or half as much again as an average modern garden.[16] Nevertheless, Twiss proposed that it should contain 42 poplars, a large almond tree, 2 rhododendrons and 20 or more roses. He estimated that, including four loads of gravel and ten of mould as well as 36 perennials and a choice of the 'best' among *all* the annuals in England, 192 sorts', for which he would provide seeds, the cost would be £21,240 *(£16 16s)*; as an afterthought, he believed the necessary bulbs would bring the total to £25,290 *(£20)*. This excluded labour costs, though 'any common gardener can lay the ground & gravel, & plant the edging and the trees'. He recommended a nursery in Mile End, in east London. On this basis, 'your little slip of ground . . . will super-eminently and super-abundantly eclipse all your neighbours in Gower Street, even unto Sommer's Town, nay perhaps as far as Pancras, may be Kentish Town!'† Despite the large number of plants, Twiss advised that 'there will be enough for your garden as it must not be crammed, but every plant must stand insulated, & have room to grow';[17] plate 28 shows the result.

Garden historians have concluded that Twiss's plan was conjectural and somewhat flawed; it would certainly have produced problems for its owner, Douce, and his gardener, since the trees would, in time, have shaded all his shrubs as well as those of Douce's neighbours. However, the total cost of the garden is a valuable estimate, when one remembers that there were hundreds if not thousands of similar houses and gardens in London. So too is Twiss's estimate that the 'garden will continually be growing more beautiful, & will not cost above 30 shill[s]. [per annum] to keep it'. That equates to £1,900 today; at the time it would have provided for only ten days per year of a

---

* See later in the chapter for further discussion of garden sizes.
† These are successive districts of London, moving northwards from what is now the university district of Bloomsbury towards the village, later suburb, of Hampstead.

jobbing gardener, suggesting that Twiss thought that his friend Douce would have to do some of the gardening himself.[18]

It is dangerous to generalize from one example, particularly when it is from London, although half the urban population of England was living there in 1801. However, contemporary descriptions and large-scale maps demonstrate clearly that small gardens were abundant in the capital. Outside London, another small Georgian town garden, though with a very different planting scheme and no record of costs, is Number 4, the Circus, Bath, in a town that was another haunt of polite society.[19] Rating (local taxation) surveys of Manchester in the late eighteenth century show terraced houses with individual back gardens, as well as large houses with gardens and orchards.[20]

But what about other English towns? The position seems to have varied a great deal from place to place. Zoë Crisp has found that, in the steelmaking town of Sheffield, in South Yorkshire, not a single house built before 1800 had a private plot, although 5 per cent shared plots with a neighbour. However, in the cotton town of Preston, in Lancashire, a quarter of houses built before 1800 had private plots and another quarter had shared plots; in Northampton, a centre in the Midlands of shoemaking and leatherworking, 42 per cent of houses built before 1800 had private, and 10 per cent shared, plots. The average size of the plots also varied: they were 205 square feet in Preston but 465 square feet in Northampton.[21]

These are small gardens or yards, similar to the terrace or patio gardens of today that are provided in many modern housing developments. Nevertheless, urban gardeners seem to have been active in towns such as these in the eighteenth century; the evidence, gathered by Ruth Duthie, lies in the activities of floral societies, whose feasts, entertainments and shows are described in local newspapers from early in the century.[22] Here florists – in this case, chiefly individuals who grew flowers for pleasure – could discuss and compare their treasured plants and compete for prizes such as cutlery or kitchen utensils, or sometimes cash.[23] Members of the Ancient Society of York Florists, founded in 1768 and claiming to be the oldest horticultural society in the world, seem to have been mainly tradesmen or skilled workers; in other areas, weavers and miners were particularly active. Different regions specialized in different flowers: across the

Pennines, in Lancashire, Cheshire and Yorkshire, the main focus of attention was on auriculas, while the miners of Durham and Northumberland favoured pinks; ranunculus, tulips and carnations were each preferred in their own areas.[24] Gardeners exhibited their work at flower shows: in 1821 there were 42 auricula and polyanthus shows, 21 tulip shows, 5 for ranunculus and 69 for carnations and pinks.[25] The Wakefield and North of England Tulip Society, established in 1836, still holds its annual show.

Small urban gardens were supplemented in many towns by detached gardens, using land set aside on the outskirts. These existed in the seventeenth century, for example in Winchester and in Southampton, long before the notion of an 'allotment' had been thought of. Land in city centres was expensive, so fields outside could be used, divided into small plots, at least until they were needed for building as the town expanded.[26] In Manchester, as the town grew from the end of the eighteenth century, 'gardens were pushed to the edges and for several decades the detached town garden was a feature' of the town.[27] Sheffield, where Crisp's research found no pre-1800 houses with private plots, had allotment gardens from as early as 1712, where the cutlers and other skilled workers and tradesmen could rent plots of 150–200 square yards (1,350–1,800 square feet) and grow not only their favoured auriculas but also a variety of vegetables. Birmingham, Nottingham, Coventry and Manchester, with other towns, had similar plots, sometimes known as guinea gardens, of which examples still survive.[28]

Lack of space did not necessarily mean lack of gardening. Window boxes proliferated, even in working-class areas. For the middle classes, gardening itself could be avoided by contracting with a nursery for a supply of plants. In the early nineteenth century, James Cochran's business – discussed in Chapter 5 – was based on hiring out flowering and scented plants for a few days for social events. But if one wanted plants all year round, others were happy to oblige. In 1839 James Mangles described in *The Floral Calendar* how

> All plants, after flowering in London, will inevitably die, unless taken the very greatest care of by a practised gardener . . . By a *contract* . . . the amateur is relieved of all this trouble and uncertainty, and he will always have before him a healthy and vigorous floration.

He recommended flowers on dining-room and drawing-room balconies and outside the bedrooms. An example was an annual contract with Mr Hopgood of Bayswater, London. It covered the provision of plants for all the balconies, for the 'Green House' (with never fewer than 70 plants), for 24 boxes and 12 vases in the back garden, and for all the beds in the front garden, and cost £51,300 (£69 10s). It seems a lot, but Mangles pointed out that the average annual wages of one gardener, together with his board, would be almost exactly the same. For those who were prepared to do their own gardening, Mangles gave recommendations for nurseries with 'reasonable' charges.[29]

The towns and cities of England in the eighteenth and nineteenth centuries were not, despite all this evidence of gardening activity, really favourable locations for gardens. Pollution from coal fires and industrial processing, together with increasing pressure on space as urban populations grew larger, severely limited what could be grown. But it is clear that there was sufficient activity, and interest, in gardening to form a basis for what was to come, mainly through the growth of the suburbs, in the next two centuries.

SUBURBAN ORIGINS

The dominant, almost universal, form of English town houses built in the eighteenth century was the terrace: rows of houses, either opening straight on to the street or with an exiguous front garden. There might, as in Bloomsbury in London, be a garden behind, accessed from a service road that, in the more affluent neighbourhoods, would provide a way in for horses and coaches. For the less affluent, the rear garden or yard might hold little more than a privy. For all households, the back road could be used to bring in the coal and take out night soil from cesspits. For the poorest, there was no rear yard at all: houses were built in rows, or 'back to back' or around a courtyard. Houses of all types could contain several households or families, renting one or more rooms with no security of tenure.

The characteristic form of English suburban living, the detached or semi-detached house, built to be occupied by a single family, was therefore revolutionary. It swept all before it and, from 1815, when the first suburb began to be built in London, until the aftermath of the Second World War, when high-rise blocks of flats were seen as a solution to housing needs, it was considered the most desirable way for people of all classes except the very rich to live. Some terraces continued to be built, but many had the essential characteristics of the suburban house, designed for single family occupation (though many later descended into multi-occupation or were divided into flats), and with gardens front and back. For the greatest defining form of the suburban house, in a sense its raison d'être, was, according to Michael Thompson, 'the garden, preferably one in front to impress the outside world with a display of neatly-tended possession of some land, and one at the back for the family to enjoy'. This led Thompson, one of the leading historians of urban and suburban housing, to conclude that the desire for a garden was at the 'roots [sic] of the demand for suburban living' and that this desire was more important, in shaping English urban growth, than the price of land, the development of transport networks, or even the Victorian 'desire for a domestic life of privacy and seclusion'. It was this demand for a private garden that produced, he wrote, that 'unlovely, sprawling artefact of which few are particularly fond'.[30]

During the first part of the nineteenth century, suburbs were synonymous with the villa, the name given to detached or semi-detached houses with their own grounds. Villa dwelling seems to have stemmed from the desire of the growing middle class to escape from the constraints of an urban existence. At first, London was the only English town or city with a middle class large enough to sustain entire communities of villas, such as those built first on the Eyre estate in St John's Wood, London, immediately after the end of the Napoleonic Wars. That suburb, as it was then considered – it is now one of the richest parts of London – was soon followed by others in and around the capital. Meanwhile, the more affluent – perhaps not constrained by the need to commute to jobs in the City – were colonizing areas further from the centre by building larger villas surrounded by more land.

Outside London, where the towns and cities were still relatively small, villas could be built on the outskirts for individual members of the middle classes. Some were very large: Thomas Goldney, an employee and then shareholder in Abraham Darby's Coalbrookdale Company – iron founders and one of the seminal businesses of the Industrial Revolution – bought 8 acres of land outside Bristol in 1748 and created a garden that included a grotto, a canal and a Newcomen engine, within a Gothic chimney, which powered a fountain and cascade.[31] The Darby family themselves had a garden at their house, Madeley Court, in Coalbrookdale,

> laid out with great Elegance and taste and ornamented with Grottoes formed of Moss Iron slags etc the Trees growing luxuriantly and Yeilding Fruit in Abundance and the Hills steep and rocky on the opposite side of the Dale, with Fish ponds and large Pools of Water; and Views of the Works intermixed.[32]

Further north along the River Severn, the physician Robert Darwin and his wife Susannah, the eldest and favourite daughter of the pottery magnate Josiah Wedgwood, built 'The Mount' between 1798 and 1800 on a 7-acre site outside Shrewsbury, Shropshire.[33] It was there that Charles Darwin, the evolutionary biologist, spent his childhood and discovered his love of botany. It became noted for the finest shrubs and flowers, with its flower gardens, hothouses, ice house, terraces above the river and numerous trees.

Similar – though usually smaller – houses seem, from the evidence of maps, to have existed around most if not all the early nineteenth-century towns, although they did not form a continuous belt but were interspersed with farms and, sometimes, houses large enough to be called 'mansions'.[34] Outside Manchester, 'the surrounding townships were the site of the country house for the merchant who wanted to demonstrate his increased status and wealth'.[35] Only one or two miles from the warehouses of the town, allowing the merchant to walk to work, such properties could still be of 10 acres or more, large enough to provide pasture for carriage and riding horses and for the cows that supplied the families with their uncontaminated milk. Later in the century, from 1879 onwards, Joseph Chamberlain, the politician, laid out Highbury, his 25-acre estate, four miles outside Birmingham. He later extended it to 100 acres and equipped it with a kitchen garden, lake, rockery of Pulhamite stone and twenty-five glasshouses, twelve of which were devoted to his passion, orchids.[36]

Too little is known about these houses and the gardens, or even small parks, in which they were set. Most were swallowed up, later in the nineteenth century or more recently, by the expanding towns and cities; they survive only as memories, sometimes through the street names given to the suburban estates of much smaller villas that often succeeded them. In Leeds, for example, according to an article on suburban development in the city, although 'individual owners of small parkland estates might attempt to stand firm against the increasing amount of building after 1870, it did not prove possible to overturn the basic sequence of the previous century. The search for seclusion of the mansion dwellers had been spoiled by the arrival of the respectable terraces.'[37] The large villas and mansions are sometimes described in local gazetteers and directories, or in sale particulars, but little information exists about their gardens. This is even the case when, as in Leicester, the grounds of a mansion – Humberstone Park – were transformed into a public amenity.

As so often in garden history, John Claudius Loudon comes to the rescue. His *Encyclopaedia of Cottage, Farm, and Villa Architecture and Furniture* of 1834 was followed by *The Suburban Gardener and Villa Companion* of 1838 and, in 1842, by *The Suburban Horticulturist: or, An Attempt to Teach the Science and Practice of the Culture and Management of the Kitchen, Fruit, & Forcing Garden to Those Who have had no Previous Knowledge or Practice in these Departments of Gardening*. The 1838 volume is particularly useful because it contains Loudon's descriptions of forty-five suburban villas, large and small, some real and some hypothetical, and his estimates of the costs of making and maintaining their gardens. His volume of 1842 shows in enormous detail, over 732 pages, what might be achieved by the suburban gardener.

Loudon's estimates of the costs of gardening in the 1830s are unique, but have apparently gone unnoticed by garden historians. No other writer has tried to describe in such detail the work, and the expense, of making a range of gardens, nor the effect on their design of different motives for their use and enjoyment, nor the costs of employing a variety of gardeners to maintain them. Most garden advice books, before and since, have ignored such topics; perhaps they don't want to frighten their readers or realize that, in times of changing prices, the figures will soon become out of date. But Loudon, ignoring these difficulties, gives us an immensely valuable picture of the middle-class garden in the first half of the nineteenth century.

He followed contemporary practice in dividing up his gardens, as was done with houses by building regulations and the local taxation system, into four classes from first to fourth 'rate'. Georgian and Victorian society could import class differences into virtually anything. The first rate had a park and farm occupying at least 50–100 acres; the second lacked a farm, but had a paddock and dairy and was at least 3 acres but often 10 or more; the third might occasionally be as large, but lacked a dairy and was more often 2–5 acres in size; the fourth, to which he devoted most attention, was from 1 perch (0.006 acres or 272 square feet) within towns, such as London, up to 5–10 acres in the countryside. He presents plans for a variety of houses and gardens within each class, paying careful attention to whether the garden is designed to save initial expense or later maintenance costs, whether it is the occupier and his family who are to do the gardening, whether it is to be used for fruit or flowers, or for a collection of ornamental plants, whether it will have a greenhouse or even 'where the object is a botanical collection'.[38] Most of his fourth-rate gardens, designed with all these different objectives in mind, are 200 feet by 60 feet, or a bit over a quarter of an acre, of which 25 per cent is taken up by the house, 18 per cent by the front garden and 57 per cent by the back garden.[39]

Loudon's basic garden (although he doesn't use that term) of such a size, designed for 'economy in the after-management', might be 'laid out and planted for 30*l*. or 40*l*.; and kept perfectly neat for 30*s*. or 40*s*. a year'.[40] In modern terms, this is an initial cost of £22,670 to £30,230 and an annual cost of £1,134 to £1,511. What did the owner or tenant get for this? There was a front garden with two rectangular grass plots, each with a large shrub in the centre and smaller ones at each corner. The back garden was

> planted with four rows of low trees, two near each walk [around the sides of a rectangle], in quincunx,* leaving a broad space in the middle,

* 'Quincunx' means an arrangement of five objects as on a playing card or on dice – one at each of the four corners and one in the centre.

about 100 ft. in length, well adapted for a party walking backwards and
forwards on in the summer season, for a dance, or for placing a tent on,
for sitting under, at the farther end.

There is no mention of flowers or vegetables, but the cost includes
paved walkways, the ornamental trees (6–8 feet tall) and grass seed,
levelling the garden, providing a refuse pit and fixing posts for a
clothes line. Loudon recommends – oddly to modern eyes – that the
surrounding walls or fences should be covered with ivy, to minimize
upkeep, while, again to reduce expense, the occupier himself could
mow the lawns or employ someone to do so. It would suit

an occupier who had no time to spare for its culture, and who did not
wish for flowers. It would not suit a lady who was fond of gardening:
but for one who was not, or had no time to attend to it, and who had
several children, this garden would be very suitable.[41]

A garden of this size could not, Loudon thought, be made profitable
through the production of vegetables; market gardens around all
towns were already producing cheaper fare to be sold by the greengrocer. But a large number of vegetables could still be grown: marrows
or pumpkins, to be used for soup or pies, together with grapevines,
figs, pears, apples, plums, cherries, currants and gooseberries. Sev-
enty feet of scarlet runner (kidney) beans would provide colour and
food for a family of six or seven – though the family must have grown
heartily sick of beans – while the front garden could be used for nastur-
tiums and a variety of herbs, with possibly dwarf fruit trees. Finally
the back garden could be used to grow many other vegetables: asparagus, sea-kale, rhubarb, artichokes, cabbages, peas, beans, turnips,
carrots, parsnips, spinach, onions, leeks, garlic, lettuce, endive, chicory,
celery, mustard and cress and radishes, grown in succession. Only the
potato is missing. The work involved in such a garden would take up
at least two hours a day from April to September, with the additional
assistance 'of the female part of the family' or could be done by a
suitable manservant. It would not be worth employing a gardener
since 'for the 15*l.* or 20*l.* a year which he must pay a hired gardener
he might purchase as much fruit and vegetables as he could grow in a
garden of the extent we have mentioned'.[42]

Variations on the basic theme are set out, like a pizza restaurant offering different toppings. A garden designed for 'exercise and recreation for the occupier and his family' would contain only lawns and flowering plants as well as bulbs, annuals and biennials. Loudon recommended eighty-nine plants for the front garden alone, while the back garden might be devoted to turf and to alpines, placed in pots; this would allow for growing 1,800 species, 'as many alpines as can be purchased . . . in any British nursery'.[43] This implied buying many rare plants, at a cost of up to £75,570 *(£100)* at the outset and possibly – if the owner would not do the work himself – employing a highly skilled gardener whose annual salary and keep would be the same amount, with a further £38,750 for fuel, pots, soil, and so on.

A garden in which flowers, fruit and vegetables were mixed would still cost about £37,780 *(£50)* to establish, with £7,550 *(£10)* or more annually for a jobbing gardener to maintain it.* Much more expensive would be a garden with a greenhouse to produce forced fruit and flowers, including peaches, grapes, melons, strawberries, cucumber, kidney beans, rhubarb and mushrooms. This would need an initial investment of £132,200 *(£175)* and also what Loudon calls a 'reserve garden', a plot of land nearby on which to sow and grow plants; the total would be £207,800 *(£275)* and the complexity of the gardening would need a skilled gardener, who would have to be provided with a house and coal for fuel, which with his wages would cost £128,500 *(£170)* annually. All for a garden of less than a quarter of an acre.

Loudon knew what he was talking about, since he had built for himself in 1825 a fourth-rate 'double detached' house (his term for a pair of semi-detached houses), which still exists, in Porchester Terrace, Bayswater – then seen as a suburb of London, now one of its most fashionable areas – and had equipped the garden of the pair of houses with rare trees and shrubs, 'being the largest plants that could be procured', at a cost of £75,570 *(£100)* as well as building greenhouses costing a total of £226,700 *(£300)*. The result is shown in plate 37. The total size of the plot seems to have been 100 feet wide by 122 feet long.[44] He also had a reserve ground at the rear, across a lane. One of

---

* Loudon actually gave a range of costs, from £45 to £55, but for clarity the average has been used in this and subsequent calculations.

his objectives was to grow as many species as possible; in 1830 he had in his greenhouse eighty-two plants – 'all the orders and tribes of hot-house plants in cultivation in Britain' – and, in the garden, 600 species of alpines. There were far more trees and shrubs: 'The greatest number of species that we ever had at one time (exclusive of varieties) was about 2000' though by 1838 this had been reduced to a few hundreds, since Loudon no longer employed a gardener.[45]

Third-rate houses were, by definition, much larger, but still considered to be suburban villas. Loudon asserts that a property of 10 acres near London could be occupied 'with propriety and dignity' if the occupier had an income of at least £1.5 million (£2,000) a year. In such a property, the garden, 'including the walls, the structures in the forcing-ground, the gardener's house, &c., may cost 1000l. more; and the tunnel, the walks, the plantations, and the ornamental buildings, from 1000l. to 3000l. additional'.[46] (In modern values, the garden would cost between £1.5 million and £3 million.*) The house itself could be built for £6.4 million (£8,500). Another third-rate house in 3 acres would cost less and could be occupied by 'a person having a clear income of from 500l. to 750l. a year, who would act as his own head gardener'.[47] (In modern values, this would equate with an income of between £378,000 and £566,000.)

Second-rate suburban gardens had to be between 3 and 10 acres or more, and presumably commensurately expensive, while the first-rate suburban gardens that Loudon describes include Wimbledon House, Surrey, with 100 acres. It was the creation of Charlotte Marryat, who, after the death of her husband in 1824, spared – Loudon said – 'no expense in enriching and adorning the place, and more particularly in procuring the rarest and most beautiful of plants'; it had 'shrubberies, a "very spacious ivy-covered summer house", a rocky cascade studded with alpine plants, a grotto, a lake with two islands, a conservatory, an orangery', with hothouses for fruit and tropical plants.[48] Mrs Marryat's achievement was recognized by her election, in 1830, as one of the first five women Fellows of the Royal

---

* Thus the garden added between a third and a half to the cost of the house. This proportion is very similar to that found for many of the large houses of the eighteenth century.

Horticultural Society.[49] Unfortunately, her garden has not survived. Another first-rater was Kenwood House, Hampstead, said to be 'the finest country residence in the suburbs of London'.[50] It still deserves that description in its role as one of the greatest art galleries in the capital, set in a Reptonian landscape only five miles from the centre of the city.

Loudon's examples show that the burgeoning middle class of the early nineteenth century had the funds and were prepared to spend significant sums of money on their villas and gardens. Some of them – he believed – were sufficiently interested in gardening to contemplate buying and then looking after hundreds or thousands of plants. Can we trust his descriptions and estimates, and what do they tell us about middle-class gardens in the first half of the nineteenth century and their influence on the gardening industry?

In 1838, aged fifty-five and five years before his death, John Claudius Loudon was one of the most successful and respected of authors, journalists and magazine proprietors and editors of his time.[51] He had earlier in his life made a fortune – £10.4 million (£15,000)[52] – through his work and then lost it through inattention to his investments; his financial situation in later life was precarious but his reputation as a gardener and garden writer remained very high. His own garden in Bayswater was renowned and, at the time he was publishing his books on villas and their gardens, he was being commissioned to design the Derby Arboretum and a number of cemeteries as well as some private gardens. Since 1830, he had been helped in his work by his wife Jane, née Webb; they are shown in plates 35 and 36. She was the author of an early work of science fiction, *The Mummy!*, published three years before; despite knowing, by her own admission, nothing about gardening when she met her husband, she soon became knowledgeable, publishing her own books and editing her husband's. This was particularly necessary as production of the work that he saw as his greatest achievement, *Arboretum et Fruticetum Britannicum* (The Trees and Shrubs of Britain) of 1838, left the couple heavily in debt in the years before and after Loudon's death.

Despite these money worries, Loudon seems to have had a very good idea of the market opportunities for his books and magazines. The first edition of his *Gardener's Magazine*, published as a quarterly in 1826, was priced at £193 *(5s)*, yet sold 4,000 copies. From 1831 it was published monthly and by 1834 was priced at £46 *(1s 2d)*; this change in price and frequency of publication, presumably to maximize revenues, suggests a strong commercial sense. Other works, including the *Encyclopaedia of Gardening* – which has over 1,000 pages – and his various books addressed to the suburban gardener, were regularly reprinted and sold well. All this indicates that Loudon knew what he was talking – and writing – about; his judgements on the costs of making and maintaining gardens are likely to have been accurate and he would not have written his books on suburban gardens if there had not been a substantial demand for them.

The evidence from maps of the areas around the towns and cities of England suggests that many substantial gardens were being laid out during the second quarter of the nineteenth century, when Loudon's books on the subject were being published. Even if the number of people who did so is unknown, there were enough who were prepared to lay out between £25,000 and £50,000 or more, in modern values, on relatively small gardens to make Loudon's ventures worthwhile. But this must have been only one aspect of the demand facing the garden industry as a whole. Loudon was a proponent of what he called the 'gardenesque' style of garden design, one important principle of which was 'distinctness, or the keeping of every particular plant perfectly isolated, and, though near to, yet never allowing it to touch, the adjoining plants'.[53] This was opposed to the prevailing 'picturesque' style, which tried to paint a picture with plants; Loudon's emphasis was on the individual plants but – very importantly – on having as many of them as possible. It was reflected poetically: 'Keats's floral language, borrowed from Hunt, seemed to imitate the gardening habits of suburbanites eager to keep up with the newest gardening trends.'[54] The typically contemporary emphasis on collecting explains Loudon's emphasis on having, for example, every species of alpine available from English nurseries, or his pride in having 2,000 species of plants in his own garden and then recommending 1,800 as suitable for a quarter-acre garden designed for recreation. It was impossible for most of his readers or clients to think of going to see these plants in their natural habitats in America or Asia; however, Loudon did not believe that those natural conditions should be imitated, but that the individuality and beauty of the particular plant should be the focus.

No advice could have been better for the nursery industry. Nurseries not only sold the plants but also provided the labour to make the gardens, together with what is now known as 'hard' landscaping – the gravel and stone paths, the fountains and their water supplies – and the tools and equipment needed to maintain the gardens. Many also provided a labour force for those unwilling to do the work themselves or to force their families to do so, either by recommending gardeners or by contracting to do the regular work that was needed; James Cochran did so in London, as did the Veitch nurseries in Exeter and London. The desire for exotic or imported plants led naturally to the greenhouse, another lucrative market, and to the conservatory where these plants could be displayed to admiring visitors. Loddiges in Hackney was one of the most celebrated examples of the nurseries that catered for this market. The first half of the nineteenth century saw a very large expansion in the number of nurseries – in and around all the towns and cities of England – and in the services they provided and the range of plants that they sold; catalogues from this time onwards to the end of the century are far larger and advertise for sale far more species and varieties than do the catalogues of today. This was the market power of the nascent middle class, which Loudon had so cleverly tapped.

The Victorian villa gardens of the middle classes became a focus for their social lives and for a new range of leisure activities in the form of garden games. Although my great-grandparents' garden in the suburb of Hampstead was not large enough for a tennis court, many others that my grandmother visited did have one and, after the invention of the game of lawn tennis in the 1870s, it spread rapidly. Its inventor, Major Wingfield, claimed the endorsement of noble clients and the *Sporting Gazette* soon predicted, correctly, that 'having won its entrée into good society . . . it will be a popular pastime in every English home which can boast a level piece of ground twenty yards by ten'.[55] It supplanted croquet, which required in its rules – if not always in practice – twice as large a playing area. Both sports could be played, with the required decorum, by both men and women and thus provide 'cover for courtship and flirtation'.[56] In the 1930s, the popular garden writer Geoffrey Henslow commented in his book *Suburban Gardens* that the 'suburban residence cannot be complete without a tennis court', although this says more about the market for his book than about the reality of most such gardens.[57] Badminton and clock-golf could be played in smaller spaces. For the less active, taking tea in the garden became a regular middle-class summer pastime. By such means, and through the ubiquitous conservatories that were another ornament of villa life, the Victorian middle classes co-opted the garden into their lifestyle. Above all, gardening was a respectable activity for their women, an extension of their domestic sphere.[58]

GARDENS FOR THE MILLIONS

When, in the 1850s, greenhouses built on Sir Joseph Paxton's principles were first advertised as 'for the millions' – by which was really meant the middle and working classes – it is unlikely that anyone realized how many millions were to be involved. The census of England and Wales in 1851 revealed that there were 3,342,000 houses in the two countries.* By the time of the census of 1931, the number of houses was 9,400,000, an increase of over 6 million in that period of eighty years, while between 1931 and 2014 the number rose even faster, to a total of 23,372,000 houses in England alone. But even these figures understate the extent of housebuilding. Old houses were demolished to make room for new. So even though the overall number of houses rose, between the 1850s and today, by about 20 million, the number that were actually built was probably at least 23.5 million.†

Not all these houses had gardens, but most did. Even around the middle of the nineteenth century, houses were still being built as courts, or back-to-backs, with exiguous or non-existent front and

---

* The summary statistics do not distinguish between the two countries of England and Wales, but the Welsh proportion was small.

† Estimates of the number of houses built in Great Britain begin in 1856 and are given separately for England and Wales from 1981. The estimate of 23.5 million is based on applying the mean proportion – 89 per cent – of houses shown in the census for England and Wales to the total of houses built for Great Britain.[59]

back yards. But the proportion of such houses fell sharply after 1875 and, by the time of the First World War, was at or close to zero. They were replaced by houses with, at least, small yards or gardens. Then, mainly after the Second World War (although 'mansion flats' had been constructed earlier), high-rise dwelling began. Yet, throughout, the overwhelming preference of renters, owners and occupiers – as distinct from architects and planners – was not for flats but for the single family home with a garden. In Birmingham just before the Second World War, householders were asked if they valued a garden: 96.3 per cent of those with gardens said yes, and 78.1 per cent of those without said they would like to have one.[60] Other surveys have continued to demonstrate the same preferences.

In recent years, the average size of gardens in England has, according to the Horticultural Trades Association, been getting smaller, as can easily be seen in modern housing developments. But in 2016 the average size of a British garden was still just over 2,000 square feet or 0.05 acre. Simple arithmetic tells us that, if that figure is applied to the 23.5 million houses built between the 1850s and today, those houses were accompanied by gardens that covered at least 1.2 million acres. That is approximately the total area of the English counties of Buckinghamshire, Berkshire and Hertfordshire combined, or about two-thirds of the area of Greater London. Even if some of those gardens replaced ones that had existed before, as they certainly did, with large country villas becoming housing developments, the numbers are still enormous. They should probably be higher, in fact, since many of the gardens built before the First World War and in the interwar period were much larger than the average today.[61]

How did this huge expansion occur and how did it affect the garden industry? Its main stimulus was demographic, the result of the growth of the population from the middle of the nineteenth century to the middle of the twentieth through birth rates that were still high compared to those today and death rates that were falling rapidly in response to increasing prosperity, better sewers and water supplies and some medical innovation. The growth in the number of houses was aided by higher incomes and more ample mortgage finance. It was carried out by the building industry, under increasing regulation from local and central government, while the new householders required help to master new gardening skills and to provide the plants, seeds and tools needed to carry out their plans and impress their neighbours.

Building begins with the supply of land for that purpose and its purchase by developers and builders. Housebuilding is a notoriously volatile industry, subject to substantial fluctuations in demand that stem in essence from the overall state of the economy and in particular from the availability of, and rates of interest on, borrowed money. Most housebuilders in the nineteenth and twentieth centuries were quite small businesses who had to borrow to buy or lease land and then finance the building of perhaps five to ten houses, hoping that they had correctly guessed the nature and extent of demand.

Demand was also complex. There were far fewer owner-occupiers than today; there is little hard evidence, but 'conventional wisdom' is that in 1914 only 10 per cent of houses were owned by their residents, although this proportion later rose rapidly to reach between 30 and 35 per cent by 1939.[62] Most families in the nineteenth century rented – sometimes a room or a floor, or a complete house – and had little or no security of tenure. Even among the more affluent, a much higher proportion than today leased or rented their property; early in the century, leases were typically of about thirty years, much shorter than the ninety-nine years that became typical by the end of the nineteenth century, so the market – and housebuilders – had to take account of the possibility of landowners repossessing the property or demanding payment to extend the lease.[63] But most renters or lessees stayed in properties for much shorter periods: in Liverpool in the 1850s half the population moved house within two years.[64] Contemporary opinion was that one should not lease or rent a house for more than three years.[65]

So there was then even greater insecurity in the housing market as a whole than in the early twenty-first century. Decisions about whether to buy, lease or rent, about obtaining a mortgage and about which house to buy, were affected as today by the purchaser's own finances, by fashion, by the proximity of shops, transport links and other facilities in a particular area. The combination of the overall state of the economy and speculative housebuilding led to booms in building in the 1870s, 1890s, 1900s, late 1920s, mid 1930s, early

1950s and late 1960s, with far fewer houses built in the intervening periods; in 1876, for example, 116,000 houses were built, while by 1886 it had fallen to 67,000, only to rise again to 140,000 in 1898. One reason was that developers and builders seem to have been consistently over-optimistic about the state of middle-class demand for housing.[66] From the end of the nineteenth to the late twentieth century, some stability came from the advent of what is now called 'social housing' and was then called 'council housing', but there were still very large variations in the overall number of houses – and therefore gardens – that were built. In addition, different towns and cities experienced building booms at different periods. The garden industry must have shared the insecurities of housebuilding, since spending on new gardens accompanied spending on new houses, even if the overall trend of demand was upward.

Until the 1850s and 1860s, there was little state intervention or regulation: landowners could sell their land for development, builders could build what, where and when they wished. They did not even have to coordinate their work with others, for example by linking up roads or providing other infrastructure such as shops or schools. When some regulations were introduced through local by-laws and in the Public Health Act of 1875, they were mainly concerned with the width of streets not the length of gardens, although in London the Metropolitan Buildings Act of 1844 had provided that every new house should have an open space at the rear of not less than 100 square feet. In other towns and cities, local regulations were mainly concerned with banning or regulating back-to-back housing. However, in Birmingham from 1845 each house had to have a private garden or yard; a study of 522 houses built there between 1878 and 1884 found that 40 per cent had front gardens.[67]

Much of the legislation was permissive: it did not require local authorities to enforce national standards, while builders inevitably did the minimum required. In general, 'by-law housing',* as it was known, led to building not of semis but of terraces of houses, with a service road between them or periodic entrances to private yards;

---

* Also known as 'bye-law housing'; the terms 'bye-law' and 'by-law' seem to have been used interchangeably.

26. Before the lawnmower: gardeners scything the grass at Hartwell House, Buckinghamshire, in the 1730s.

27. Horsepower in the garden: an early lawnmower, *c.*1910.

28. 'All the annuals in England': Francis Douce's garden in Gower Street, London, in 1791.

29. Essential equipment for the Victorian garden: nineteenth-century mantraps.

30. 'Better men than we': gardeners at Wentworth Castle, Yorkshire, in the 1890s.

31. 'A labour of love': William Coleman, a devoted head gardener, in 1875.

32. Men, boys and horses: the garden labour force at Wrest Park, Bedfordshire, in 1903.

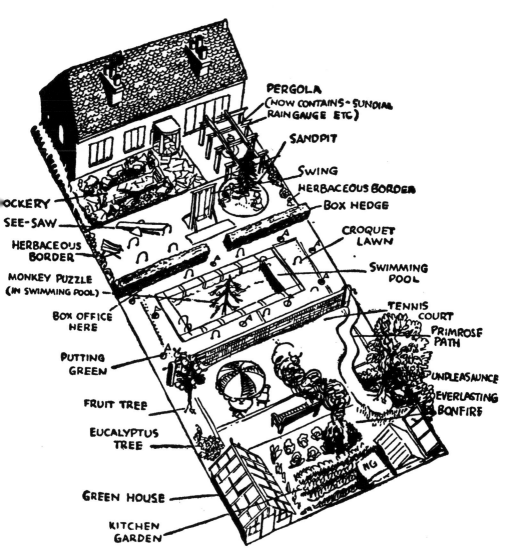

PERGOLA
(NOW CONTAINS - SUNDIAL
RAIN GAUGE ETC)

SANDPIT

SWING

HERBACEOUS BORDER

BOX HEDGE

CROQUET LAWN

SWIMMING POOL

TENNIS COURT

PRIMROSE PATH

PUNDLEASAUNCE

EVERLASTING BONFIRE

OCKERY

SEE-SAW

HERBACEOUS BORDER

MONKEY PUZZLE
(IN SWIMMING POOL)

BOX OFFICE HERE

PUTTING GREEN

FRUIT TREE

EUCALYPTUS TREE

GREEN HOUSE

KITCHEN GARDEN

33. 'Planning the Modest Plot'. W. C. Sellar and R. J. Yeatman, the authors of *1066 and All That*, comment in *Garden Rubbish* on fashions in 1936.

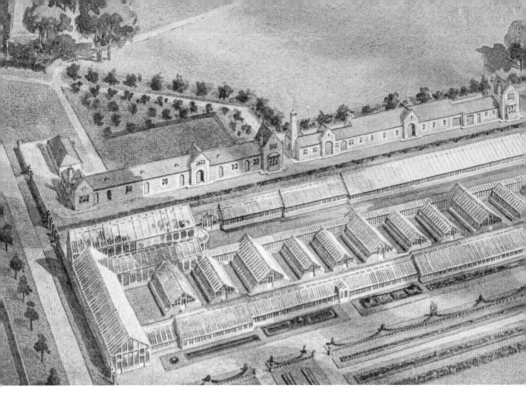

34. 'The most perfect garden in Europe, of its kind': the Royal Kitchen Garden, Frogmore, in the 1840s.

35. and 36. The nineteenth-century encyclopaedists of gardening: Jane Loudon and John Claudius Loudon.

37. Two thousand species of plants: John Claudius Loudon's villa garden at Porchester Terrace in Bayswater, London, in the 1830s.

they were set out in a rectangular grid and with a density of up to forty houses per acre. The result – still to be seen in some London suburbs and around the country – was often criticized as monotonous, despite a gradual improvement in the standard of what was built.[68]

Meanwhile, the building of middle-class villas or mansions continued in areas such as the home counties around London and, for the rich entrepreneurs of northern England, in areas such as the Lake District; Thomas Mawson made his reputation as a garden designer in catering for the latter. But the same process took place all over England, as the grounds of earlier villas were developed for housing and their owners, or their children, moved further out.[69] The contrast between these spacious mini-estates, increasingly shielded from view by gates, trees and hedges, and the terrace by-law housing remained acute.

Change came first with local authority housing, where early schemes from the 1890s adopted densities of twenty to twenty-four houses per acre and thus larger gardens than in the by-law houses. Much more influential, however, was the ideology of the garden city movement. The first Housing and Town Planning Act, of 1909, put the emphasis 'on separation, on space rather than on what filled it, on trees, grass and gardens more than on shops, factories and pavements'.[70] It followed the precepts of Ebenezer Howard and the lead set by the philanthropist and chocolate king Joseph Rowntree, who built New Earswick, a model village for his workers, on the outskirts of York. As the *Municipal Journal* reported in 1906:

> The average size of the gardens is 350 square yards – a size determined upon after careful consideration of the amount a man can easily and profitably work by spade cultivation in his leisure time. The gardens have proved a great source of health and enjoyment to the villagers. Prizes are awarded in competition for the best-kept gardens, and we find many well-cultivated fruit, vegetable and flower gardens.[71]

Another chocolate philanthropist, George Cadbury, designed his new settlement at Bournville, south of Birmingham, with the house taking up a maximum of a quarter of each site, leaving plenty of room for a garden.[72]

The density of housing established at New Earswick was explained – virtually in the same words – by its architect, the influential town planner Raymond Unwin, in 1909. As he wrote:

> Twelve houses to the net acre of building land, excluding all roads, has been proved to be about the right number to give gardens of sufficient size to be of commercial value to the tenants – large enough, that is, to be worth cultivating seriously for the sake of the profits, and not too large to be worked by an ordinary labourer and his family.[73]

TWELVE HOUSES PER ACRE

During the First World War, Unwin had great influence on the Local Government Board and his views were endorsed in 1918 by the Tudor Walters Report on working-class housing.[74] As a result, the approved standard for construction became eight houses per acre in rural areas, twelve per acre in urban, a huge change from the 'by-law' terraces of about forty per acre. Although there was some increase in the size and number of rooms, the basic pattern remained: a living room, possibly a parlour, a scullery and washroom and three bedrooms, reflecting the massive fall in the number of children per family that took place from the end of the nineteenth century onwards. So the main beneficiary of these lower densities was the garden: plots in typical inter-war houses were – and still are, for most of those houses still stand – 80–200 feet long and 20–40 feet wide (between 1,600 and 8,000 square feet) with larger areas for higher-priced homes.[75]

Providing gardens for people to grow vegetables was not the only motive for establishing a standard density of housing that has endured ever since.[76] The British government, at the end of the First World War, was seriously worried by the threat of revolution – which had swept through Russia and Germany – and decided that one way to stave it off would be to provide 500,000 'homes fit for heroes'.[77] The homes had gardens for the same reason: Neville Chamberlain argued in 1920 that 'every spadeful of manure dug in, every fruit tree planted' would convert a potential revolutionary into an upstanding citizen.[78] An ambitious programme of housebuilding followed, bolstered by further regulation of housing standards and, initially, financial incentives for both council and private housing.*

Public and private builders together constructed 4 million houses in England between the wars and, even if providing gardens for the masses was not the only reason, that was certainly the result. Of course, not all households had gardens, even after this building boom. A survey of urban gardens and allotments conducted by the Ministry of Agriculture in 1944 found that in 1939 only 42 per cent of urban or suburban working-class households had a garden, compared with 69 per cent of the middle classes. More people took up gardening on allotments as a result of the 'Dig for Victory' campaign, but even so only a minority of households – even if they had a garden – could be persuaded to grow fruit and vegetables as a contribution to the war effort, although many continued to grow flowers.[79] So the hopes of Unwin and the garden city movement that providing plots large enough to supply household needs would persuade people to use them for that purpose do not seem to have been fulfilled.

However, after the Second World War, the new towns were planned on the same principles and the area of English land devoted to gardens continued to grow. The fashion for building high-rise flats from the 1950s did not make much difference to the onward march of gardens, since many of them replaced the by-law or other cramped houses of the inner suburbs, which had only had small plots.

The low-density housing recommended by the Tudor Walters Report was not mandatory in private housebuilding.[80] It is therefore surprising, but also significant, that developers and builders did adopt that standard. The likelihood is that they did so because that is what their potential buyers – and more and more of the houses were being built for sale to owner-occupiers – now wanted and were prepared to pay for. Low interest rates and easily available mortgages (with interest on the borrowing attracting tax relief) were helped by the expansion of building societies. This made it possible for a wide swathe of the middle classes and upper working classes to buy their houses, as their main form of saving. Their demand was for semi-detached houses

---

* Subsidies for private housebuilding declined in the 1920s and were virtually nonexistent in the 1930s.

with sizeable front and back gardens and that is what the builders provided, together with even lower-density developments for the wealthier buyers, often with outbuildings and, later, garages. The market was speaking and it wanted gardens.[81]

Even today, most urban developments are still built at a density of around twelve houses per acre (thirty per hectare) and many local authorities are identifying land for development to supply, at that density, the very large number of houses that the government now wants them to authorize. But the density is gradually creeping upwards, one of several factors that explain why the average size of British gardens has recently been falling. New houses have been built in the large gardens of many suburban villas and even in those of the inter-war houses, on relatively cramped sites. Houses have tended to get bigger and, therefore, to take up more of their plots, although this tendency was halted or even reversed in the 1980s.[82] Above all, there is the car. The lower-priced suburban house of the inter-war period rarely had space for a garage, although higher-priced houses might have one. Nowadays planners require at least two car spaces for each house – partly because internal garages are quickly turned into rooms – while hundreds of thousands of front gardens have been paved over as car parks. There were 32,000 private cars in the UK in 1907 and 2 million in 1939; now there are over 37.5 million registered in Great Britain.[83] The car is probably the main reason why English gardens are becoming smaller, although developers tend to blame higher prices for land and less demand from owners; there is little evidence for the latter.

Since the second half of the nineteenth century, more and more land in England has been devoted to gardens. That means more and more shrubs, trees, flowering plants, bulbs, seeds and tools. It means more nurseries and garden centres, more flower shows and, above all, more gardeners.

THE AGE OF DO-IT-YOURSELF GARDENING

John Claudius Loudon expected that most of the owners of the fourth-rate villas that he described would do at least some of the gardening, or get a manservant to do it. He, and his wife Jane, also thought that wives and children would be enlisted to help. In practice, it took a long time for these expectations to become the norm. Villa gardens needed paid gardeners.

As middle-class incomes rose, as new villa and suburban gardens and public parks were laid out, and as gardens became more elaborate and labour-intensive, with greenhouses and planting such as carpet-bedding and other fashions, so increasing numbers of gardeners were employed, as described in Chapter 6. But, even so, most people did not have a paid gardener; most gardens were, and are, tilled by their owners, as Loudon hoped and recommended. They might employ a gardener intermittently to do the hard digging or trim the trees or, of course, to undertake expensive hard landscaping and the installation of garden features such as ponds, fountains, sheds and pergolas. But gardening is an unusual economic activity in which, overwhelmingly, the consumer is also the producer and in which a great deal of the pleasure of consumption lies in the activity of gardening itself.

Gardening has long been seen as a highly moral activity, capable of keeping the working classes out of the public house and providing healthy exercise for the middle-class male after a hard week at work.[84] In 1919 the Conservative MP for Chelmsford was reconciled to spending money on housing by the thought that good garden plots would ensure that when the man of the house got home at night 'he will find not only a healthy family, but healthy occupation outside where they can go and work together in the garden'.[85] Accounts abound of the office worker of the late nineteenth or early twentieth century devoting his evenings and weekends in his suburban villa to gardening, with the state of his front garden one of the most significant indicators of his social status and pride in ownership.[86] The front garden was 'the most public and visible part of the house', as the parlour and the front step had been in the earlier terraces opening directly on to the street.[87] Gardens were preferred to the street as places for children to play and the Tudor Walters Report recommended house designs that allowed the housewife to keep an eye on the children while cooking or cleaning. Reminiscences of the Stoneleigh estate in south-west London, and many others, are of people spending most of their weekends in the garden.[88]

With the fulfilment of desire for a garden came responsibility: it needs to be nurtured and kept tidy. On council estates, in the 1920s and 1930s, inspectors were sent round to make sure that front gardens, in particular, were in good shape. Peer pressure was also important: no one likes to think of weed seeds blowing on to the garden from a neighbour's plot. There were pressures to conform with prevalent styles, such as a lawn in the front garden with a circular bed and a rosebush or small tree in the middle. But there was, in England, some scope for individual taste and variation, even in the regimented suburb. W. C. Sellar and R. J. Yeatman, authors of *1066 and All That*, satirized the possible result in plate 33. In the United States, by contrast – and despite the famed individualism of its culture – there was greater uniformity. This is because while the English suburb is composed of detached or semi-detached houses, each surrounded by a hedge or fence, the American suburb – from the outset – has been a place of detached houses set in the midst of a grass lawn, with no barriers between one house and the next. The garden writer Frank Jesup Scott even brought religion into it: 'It is unchristian to hedge from the sight of others the beauties of Nature which it has been our good fortune to create.'[89] By 1992, four-fifths of the 30 million acres of space around American detached homes was taken up with lawns,[90] and there is enormous pressure on homeowners to keep them watered, weeded and manicured.

Lawn care is only one aspect of the huge amount that the English – and the Americans, the French, the Germans and virtually every other nation – spend on this pastime. The Horticultural Trades Association, which represents the nursery and garden centre industries, estimates that annual retail expenditure, by the 18.7 million households in Britain that have a private garden of some description, is now about £5.5 billion annually. Dividing one figure by the other shows that the average household spends £29.41 each year on its plants, statuary or bird food (on which we spend about £337 million[91]). This figure fits with evidence from surveys conducted by the Horticultural Trades Association, but it may surprise anyone who has stood in a checkout queue at a garden centre. We probably spend much more in reality, as this figure excludes expenditure on paid gardeners, on landscaping and paving, on mowers and tools other than those bought at a garden centre and on materials purchased at build-ers' merchants.[92] The recent study by Oxford Economics of the entire industry in 2018, conducted at the behest of the Royal Horticultural Society and other bodies, suggests a figure that is about twice as large, implying average annual expenditure of well over £50.[93]

The average also conceals a very wide variation in spending; sur-veys show that we spend more money on gardening as we get older and as we get richer, and of course more on large than on small plots. I tried a personal experiment while writing this book: I recorded all my expenditure on a half-acre garden in rural Buckinghamshire for a year. It included a small amount of landscaping and payments for four hours a week to a jobbing gardener and for mowing the lawn, but not the value of the labour of myself and my wife. The total was £7,334, of which only about £500 was spent at garden centres, show-ing how much is omitted from the £5.5 billion total; as suggested by Oxford Economics, that figure probably needs to be multiplied sev-eral times to get a true indication of how much we spend on gardens, even without taking account of the value of our time or of the land that we use.

We don't know how much was spent in earlier periods. It would be possible to extrapolate backwards on the assumption that gardening took up the same proportion of our average incomes as it does today, but it is difficult to know whether this assumption is justified. On the one hand, expenditure on leisure tends to increase as a proportion of our income as that income rises; on the other hand, we now have many more opportunities for spending our money on different leisure activities than did the villa or suburban gardener of one or two cen-turies ago. The figure quoted by the Horticultural Trades Association of £29.41 today equates to 4 shillings (or up to 8 shillings if the HTA figure is an underestimate) in 1939, when Orwell referred to the 'five-to-ten-pound-a-weekers' in the suburbs, so very little even of a lower-middle-class income would have been spent at nurseries or embryo garden centres. It would have covered a few packets of seeds or a few annuals. But that sum doesn't include, of course, the value of seeds exchanged with a neighbour, or cuttings taken from his or her plants, nor the value of the time spent lovingly tending the roses.

How did the suburban gardener know how to spend his money and his time? Although it is sometimes asserted that the English have an innate love of gardening, it seems far-fetched to suppose that they have an instinctive ability to plant a shrub or prune a fruit tree. Nor can it be assumed that the suburban gardeners of the late nineteenth and early twentieth centuries had acquired a knowledge of gardening in some rural idyll of a cottage garden, knowledge that they then transferred to the suburbs. Although there continued to be some small movement of people from agriculture into the urban areas, England was already so urbanized by the middle of the nineteenth century that most suburban residents would have been born and brought up in towns and cities. Ebenezer Howard, the founder of the garden city movement, realized this when he praised the idea of attaching half an acre of land to each government-funded elementary school to 'give the young an interest in horticulture'.[94] Working-class gardens in rural areas in the nineteenth century were also not a good model. Flowers were grown, as Margaret Willes, a historian of working-class gardening, has shown, but the contents of cottage gardens – which gave far more space to vegetables, or even to keeping a pig – were very unlike those of suburbia.[95]

New gardeners learn, of course, from their parents and from friends and neighbours; gardeners are never shy of offering advice. They exchange seeds, lend tools and transmit gardening lore – and probably myth. One source of information (as well as another kind of spending on gardens that doesn't count towards the £5.5 billion) is the printed word – books and magazines. Books on gardening date back to the Middle Ages; Blanche Henrey's bibliography of British books on botany and horticulture lists 376 publications, counting each edition separately, for the sixteenth and seventeenth centuries, while many more were published in the eighteenth.[96] The catalogue of the British Library, which by law receives and preserves a copy of every book published in the British Isles, lists 137 books that appeared in the 1820s with the word 'garden' or 'gardening' in the title (excluding a number about Covent Garden in London). The number mounts from decade to decade with scarcely a check: 296 were published in the 1890s, and 539 between 1900 and 1909; by the 1970s it was 1,462 and, between 2000 and 2009, it had risen to 2,508. The total from 1820 to the present day is 15,652. It is an imperfect indicator because it includes some books published overseas and purchased by the library and some on entirely different topics, such as the Garden of Eden. Above all, we have only fragmentary knowledge of how many copies were sold. One of the earliest, John Evelyn's *Sylva* of 1664, sold more than 1,000 copies in less than four years and many subsequent editions were published. John Claudius Loudon's huge *Encyclopaedia* went through eight editions between 1822 and 1878; in 1846, Jane Loudon's *Ladies' Companion to the Flower Garden* sold more than 20,000 copies.[97] Some recent gardening manuals sell hundreds of thousands of copies.

Another indication of the appetite of gardeners for knowledge is the publication of magazines.[98] *Curtis's Botanical Magazine*, now published by the Royal Botanic Gardens at Kew, first appeared in 1787 and sold over 3,000 copies of its early issues, at £71 *(1s)* each. It was followed by the *Botanical Register* in 1815, by Loddiges' *Botanical Cabinet* in 1817 and Robert Sweet's *Flower Garden* in 1828, each with hand-coloured printed plates that were their principal attraction. John Claudius Loudon's *Gardener's Magazine*, initially priced at £193 *(5s)* a copy, gave him an annual income of between £382,000 *(£500)* and £573,000 *(£750)*.[99] This didn't last and in the 1830s, despite the fact that the price per copy had fallen to £46 *(1s 2d)* by 1834, cheaper rivals, such as the *Florist's Magazine*, cut into his market; helped by reductions in taxes on paper and advertising, its owners were able to sell over 10,000 copies at £19 *(6d)* each.[100] This was also the price of Joseph Paxton and John Lindley's *Gardeners' Chronicle*, probably the most successful and longest-lasting of the nineteenth-century magazines, which began publication in 1841. Some books, such as Jane Loudon's various volumes under the title of the *Ladies' Flower-Garden* (1841–8), began life as monthly issues, later being sold as bound volumes. Their market was clearly the new suburban villa: 'a lady, with the assistance of a common labourer to level and prepare the ground, may turn a barren waste into a flower garden with her own hands'.[101] Her book *The Ladies' Companion to the Flower Garden* stayed in print for over thirty years.[102] However they were published, garden books and periodicals seem to have been lucrative. Later in the century, William Robinson, the proponent of the naturalist style of gardening, is thought to have made a large fortune from the *Gardening Illustrated* magazine. By that time, every popular newspaper had gardening columns, dispensing advice and describing gardens to be emulated. When *Ideal Home* was launched in 1919, it set out its stall: 'readers will find this journal of very decided assistance in not only the Ideal Home, but likewise the Ideal Garden without which the former would, of course, be incomplete'.[103] There were even various series of cigarette cards in the inter-war years devoted to suburban garden designs. More recently, in the 1960s, the five principal gardening periodicals are said to have sold, usually on a monthly basis, 2 million copies between them.[104]

Many of the specialist journals founded in the nineteenth century were aimed at the large garden owners – who could afford them – although they might be passed on to their head gardeners and, from them, to the journeymen and apprentices; Loudon and other garden writers urged employers to provide a library of books and magazines as a form of self-education for gardeners. But they were succeeded, in the late nineteenth and throughout the twentieth century, by a plethora of books and magazines designed for the smaller garden, although the definition of that varied. F. M. Wells, in *The Suburban Garden and What to Grow in It*, published in 1901, was thinking of 'slightly more than half an acre of freshly cleared ground', while Brigadier C. E. Lucas Phillips, whose *The Small Garden* was first published in 1952 but is still in print, had 'borne in mind a limit of about an acre, but with special consideration for suburban gardens of less than half that size'.[105] Few books were aimed explicitly at the more typical, smaller, suburban garden, whose owners were expected to adapt the advice for their limited space, but several warned against the tendency to try to cram into the suburban back garden all the features – rockeries, pools, statues and arbours – of the larger villa gardens. A model garden designed as a suburban plot at the Chelsea Flower Show in the early 1930s adopted a more restrained approach.[106]

Then, in the twentieth century, came the radio, the television and finally the internet. The BBC had begun to broadcast talks on gardening from soon after its inception in 1922, but it was Cecil Middleton, who began broadcasting in 1931 and was given his own programme, *In Your Garden*, in 1934, who made the real breakthrough. His appeal was summed up by a friendly critic: '[Mr Middleton] will assume that your soil is poor and your pocket poor. All he asks is that your hopes are high and your Saturday afternoons at his service.' His programme continued until his death in 1945 and, at its peak, had 3.5 million listeners.[107] He was even able to become, in 1937, the first television gardener, with a special garden built for him at Alexandra Palace in north London.

He was the first of many. Gardening television programmes abound, several of them with spin-off gardening magazines. There are televised visits to gardens, demonstrations of how to take cuttings and plant shrubs; there are garden makeovers with decorative presenters; there are commemorations of past gardeners, such as the plethora of programmes about Capability Brown for his 300th anniversary in 2016. The gardens of various pundits are displayed weekly on television and then become visitor attractions. All this is supplemented by a huge range of websites, led in Britain by the Royal Horticultural Society, which has about half a million members. It freely dispenses advice, leads gardeners to plant species to suit their soils and conditions and enables them to find nurseries at which to buy rare varieties. The website of the National Garden Scheme – supplementing the much-loved 'yellow books' – makes it easy to visit gardens great and small, while the Parks & Gardens website reproduces the listings of more than a thousand historic gardens.[108]

Garden shows – several of them, such as the Chelsea Flower Show, also televised – attract hundreds of thousands of visitors, as they did in the nineteenth century, when there were international, national, regional and local shows, down to the village horticultural societies that still flourish. Garden visiting, also, has a long tradition in England – the first guidebooks to Stowe, in Buckinghamshire, for example, were produced in the mid eighteenth century. A 'Lady of Distinction' pointed out in 1775 that garden visiting in England was the female equivalent of the Grand Tour of Europe enjoyed by young men.[109] Today, the gardens of the National Trust are as popular, if not more so, as its houses with its 5.5 million members and many individual visitors. In all of them one can see garden lovers, like their predecessors for three centuries, discovering new species, imagining improvements to their own gardens and tut-tutting over the weeds.

Garden books and magazines, garden visiting and chatting with neighbours over the garden fence all lead to spending money, on tools, fertilizers, weedkillers, water features, gnomes, sundials or bird baths. Plants are the least of it, although they are still bought in very large quantities – mostly now imported from Holland and other European countries – which is why garden centres have expanded from their eighteenth-century origins, when visitors might receive a cup of tea after inspecting the shrubs, to the huge emporia selling anything from greeting cards to scratching posts for cats, with ever-expanding cafés and children's play areas. Bird seed is a booming product. All this is why the British spend at least £11 billion a year on their gardens.

Television commentators talk regularly of football as Britain's national sport. It is dwarfed, in fact, by angling, which is the most popular of pastimes except, that is, for gardening, which engages far more people, year in and year out, often from their twenties and thirties to the day of their death. Conversations about the weather, another major pastime, segue rapidly into debate about the effect of wind, rain or drought on our gardens. There are now millions of them, created in the main since the philanthropic notions of Joseph Rowntree at New Earswick began to ripple across the land. The collective decision to build the suburbs at the rate of twelve houses per acre has changed the face of England more decisively than even the landscape parks of the eighteenth century or the inner cities of the nineteenth. It gave gardens to the people.

9
Kitchen Gardens

'To trim up their house, and to furnish their pot'[1]

In the week before Christmas 1783, King George III and his household of about 250 people dined well on the produce of the royal kitchen gardens. Apart from the meat, fish and bread that made up most of their diet, they consumed 60,000 asparagus spears, 20,000 French beans, 32 baskets of pears, 13 of apples and 6 of grapes, 236 lemons and 27 'salads' (lettuce) – all of them out of season. There was also a range of root vegetables, including 419 lb of potatoes and 17 bunches of carrots, together with 112 cabbages, 16 heads of celery, and chestnuts and currants. Oddly, the royal chaplains seem to have been largely responsible for consuming 122 heads of endive. On Christmas Day itself, there was a special treat of a pineapple and 18 home-grown oranges for 'their majesties' table'.[2]

Around the country, the gardeners in the walled kitchen gardens of the English aristocracy all had the same task, to feed their master and mistress with whatever was requested. The unusual feature of the royal Christmas feast was that, despite there being four kitchen gardens, at Kensington, Richmond, Hampton Court and Kew, there was only one pineapple – since many head gardeners prided themselves on producing large numbers of the fruit. The Lord Steward's list of the produce of the gardens supplied to the kitchens, thirty-five different types of fruit and vegetables, demonstrates the extent to which, during the eighteenth century or even earlier, skilled gardeners had mastered the English weather and the technology required to bring a wide range of food plants to the table for many if not all months of the year. By the nineteenth century, the extensive use of glasshouses meant that, as an inquiry into the state of the royal kitchen gardens was told in the 1830s, a head gardener should be able to supply any fruit or vegetable required by his master and mistress, or by their cook, at any time of the year. That was on top of providing flowers to grace tables, reception rooms and breakfast trays and, during the autumn and winter hunting season, buttonholes for each participating member of the family and their guests.

It was a difficult feat to pull off: the Royal Gardens Inquiry of the 1830s documented everything that could go wrong. But when it worked, as it still does at the Rothschild family garden at Eythrope in Buckinghamshire,[3] at Knightshayes in Devon, Heligan in Cornwall, Holkham in Norfolk and at a few others around the country, it was a marvel of technology and skill, the product of centuries of experimentation and of knowledge exchanged between head gardeners, who transmitted it to the apprentices and journeymen whom they trained and nurtured if also sometimes bullied. All that skill deserves to be celebrated, though given that it was more about what could be done, for show, rather than what was strictly necessary, its impact and importance can be exaggerated.

What was the point of the kitchen garden? Did it supply a vital part of the food of both the rich and the poor? How important were fruit and vegetables in the English diet? These questions will be considered in greater detail below. As so often, much depended on who you were and how much money you had.

EXPENSE NO OBJECT

The walled kitchen gardens of the English upper classes inspire almost as much adulation among garden historians and garden visitors as the landscapes of Capability Brown. The serried ranks of vegetables, with never a weed to be seen, the walls covered with carefully trained fruit trees, the Gothic greenhouses, are indeed beautiful and, at the same time, utilitarian. But they were also as important a means of displaying the wealth and taste of their owners as the statues, temples, cascades or artfully placed clumps of trees. They, like the pleasure gardens or the lakes, were there to be seen, to be viewed and to incite envy and emulation as part of the guests' tour of the estate.

They were also, like the paintings that adorned the walls of the stately homes that they served, a sign of luxury, clearly not justified by the monetary value of the fruit and vegetables that they produced. They were status symbols, the embodiment of Thorstein Veblen's 'conspicuous consumption'.[4] In a recent National Trust book on walled gardens, Jules Hudson states how from 'their inception walled gardens were a wholly practical addition, relied upon to produce enough fruit and vegetables to support large or small households throughout the year. They were, in effect, the private supermarkets of those who built and worked them; a vital food source expected to repay the investment required to build them.'[5] Nothing could be further from the truth. As Loudon realized in the 1820s, relying solely on produce from the kitchen gardens, particularly of the great estates, was a wildly impractical solution to the feeding of a whole household.[6]

Although fruit and vegetables have been grown in English gardens since Roman times, and presumably before, the great age of the kitchen garden did not begin until the late seventeenth century and lasted until the Second World War. A historian of the kitchen garden, Susan Campbell, has described in a number of books and articles – partly through the medium of a fictional country house, Charleston Kedding – the huge variety of buildings and other equipment that made up the kitchen gardens of the eighteenth and nineteenth centuries: the walls, pineries, vineries – even pinery-vineries (glasshouses for both pineapples and grapes) – mushroom houses, potting sheds, beds, borders, glasses, bells, frames, dipping ponds, seed rooms, orchards, nutteries, ice houses, bee hives, fruit houses, carrot and potato clamps, racks, baskets, root and fruit cellars, and much else besides. That was on top of the many spaces in the country house devoted to the preparation and preservation of fruit and vegetables or to the arrangement of flowers for tables and rooms.[7] Some of those gardens have been recreated in the twentieth, as in the so-called Lost Gardens of Heligan in Cornwall and a number restored by the National Trust. Some, for example at Gravetye Manor in West Sussex or Le Manoir aux Quat'Saisons in Oxfordshire, have found a new role and exist to supply a hotel and restaurant with fresh produce. Restoring old walled kitchen gardens has become a hobby – though only for those with a great deal of money – and there is a website devoted to it; fifty are open to visitors.[8] Where they exist, as at Heligan, the pineapple pits, requiring 90 tons of manure a year, stay in the minds of most visitors.

But pineapple pits were only the crowning glory. The archetypal kitchen garden is of 3 to 7 acres, though some were much larger, usually a square or rectangle of ground surrounded by brick walls 12 feet or more in height. Some – a good example is that of Capability Brown's friend the Earl of Coventry at Croome in Worcestershire – had another wall dividing the area in two: it is a so-called 'hot wall', with internal steam pipes like a vertical hypocaust, powered by underground boilers, that heated the wall and the plants grown against it. Plate 24 shows how such walls worked, while plate 25 shows a later example, at Weston Park in Staffordshire. The inside of the surrounding walls provided shelter from the wind; greenhouses were usually placed alongside the northern wall, facing south or southeast, to make the most of the sun, while the other walls were covered with espalier fruit trees. Much of the ground within the walls was used for growing vegetables (and flowers to be cut to decorate the house) in rectangular beds often arranged around a central 'dipping' pond from which the plants could be laboriously watered. Some of the beds were used for the cold frames, propagating beds and hotbeds for pineapples and melons. Gardeners moved plants between beds and frames as they grew from seed to fruiting; one of the main objectives was to advance or retard the crop so as to extend the growing season for as long as possible.

The kitchen garden extended further than its walls. A house for the head gardener – increasingly large and elaborate as the decades wore on – was accompanied by bothies for the apprentices and journeymen, as described in Chapter 6, usually set against the cold and unproductive outer wall to the north. There were fruit stores, boiler houses and sheds. 'Slip' gardens, patches of land outside the walls but obtaining some protection from them, could be used for transplanted seedlings and young plants. Further out still would be potato grounds. Initially many kitchen gardens were placed close to the country house, but as they became larger and as the smells from the manure used in them became more offensive, they were moved away, sometimes as much as a mile or more, although usually less, since they were always seen as something to be shown off to visitors and guests, part of the carriage or walking tour of the grounds. As such, the garden had to be kept in pristine condition throughout the year.

To many people at the time and since, the kitchen garden represented the acme of the skill of the head gardeners who, for several centuries, conjured up perfect fruit and vegetables, extending the growing season almost throughout the year. They provided for the English upper classes on their country estates the range and quality of fresh produce that can, now, only be found by transporting it by air over vast distances from polytunnels in far-flung parts of Europe and beyond. The skill that was expended in grafting and propagating, the elaborate constructions of hot walls and innumerable varieties of greenhouse, the experimentation with new varieties, have helped to shape the nursery industry and, even if slowly, to enable fruit and vegetables to play a larger part in the English diet. The walled gardens that existed in their thousands are also – where they survive – beautiful in themselves, symmetrical, carefully tended, marvels of functional design.

Nothing exemplifies this and the luxury consumption embodied in growing fruit and vegetables more than 'perhaps the most perfect Garden in Europe, of its kind', as the *Gardeners' Chronicle* described Queen Victoria's new kitchen garden at Frogmore, built in the 1840s.[9]

It had long been been clear that the system of kitchen gardens that fed the royal family and their household was expensive and inefficient; Joseph Carpenter and Henry Wise complained about it as far back as 1717.[10] There were too many royal kitchen gardens and too little coordination between them. Thus concluded the Inquiry into the Management, Superintendence and Expenditure of the Royal Gardens, which, in 1837, finally got to grips with the situation. Just as they so often are today, management consultants were brought in: the work of the inquiry was done by Dr Lindley, Professor of Botany at University College, London, and, at the time, Secretary of the (later, Royal) Horticultural Society, assisted by Messrs Paxton and Wilson. Joseph (later Sir Joseph) Paxton was superintendent of the

Duke of Devonshire's gardens at Chatsworth and Chiswick (and later the architect of the Crystal Palace) while John Wilson was the Earl of Surrey's gardener at Worksop Manor, Nottinghamshire. Their report was damning.[11]

Lindley and his colleagues surveyed the condition of the royal gardens at Windsor, Hampton Court, Buckingham Palace, Kew and Kensington. They estimated that the current expenditure of the gardens (exclusive of the Botanic Gardens at Kew) was about £7.4 million *(£10,000)* per annum, of which just over £2 million *(£2,750)* was for annual repairs. Despite this, over £583,000 *(£750)* per annum had to be paid for fruit and vegetables purchased elsewhere. According to the report, 'None of the Gardens were in communication with each other; there was no unity of purpose among them; nor the least arrangement as what they were respectively to supply. It did not appear that they had ever received orders upon such a point.'

The equipment of the gardens, taken together, was breathtaking. There were in all

> nearly 900 yards in length of forcing houses of all descriptions ... more than 6,200 yards in length of walls for the cultivation of Pears, Peaches, Cherries, Apricots, Plums, Apples etc ... 429 sashes of frame lights, 1,260 feet in length of brick pits ... 30,000 square feet of Glass roofs ... and 64 acres of land.

Despite this, the gardens did not 'in the year 1837 furnish a Strawberry or Grape in the months of January or February, and scarcely even in March'. The same failure was true of vegetables, since the gardens could supply

> only 15 cucumbers in the two first months of the year, two hundreds of kidney beans in January and the same number in February, scarcely any early potatoes and hardly a Bunch of Asparagus per diem before the month of May ... The Royal Table is not better supplied with fresh fruit than that of most private gentlemen.

Conversely, there were times of plenty when the household – 241 people, including 160 servants, all entitled to be fed at royal expense – was overwhelmed with fresh fruit, vegetables and salads.

The experts recommended a total reorganization, in which Kew would supply fruit and vegetables in the spring and summer, Windsor, with its peach house, vineries and pineapple pits, in the autumn and winter. The other kitchen gardens should be closed. They thought this would reduce annual expenditure from £7.4 million *(£10,000)* to £4.7 million *(£6,400)*, although fruit and vegetables amounting to £1.1 million *(£1,500)* would still have to be bought on the open market. After several years of discussion, the main recommendations were accepted, but it was decided to concentrate the work of the kitchen gardens on an entirely new site at Frogmore, near Windsor. This would be paid for by leasing out the kitchen garden at Kensington for building plots, producing an estimated income of £740,000 *(£1,000)* per annum, since 'valued as a site for Villas it is estimated to be worth, with the Buildings, Walls and Materials, at present standing upon it', about £22.2 million *(£30,000)*. This would finance a new, 20-acre, garden

the execution of which, according to Estimates very carefully formed, would involve an outlay of not less than £29,781 [£22 million today] but the net expense of which after deducting the value of old materials in the Gardens given up and modifying some of its details might, it is believed, be reduced to about £25,000 [£18.5 million today]'.[12]

Remember that this investment was to supply fewer than 250 people; £22 million was to be spent at the outset in order to continue spending at an estimated cost of £74,000 per head each year.*

Sites 'for Villas' ultimately decided the fate of many gardens around the growing towns and cities of Britain. The villas in this case would ultimately become some of the most expensive houses in the country, the mansions and embassies of Kensington Palace Gardens. The new kitchen garden was built on an equally lavish scale, with its 'garden walls, Forcing Houses, Conservatories, Pine, Melon and other Pits, Mushroom Houses, Gardeners' and Under Gardeners' Houses and Shed Buildings'.[13] *The Times* reported on 11 August 1841 that

A brick wall, 12 feet high, and nearly 2000 yards in length, to extend completely round the gardens, has just been contracted by the [Office

---

* This may be an overestimate, since it was presumably necessary to supply produce for state banquets and other occasions.

of] Woods and Forests to be erected by Mr Chadwick (the well-known extensive railroad contractor) of the best stock bricks and to have a feather-edge York coping, with four cart entrances. A handsome terrace, with a wall in front, 800 feet long, will be constructed within the gardens (for the promenade of the Royal Family) extending nearly the length of one side, with several flights of stone steps leading from the gardens.

Construction started in late 1841 and did not go entirely smoothly. The eminent civil engineer William Cubitt had to be enlisted to advise on the water supply, a bugbear of many gardens. The water tank that he subsequently recommended was badly built and, after disputes with the contractors, had to be replaced. On 25 November 1843 a letter from the Board of Green Cloth,* which administered the royal household, admitted that 'the outlay has already largely exceeded the amount of the estimate' and advised that no further unavoidable expense should be incurred.[14]

One item that may have fallen into this category reflects the close interest that Queen Victoria and Prince Albert were taking in the scheme:

> The Bells required at the Entrances are for the convenience of admitting the workmen and others generally as well as to prevent injury to the doors from persons kicking violently against them as at present for admission; besides which, the want of them has recently been complained of by Prince Albert who (with Her Majesty) frequently visits the Gardens without any previous notice: Upon a late occasion His Royal Highness was obliged to return, being unable to get into the Gardens, and having no means of making known his intention to enter.[15]

John Phipps, the contractor, later explained that extra expense had been caused by changing the location of the principal entrance 'by order of His Royal Highness Prince Albert', who had also ordered extra rooms in one of the gardeners' houses and an additional water closet for the use of the royal attendants on their visits.[16]

By late 1843 the estimate for the cost was £29.3 million (£41,028

---

* So named after the cloth of green baize covering the table at which members of the board sat.

10s 5d). A familiar litany of explanations was offered: the location of the greenhouses had been changed and metal-framed forcing houses – costing £6 million *(£8,399)* – had been bought instead of ones with wooden frames; only half the necessary greenhouses had been included in the tender document so all the estimates had been based on the wrong quantities; extra propagating, vinery and forcing houses (including houses for growing melons) were required. The final cost, according to the *Gardeners' Chronicle* of 31 March 1849, was £32.5 million *(£44,962 6s 1d)*, nearly twice the original estimate. This was significantly more than the cost of the Great Stove at Chatsworth, Paxton's first great building, and about half that of his second, the Crystal Palace.

But the resulting garden was, everyone agreed, splendid, as plate 34 indeed shows. Hothouses extended for 840 feet and a line of pits, 1,665 feet long, provided for early asparagus, along with cucumbers, melons, pineapples, French beans and grapes. Thirteen acres of vegetables were extended in 1848 to 21, the vegetables screened by dwarf ornamental fruit trees. A substantial house was provided for both the head gardener and his deputy. At the centre of the garden was a fountain with a basin 30 feet across and made, like the fountains in Trafalgar Square, of polished Peterhead granite. Pipes to every part of the garden made watering easy. Like other kitchen gardens, it was an object to be displayed, and it was for this reason that Prince Albert had requested that two reception rooms be built in one of the gardeners' houses for the use of the queen and the prince when they took visitors there.

Frogmore was built on a particularly lavish scale. It was 27 acres in size, later extended to 30, when a household of 241 would normally have been expected to need about 17 acres, though perhaps the extra 10 acres were required to provide for guests and banquets; it is also not clear whether flowers – needed on an increasingly lavish scale – were included within the rule-of-thumb of an acre for fourteen people. The greenhouses, of which Sir Robert Peel ordered a smaller replica, were especially splendid, but other Victorian houses, such as Waddesdon or Tatton Park, had similar ranges. The glasshouses at Waddesdon cost £22 million in modern values later in the century. Another indication of the possible costs of establishing walled gardens is that the current owners of the garden at Croome have spent £5 million on restoring, so far, only a quarter of its acreage; truly a labour of love.[17] And this was replicated all round the country.

But, in addition to the building costs, the expense of running the kitchen gardens of England was enormous. The new royal kitchen garden was expected to cost £4.8 million a year to run in modern terms and even then would not provide all the produce needed to feed the household. A detailed survey of Oxfordshire, using nineteenth-century Ordnance Survey maps, has identified 253 walled kitchen gardens in that one county alone, although the minimum-size criterion it applied – 9,720 square feet (or about a quarter of an acre) – is quite modest.[18] Even so, a kitchen garden of that size would have required at least one full-time gardener, while very many were much larger: Blenheim, the largest in the county but still only a third of the size of Frogmore, would have had at least twelve gardeners in its 7-acre kitchen garden – now mainly a children's playground.* Many more would have been needed for the 20 acres of the Duke of Portland's garden at Welbeck Abbey, Nottinghamshire.[19] It was surrounded by high walls with recesses in which braziers could be placed to hasten the ripening of fruit. One of the walls, for peaches, measured over 1,000 feet in length.

The accounts of other stately homes suggest that the garden workforce was often split roughly equally between the kitchen and pleasure gardens, although the kitchen gardener was subordinate to the head gardener. At Wrest Park in Bedfordshire in the early eighteenth century, John Dewell Senior was paid £37,200 (£20) per annum as the gardener in the Great Garden, while his son, John Junior, received £35,340 (£19) as the kitchen gardener. Between them, they led a staff of twenty men and eight boys; including the two Dewells, annual wages in the gardens amounted to at least £740,000 (£397 16s).[20] Their successors, the men and boys – and the horses – on whom such sums were spent even as late as 1903, are shown in plate 32. This seems to have been on top of the yearly expenses of the garden, which, it was noted for 1773–80, had been a 'fix'd' sum of £440,000 (£300)

---

* It is a sign of the impact of mechanization that Blenheim now has six full-time gardeners to look after all its kitchen and pleasure gardens.

for many years.[21] It is reasonable to think that the kitchen garden might have accounted for half of these wages and expenses, or about £600,000 a year.

As Wrest shows, although wages were probably the major cost of the kitchen garden, there were many other significant expenses. At Shardeloes, near Amersham in Buckinghamshire, owned by the Drake family, the bills for the garden in 1755 came to over £267,000 (£154 11s 9d), but in addition the gardeners bought 1,961 bushels of tan for £14,610 (£8 9s 5d).[22] Tan or tanner's bark was a vital heat source for growing pineapples in hotbeds and conservatories, since it retained a constant heat of 75–85 degrees Fahrenheit (24–30 degrees Celsius) for as much as a year, much longer than horse dung would have done.[23] John Cowell had commented in *The Curious and Profitable Gardener*, published in 1730, that 'Stoves and Glasscases for the culture of the Pine [pineapples] ... are now found in almost every curious garden'[24] and indeed the Shardeloes' accounts contain an estimate for a garden building costing £52,000 (£29 5s 10d) as well as an uncosted design for a hothouse with a tan pit and fireplace.[25] This is likely to have followed the pattern of a stove built at Croydon in 1723 that combined heat from tan and coal. Susan Campbell believes that this design 'ushered in the great era of glasshouse horticulture in England'. By 1730 'the Nobility and Gentry' of England were in the grip of what can only be described as 'Pineapple Fever'.[26] In spending £15,000 or more in one year on their 'pines', the Drakes were clearly addicted. So too must have been the anonymous author of an article in *Floral World* a century later, in 1868; he calculated that his pine pits covered 500 square feet and contained 139 plants. The cost of fuel, tan, manure and general maintenance, together with two hours of labour every day, meant that each pineapple cost £569 (£1) to produce.[27]

The country houses of the nineteenth century continued to be built with walled kitchen gardens on the same scale as their predecessors, even if not as lavishly as at Frogmore, and employed large numbers of gardeners. The First World War removed many of them to the trenches, never to return, but work in the kitchen gardens revived in the 1920s and 1930s; the myth that the war had ended the era of the kitchen garden still prevails. Arthur Hooper's memoir of his time as a garden boy, labourer and foreman between 1922 and 1939, published in 2000 as *Life in the Gardeners' Bothy*, makes their survival clear. In 1926, at the age of eighteen, he was put in charge of seven glasshouses at Fonthill House in Wiltshire, the home of Hugh Morrison, MP. The kitchen garden was there to be viewed and he was told how important it was to keep all the door handles perfectly clean, 'as Mr Morrison's family or their guests would be most annoyed if they soiled their hands or gloves when walking through the greenhouses'.[28] The family went to Scotland every August and 'our garden had to supply vegetables, fruit, cut flowers and flowering plants to the house in Scotland for the ten weeks the family were there'.[29] They had to be transported at least five hundred miles.

When Hooper moved to Gatton Park, Surrey, which was owned by Sir Jeremiah Colman of the mustard firm, the extensive kitchen garden was staffed by six men and there were seventeen glasshouses for growing peaches, grapes, melons, tomatoes, chrysanthemums and alpines as well as twelve orchid houses. An important part of Arthur Hooper's duties was to supply and arrange flowers and pot plants throughout the mansion. In 1930 he moved to Heathfield Park, Worcester, where there were twenty-two gardeners, including four in the 2-acre kitchen garden, while at Highbury Hall, Cambridge, his next job, there were thirty-seven gardeners of whom seven worked in the kitchen garden. In general, Hooper's employers weren't aristocrats – one was a wealthy tea merchant – but they lived as if they were, with house parties, shooting and hunting, their houses embellished with lavish floral decorations and pristine glasshouses.

A modern example of kitchen gardening is Lord Rothschild's family garden at Eythrope. Part of a park and garden of 60 acres, near the much larger Waddesdon Manor in Buckinghamshire, it was created in the late nineteenth century by Miss Alice de Rothschild around a daytime house; having been told that, after suffering from rheumatic fever, she should avoid the proximity of water at night, she never slept there but returned every evening to the main house at Waddesdon.[30] The scale of her gardening in general is shown by the fact that not only did she and her brother, Baron Ferdinand, employ 160 gardeners at Waddesdon but from 1888 onwards she created the Villa Victoria, a house and garden of 135 hectares (333 acres) near

Grasse in the south of France; even after the First World War, the villa garden cost over £80 million *(£500,000)* a year to run, partly because it had 100 gardeners, who – one guest mischievously speculated – were positioned every 20 yards around the garden to catch or pick up any leaf that fell.[31] Each year they planted a total of 55,000 daisies, 25,000 pansies, 10,000 wallflowers, 5,000 forget-me-nots and 23,000 bulbs of tulips and narcissi. The Rothschild family archives describe it as 'possibly one of the most ostentatious and extravagant gardens ever seen'. It was visited frequently, in her last years, by Queen Victoria, after whom the villa was named, who in private always referred to Miss Alice as 'The All-Powerful'.[32]

By these standards, Eythrope's kitchen garden – revived by the current Lord Rothschild – is positively modest, at 4 acres and including five greenhouses, originally with wooden frames but now recreated in aluminium; yet even its modern designer, Lady Mary Keen, acknowledges that 'as an exercise in the conservation of old methods and traditional techniques, it cannot, even with a Rothschild purse, be continued in the same way and on the same scale for ever'. Meanwhile, it maintains the tradition of the great kitchen gardens, supplying fruit, vegetables and flowers to the family and a number of relatives throughout the year, whether in Buckinghamshire or in London, all 'home-grown and grown to perfection'.[33] Such attention to detail would have satisfied Miss Alice: whereas her brother 'would not . . . have been averse to making up the supplies he wanted from any convenient outside source, [she] had other views'.[34]

But such year-round abundance – Keen's book is rightly called *Paradise and Plenty* – does not come cheap. The biographies of head gardeners published in the *Gardeners' Chronicle*, and discussed in Chapter 6, together with the accounts of kitchen gardens from the seventeenth century onwards, show that for at least two centuries the English upper classes were prepared to pay whatever was necessary for perfection. They demanded fruit, vegetables and flowers if not all throughout the year at least for most of the year, grown in a garden always ready to be shown off to visitors. It would be foolish to attempt an estimate of the total cost, either of building or maintaining the kitchen gardens; we do not even know how many walled kitchen gardens there were in their heyday, although there is no reason why the

250 in Oxfordshire should not have been replicated in the other forty-seven English counties. Nor were all of them, of course, run on the scale of Wrest Park, Blenheim, Waddesdon or even Shardeloes. But even if they were each staffed by an average of two gardeners and at an average cost of £50,000 a year in modern values, the annual total for England would have been £600 million; it could well have been much more, every year for a period of nearly two and a half centuries.

For a long time, the English upper classes – as well as their peers in Scotland, Wales and Ireland – were prepared to pay the price. They revelled in their tomatoes picked at Christmas or their out-of-season flowers: King George V still demanded, in the years after the First World War, that a fresh gardenia should be placed every day throughout the year on his breakfast table so that he could insert it in his buttonhole.[35] They competed in displays of fruit and vegetables at the shows of the Royal Horticultural Society, happily taking the credit for the work done by legions of gardeners, from the head to the lowliest apprentice. They demanded that bothy boys slept by the furnaces of the greenhouses to keep them stoked twenty-four hours a day, while the under-gardeners opened and closed windows to keep the fruit at its optimum temperature. They tolerated huge overproduction: at Eythrope, for example, the cook might want forty beetroot, which all had to be of the same size, and therefore at least eighty were pulled.

Ultimately, but not until the twentieth century and sometimes only after the Second World War, reason prevailed. Even the wealthiest, such as the royal family, had long bought some fruit and vegetables in the market; James Cochran bought pineapples in Covent Garden for Lord Salisbury in the early nineteenth century, for instance. The railways probably delayed the demise of the kitchen garden, since fresh fruit, vegetables and flowers could be shipped to wherever in the country the aristocratic family were staying. As budgets tightened, surplus production was increasingly sold to local markets, although this had actually occurred since medieval times. Finally, inexorably rising labour costs and the increased scale and sophistication of the horticultural industry and, ultimately, of the supermarkets supplied by air from around the world, combined to persuade almost every country house owner that it would be better to buy their fruit and vegetables from an emporium such as Fortnum & Mason in Piccadilly;

it is one of a small number of Royal Warrant holders who currently supply the royal household with the produce that would once have come from Frogmore, now sadly an open field. Even the rich have to make some choices and the luxury of country house parties, supplied with perfect food from the immaculate kitchen garden, came to seem less attractive than the lures of other ways of spending their money – foreign travel, a fleet of cars, a yacht or private jet – and for their food, of course, the Michelin-starred restaurant.

FOOD FOR THE MIDDLING SORT

The middle classes came much earlier to the same conclusion. They had the funds and they increasingly wanted to consume more fruit and vegetables, but this did not mean that they grew their own. Some did, perhaps because they liked the taste of freshly picked carrots or lettuces from their own back garden, but most, from the seventeenth century onwards, took advantage of – and supplied the demand for – what the English call 'market gardening' and the Americans 'truck farming'.

John Claudius Loudon was, as usual, reflecting reality when he advised in 1838 against trying to use a villa garden to grow one's own fruit and vegetables. It was not worthwhile for an owner because 'for the 15l. or 20l. a year which he must pay a hired gardener he might purchase as much fruit and vegetables as he could grow'.[36] William Cobbett, the radical agitator and, later, Member of Parliament, who was also a keen gardener and garden writer, seems to have disagreed. His book *The English Gardener*, published in 1829, was adapted from his earlier book *The American Gardener* of 1821, which he had written during two years of self-imposed exile in the United States. Most of the English book is taken up with advice on growing fruit and vegetables; there is no discussion of their costs and he seems to have thought that a kitchen garden 'suitable for the average household is an area approaching an acre' but says nothing about the number of gardeners.[37]

Loudon, on the other hand, had evidence for his view, at least for London residents: in the revised, 1825 edition of his *Encyclopaedia of Gardening* he had tabulated 'the average prices of vegetable productions, of the best quality, by retail, in Covent-garden Market, throughout the year'.[38] His list would not disgrace a modern supermarket: it comprises fifty-six kinds of fruit and vegetable, including the usual root vegetables, as well as asparagus, artichokes, capsicums, truffles, cucumbers, several different berries, pineapples, oranges, lemons, apricots, nectarines and even pomegranates. Even more impressive are the seasons during which they were available: kidney beans and cucumbers could be bought from January to September; oranges every month except September and October; lettuce every month except February and March; and pineapples throughout the year. Few private gardens, except perhaps the best-organized and most lavishly financed walled kitchen gardens of the great estates, could have provided produce in such profusion.

Market gardening around London had existed for at least four hundred years, and probably much longer, before Loudon wrote; there had been a regular market near St Paul's Cathedral in London, at which the surplus from noble and monastic gardens was sold, since before 1345.[39] One such garden may have been that of the Earl of Lincoln, now Lincoln's Inn Fields, which in 1296 was producing apples, pears, large nuts and cherries sufficient for the earl's table, together with onions, garlic, leeks and beans.[40] The market near St Paul's disturbed the priests 'singing Matins and Mass in the church of St. Austin' because of the 'scurrility, clamour and nuisance of the gardeners and their servants there selling pulse, cherries, vegetables and other wares', while sellers of fruit, leeks, onions and garlic congregated elsewhere in the City.[41]

An embryo Company of Gardeners seems to have existed in the City of London since the fourteenth century or possibly earlier, but a Company of Fruiterers is recorded as early as 1292. They were two of the guilds or companies that regulated – and tried to monopolize – the trades and crafts of London and many other medieval cities.[42] Both companies received royal charters from James I, in 1605 and 1606 respectively, and by 1622 the membership of the Company of Gardeners had reached 500. They rejected applications for membership by the gardeners of Fulham, Kensington and Chelsea on the grounds that they 'pretend themselves to be Market Gardeners and to furnish the market with eatables' but were really only farmers who had not served an apprenticeship in gardening.[43] By 1649, 'members of the Company maintained large market gardens outside the City on the south bank [of the Thames] near Tower Bridge, and employed 1,500 men, women and children in addition to four hundred apprentices, the land being divided amongst the members with a maximum of ten acres for each'.[44] In 1661–2 the company had members in fifty different parishes.[45] The scale of these memberships and landholdings indicates the extent of the market for their produce, although it is not possible to distinguish between private gardeners, nurserymen and market gardeners.

The trade extended far beyond London: in Yorkshire in the late sixteenth and early seventeenth centuries, a range of fruit and vegetables – including walnuts, filberts (hazelnuts), pears, plums, damsons, artichokes, parsnips, carrots, turnips and hops – were all subject to tithe (a tax paid to the Church on the value of goods), showing that they had been produced for sale. Globe artichokes were particularly popular because eating them was thought to produce male children.[46] Malcolm Thick has pieced together the evidence from cookery books and other sources in the late seventeenth century and concludes that roots, cabbages and beans were coming into London 'in enormous quantities after the Civil War' to be eaten by both rich and poor, but that what he calls 'superior vegetables' were also available for 'polite' London society, in other words for the middle and upper classes. Asparagus, artichokes and lettuce were, he says, old favourites, together with cauliflower and mushrooms, with broccoli increasing in favour in the early eighteenth century.[47] Out-of-season vegetables were particularly favoured and Stephen Switzer describes how the cauliflower was available for six or seven months of the year, 'furnishing the tables of the curious ... and by good management mock[ing] the severity of our unsteady climate'.[48]

By 1675, the Neat House gardens, established on the Thames west of the modern Vauxhall Bridge around 1600,[49] covered over 107 acres and the area was let out in gardens of 3–18 acres;* they were laid out with hedges between them and also internally, marking out

---

* These market gardens were in a different part of London from those of the Company of Gardeners.

four regular plots or 'quarters'. Some had walls for shelter against which plants could be grown. One inventory shows asparagus in one quarter, cauliflowers and colewort in a second, cauliflowers under glass in a third and radishes and carrots in the fourth. There was a 'bank near the Willow walk with wide rows of artichokes planted with Colley flower plants'.[50] Reed mats were erected to protect plants from the wind. Water could be drawn from shallow wells by a bucket suspended from a counterweighted pole, a form of technology dating back to ancient Egypt. In short, land was used intensively and efficiently, particularly with the aid of large quantities of manure and night soil. As a sideline, the gardens were opened to the public and supplied beer, food and entertainment.[51]

Over a century later, in 1792, the writer Daniel Lysons surveyed the parishes within twelve miles of London and concluded that within them 5,000 acres were cultivated for vegetables, 1,700 for fruit and another 1,700 for potatoes; another writer, John Middleton, estimated in 1798 that market gardens around London occupied 10,000 acres.[52] In other words, long before Loudon was writing, market gardening and the production of fruit and vegetables, predominantly for a middle-class market, was a major economic activity around London and probably many other towns and cities. They certainly existed around the growing industrial town of Manchester.[53] So the middle classes did not need to grow their own.

The agricultural historian Joan Thirsk has suggested that the growth of commercial horticulture slowed between 1750 and the 1870s, as farmers – reacting to the demand for bread from an increasing population – could make more money growing their traditional grain crops.[54] But other indications, such as the overcrowding of London's main fruit, vegetable and flower market at Covent Garden, suggest that there was still plenty of demand for its wares. The area had been owned in the medieval period by Westminster Abbey and seems to have been used for arable crops and, possibly, for gardens bordering what is now the Strand. The dissolution of the monasteries and later land sales brought it into the possession of the earls, and later dukes, of Bedford, who first developed it for luxury housing in the 1630s. The famous architect Inigo Jones was employed to build houses around a piazza – a novel concept at the time – together with a neo-classical church which, rebuilt after a fire in the early nineteenth century, still stands. In a familiar pattern, aristocratic houses became, later in the seventeenth century, bath-houses, coffee houses and brothels; the central open space gradually filled up during the day with fruit, vegetable and flower sellers, their wares supplied from market gardens along the Thames. The produce was landed at wharves south of the Strand and carried – much of it in towers of tiered baskets on their heads – by market porters up the sloping narrow streets to Covent Garden. There, from early morning, it was sold to retail greengrocers whose carts or vans came in from all around London. The flower girls, who sold posies to party- and opera-goers as they staggered home in the early hours of the morning, were later immortalized in the shape of Eliza Dolittle in Bernard Shaw's *Pygmalion* in 1913, followed by its musical adaptation, *My Fair Lady*, in 1956.[55]

The chaos of market stalls was replaced, in 1830, by the neo-classical market building initially occupied by greengrocers and florists and, increasingly, by wholesalers and importers, while market gardeners continued to pitch stalls in the piazza under large umbrellas. The streets around filled up with traffic generated by the market and were lined with shops and offices connected to the fruit, vegetable and flower trades, particularly as the importing of luxury foods expanded. The Floral Hall was built in 1860 at a cost of about £19 million (*£30,000*), an indication of the value of the market to the dukes of Bedford, and another new flower market followed in 1870. The journalist Henry Mayhew, writing in the 1840s, his articles later compiled as *London Labour and the London Poor*, estimated that on a Saturday morning 2,000 donkey barrows and 3,000 women with head baskets or shallow baskets of flowers or fruit could be seen, together with innumerable costermongers, porters, salesmen and carters.[56] It was said to be the largest fruit and vegetable market in the world, first thought to be overcrowded in the 1760s but not moved out of central London until the 1960s. By then, according to Ronald Webber, 'Over a million tons of fruit and vegetables were being handled every year with a total turnover of £67 million representing some sixty million packages, about half of which contained imported produce. Flowers added another £10 million and a million packages to those figures.'[57]

Although Covent Garden continued to thrive and market gardens were established around many towns, supplying local markets and greengrocers, British farmers could make more money from grain, as the population grew rapidly and the 'landed interest' in Parliament secured tariffs on imported grains through the contentious Corn Laws, introduced to keep prices high after the Napoleonic Wars. A long political campaign by free traders, principally the increasingly self-confident 'manufacturing interest' but assisted by the Chartists and others who championed the working poor, brought the repeal of the Corn Laws in 1846; this initially did little to achieve the campaign's objective, to reduce the price of bread, and it was not actually until the 1870s that imports from Russia and the Americas brought a collapse in wheat prices and a depression in British agriculture and land values that lasted nearly until the onset of the First World War. Many farmers and landowners hung on to the old ways, or converted arable land to pasture, but others turned to market gardening in what Thirsk calls a 'spectacular development' from the 1880s onwards, with huge investments in glasshouses, mushroom beds, orchards and fields of soft fruit. This form of horticulture seems to have doubled in size between 1870 and 1910 and market gardens themselves became much larger; the 10 acres or fewer of earlier generations were replaced by enterprises of 100 acres or more, such as the Tiptree Farm of Arthur Charles Wilkin, still famous for its soft fruits and jams. New areas of the country took up the trade, aided by 'scientific advances in food preservation, refrigeration, and improved transport'.[58] There was also 'spectacular new growth' in the flower and plant business.[59] All this is evidence of growing demand, much faster than the growth of population at the time, although Thirsk cautions that 'it did not add greatly to the number of foodstuffs offered, nor did it yet broaden English perceptions of fruit and vegetables in diet to match those of the French'.[60] Some of the demand was satisfied by imports from the rest of Europe, aided after the middle of the nineteenth century by the greater regularity of supply made possible by steamship services and, later, by the development of refrigerated shipping. These were the precursors of the air freight that stocks British supermarkets with fruit and vegetables all year round today.

It was probably not until the Mediterranean-style recipes of

Elizabeth David entranced the English middle classes in the 1960s, as they emerged from the austerity and rationing of the Second World War and its aftermath, that vegetables ceased to be boiled to death. The nutritional impact of prospering horticulture had, until then, been literally diluted. But in her *A Book of Mediterranean Food*, first published in 1950, only eleven pages out of 200 in the 1955 edition are devoted to vegetables; it is clear that she thought that many of those she mentioned, such as courgettes, fennel and celeriac, would be unfamiliar to English readers and might be difficult to obtain. In *French Country Cooking*, first published in 1951, she tries to reassure readers 'who may be alarmed by the quantities of garlic used' and, of only fourteen pages of vegetable recipes (in a book of 207 pages), several are devoted to potatoes.[61] So vegetables were still slow to enter the English middle-class diet, peculiar as this may seem to modern readers brought up to believe in the benefits of the 'Mediterranean' diet. But that lack of interest in vegetables, combined with the growth of market gardens, suggests that there was little reason for middle-class gardeners to grow their own, except possibly during the war in response to calls of 'Dig for Victory'. It was not until the 1970s and 1980s that the English middle classes took up allotments, which had languished after the Second World War, in search of better-tasting and chemical-free fruit and vegetables, even if each carrot or cauliflower cost them – particularly in their own time and energy – far more than they would have spent at the greengrocer or supermarket.

THE PEOPLE'S FOOD

How important were fruit and vegetables to most of the people of England? In 1695, England's first social scientist, Gregory King, estimated that fruit and vegetables made up only 5.75 per cent of the national average expenditure on food.[62] Two centuries later, on the eve of the First World War, another statistician, R. H. Rew, made another estimate: those two food groups accounted for 12 per cent of the value of food consumed in the United Kingdom.[63] Most recently, government statistics suggest that expenditure on fruit and vegetables is about 16 per cent of spending on food and non-alcoholic drinks.[64] There seems, therefore, to have been a gradual rise in such spending over the centuries, speeding up in the twentieth century. However, even in 2015, 7.3 per cent of adults and 5.9 per cent of children did not include any fruit or vegetables in their diet and the amount of spending on those items still rises strongly with income, with the top 5 per cent of the population by income spending 46 per cent more on them than the bottom 5 per cent.

These figures all suggest that fruit and vegetables made up a small proportion of spending on our diet in past centuries, when the population was much poorer than today. In addition, since we know that the upper and middle classes ate at least some fruit and vegetables in the past, such low numbers as averages for the whole population imply that the working classes ate very little indeed. These foods were, and are, luxuries in the sense that they will be purchased only when some income is left over after necessities have been bought.

More evidence comes from the budgets of households, predominantly in rural England – where access to gardens was most likely – that were collected by David Davies and Frederick Eden during years of high food prices between 1787 and 1796.[65] Nearly all the households were headed by agricultural workers, with a few engaged in mining, textiles or other manufacturing pursuits. Only about one-third of the families had gardens and most of their food, 'flour, bread, meat, beer, milk, butter and even some of the vegetables', was 'purchased from tradesmen, peddlers or at market'.[66] Fruit was sold by half of a small sample of village shops between 1660 and 1690, although this probably included dried fruit such as currants and raisins; further small samples showed a slight decline in availability during the eighteenth century.[67] Potatoes and other vegetables made up only 7.9 per cent of annual food expenditures among the families in the more industrialized north of England and a tiny 2.6 per cent of spending by families in the south. Spending on them was even less in families with lower incomes and more children.

Little changed during the course of the nineteenth century. It was stated in 1867 that working people, from both urban and rural areas, ate vegetables no more than once a week – partly because cabbages, for example, were too cheap to justify the costs of transport and were therefore not sold by greengrocers – and fruit, of the lowest quality, only in times of glut and low prices.[68] Government inquiries in the 1880s found that where vegetables were eaten, they had normally been cut up and sold in a mixture for making vegetable pottage or soup – we now know that they lost most of their nutritional value in the process.[69] A 1904 survey shows 'very limited purchases of fruit and vegetables'.[70] One reason was a long-standing belief, which persisted throughout the nineteenth century, that fruit and vegetables were bad for children and could cause 'summer diarrhoea', which might prove fatal. *A Treatise on Food and Dietetics*, published in 1875, judged that 'in a raw state the apple must not be looked upon as easy of digestion' while cherries and plums might produce 'disorder of the bowels'.[71] Newly introduced vegetables, such as the tomato, were initially thought to be poisonous, as potatoes had been in earlier centuries, so that tomatoes did not become a major ingredient of the British diet until the twentieth century.

It is possible that the low expenditure on fruit and vegetables shown in household expenditure surveys might be misleading, if it was common for people to grow their own in gardens or allotments. Evidence on the extent of 'self-provisioning', as it is called, is very difficult to obtain since, even when surveys noted the existence of gardens, they rarely recorded or valued the produce. Jeremy Burchardt, the historian of allotments, believes that 'contrary to what might have been expected, garden provision was grossly inadequate in rural areas by the early nineteenth century'.[72] The Davies and Eden studies conducted between 1787 and 1796 have been analysed to show that about a third of households in rural areas had a garden or an allotment or grew their own potatoes, although another survey in 1835–46 put the figure at only 17 per cent. This certainly changed: by the end of the nineteenth century, a survey showed that 42 per cent of agricultural labourers had gardens or allotments, while another in 1912 put the figure as high as 81 per cent. The problem lies in estimating how much a garden might produce, what it produced and what it contributed to the diet and to nutritional status.

This is a contentious topic and has been so since the end of the eighteenth century when the provision of allotments for the poor was first proposed. They were seen as compensation for the loss of common rights resulting from the enclosure of the open fields, although recent research has shown that as few as 15 per cent of agricultural labourers in the Midlands actually had those common rights before enclosure.[73] Then, during the nineteenth century, it was thought that allotments could help to feed the labourer's family and compensate for the low wages and irregular employment that were so common at the time. This would have the incidental effect of reducing demand for poor relief; meanwhile, the labour of cultivating an allotment could both employ the whole family – thus cementing the family unit – and keep the labourer out of the pub. It was pointed out, correctly, that through intensive cultivation the allotment could produce much higher yields per acre than could commercial horticulture or agriculture. Similar arguments were employed into the twentieth century and underpinned the provision of gardens for the working classes in the garden cities and the suburbs of the inter-war period. Raymond Unwin, who, as discussed in the previous chapter, was responsible for the density of housing in those suburbs, argued that their gardens were of a size to be cultivated by a single man and to feed a family.[74]

Despite this continuity of motive, the size of allotments and gardens changed a great deal. Between 1793 and 1829, the average size of an allotment on the fifty-four sites that existed by 1830, most of them in Wiltshire, was just over half an acre; this was clearly large enough to keep a pig and some chickens and some allotment holders even had a cow. But later the average size of allotments fell to 0.38 acres between 1830 and 1849, 0.28 acres between 1850 and 1873, 0.24 acres in 1873 and then down to the modern standard of 0.06 acres.[75] The standard suburban garden of 100 feet by 30 feet was even smaller, at 0.004 acres. As the size fell, so must have the capacity of the garden or allotment to provide enough fruit and vegetables for a family. Burchardt has calculated that an allotment of 0.25 acres could increase a labourer's income by 11 per cent, or up to 21 per cent if he kept a pig – possibly more in low-wage areas – which might have been possible on the larger allotments of earlier decades. However, B. S. Rowntree and M. Kendall used their very detailed study of the budgets of forty-two labourers in 1913 to suggest a lower figure: 'But the value of the garden to the labourer, important though it is with present money wages, must not be overrated. Less than a twelfth of the food consumed by these families was derived from their gardens.

And it must be remembered that from a garden of the usual size produce cannot be obtained the whole year round.'[76]

There is no doubt that holders valued their allotments. Allotment numbers rose steadily, presumably responding to demand: there were said to be 100,000 allotment plots in England in the 1840s, 344,172 in 1887 and between 450,000 and 600,000 by the time of the First World War. They played a role in keeping the population fed during both wars. But although it was said that by 1914 any agricultural labourer who wanted an allotment could have one, farm workers were just a small part of the population – only about one-tenth by 1911 – and all allotment holders were a small proportion of the 6 million or so households at the time.[77]

Middle-class observers were censorious about the diet of the poor. This has been true for millennia: James Davidson's *Courtesans and Fishcakes: The Consuming Passions of Classical Athens* chronicles hostility to the consumption of fish in ancient Greece, expressed in much the same derogatory terms that are used today about takeaway and 'junk' foods.[78] The upper classes, of course, were exempt from such strictures: Lady Astor, one of the first female Members of Parliament, is once reputed to have given a speech to a group of women extolling the merits of fish-head soup; she was lost for words when one asked, 'If you want us to use the heads, who's eating the rest of the fish?' Numerous Victorian ladies, doing good by visiting slums, recommended nutritious broths and complained of the consumption of white bread and bacon. Complaints of this kind confirm that healthy foodstuffs such as vegetables made up a only small part of the diet of the poor, but also show a failure to understand why.

So while George III was enjoying asparagus and oranges, his subjects were eating hardly any fruit and vegetables, let alone exotic varieties such as these. Understanding the reasons for this difference requires an exploration of the nutritional state of the English working classes then and in succeeding centuries.

George III and his household, with his aristocratic friends, dined well; the results can be seen in numerous portraits of the Georgian monarchy and upper classes. George himself is said to have been 6 feet 6 inches tall, some 3 inches taller than his American counterpart George Washington; there is always a tendency to exaggerate the height of our leaders, so these estimates may not be reliable. But the stature and corpulence of the royal and aristocratic males, particularly George III's son the future Prince Regent in his tight-fitting breeches, are clear to see in portraits or in statues of them clad as Roman emperors. Even if we lop a few hyperbolic inches off, George and his friends would still have towered over his subjects. The average height of male adults was only 5 foot 5 inches – compared to today's 5 foot 10 inches – and the poorest boys were no more than 4 foot 8 inches at the age of fifteen.[79] Every member of the working classes would have looked very short and very thin, even by the standards of underdeveloped societies around the world today. Many would have been deformed by lack of essential vitamins and minerals; they suffered from diseases such as scurvy or rickets. The average length of life – although this included many infant and child deaths – was still less than forty years and was not to rise above fifty years until the early twentieth century; in many of the growing towns, it was much less.

The primary cause was lack of food as a result of low wages, though poor housing, childhood disease combined with poor medical care, and forcing young children to work, all played their part. Looked at as a whole, the height, weight and mortality of the population indicate what modern scientists call 'nutritional status' and how it is compromised by the combined effects of too little, poor-quality food, diseases, cold houses, inadequate clothing and hard physical labour. The English working-class population was in this deprived state because its energy intake, principally from food, could not match the energy demands of living and working. The result was that many people were stunted and emaciated, to the extent that they were unable to work a full day or week. The energy demands of different jobs vary widely, but heavy physical labour – of the kind required in farming or mining – is particularly taxing. For many people, this led to a vicious circle in which low energy supply produced low output, which in turn produced low wages and thus low food consumption.[80]

Historians argue at length about how bad the situation was. One rule of thumb is that an average of 2,000 calories per head per day produces an 'adequately-fed population'.[81] Complex calculations of the food supply in Britain – from its farms and from the imports from around the world that began in the eighteenth century – suggest that it met this standard; the average available number of calories per head per day was just over 2,200 in the eighteenth century, rising to 2,500 by the middle of the nineteenth century and nearly 3,000 by its end. If these figures are correct, the British population was one of the best fed in the world at the time.

But that is not the complete picture. These are averages and they conceal beneath them – in what were throughout these years very unequal societies – a range of actual consumption of food depending on whether you were rich or poor, male or female, adult or child, and even whether you lived in the north or the south. Heavy manual labour needs more energy, and therefore food, than sedentary occupations such as office work. The household surveys show that 66 per cent of food expenditure in the south went on cereals – bread, flour and meal – while in the north it was 50 per cent; southerners spent more on wheaten bread, northerners on oatmeal and potatoes.[82] The averages suggest, however, that perhaps four out of ten households, with lower incomes than the average, would not have met the adequacy standard and some people, with the lowest incomes, would have been severely malnourished. There is also plenty of evidence that the food that was available to a family would go disproportionately to the adult male – the breadwinner on whom the family depended and who might well be carrying out hard physical work; children, and particularly women, would eat bread and gruel so that the man might have his bit of bacon. Even so, having more children meant less food to go around. So there was a great deal of malnutrition despite statistics indicating the relatively high amount of food that was generally available.

The composition of foods is also important. Household surveys have been used to calculate how many vitamins and minerals were present in the diet of the working classes in town and country during the eighteenth and nineteenth centuries. The results are complicated, but suggest that more than half the population of agricultural labourers before 1850 lacked the modern recommended levels of calcium, vitamins A, C, D and the B vitamins cobalamin, riboflavin and niacin (B_{12}, B_2 and B_3). Matters improved slightly in the later nineteenth century, but deficiencies of calcium, vitamin C, cobalamin and riboflavin – all vital for child growth and general health – persisted. The same deficiencies were found in a survey of urban households in 1904.[83] The historians who have analysed that survey conclude that 'the array of nutrients in which deficiencies show more strongly, that is, calcium, riboflavin and vitamins A, C and D, are those one might expect from a diet in which fruit and vegetable play a minor role. The UK Food Standards Agency today recommends that fruit and vegetables constitute about 30 per cent of a healthy diet, with starchy food, bread, potatoes and rice, for instance, also at about 30 per cent. By weight, in the average 1904 British household, fruit and vegetables constitute about 6 per cent, while starchy food occupies just below 60 per cent.'[84]

Although there certainly were prejudices against eating fruit and vegetables, it is more likely that low consumption of them by the working classes stemmed – as it still does today – from limited choices made under severe constraints. As the Ministry of Health noted in 1926, 'the high cost of milk and green vegetables partly explained the very low levels of consumption' at the time.[85] Families with low incomes had to choose foods that gave the highest level of energy per pound or penny spent, or which filled up hungry stomachs. They also had to think about the costs of cooking when fuel was expensive. They had to ensure that the breadwinners, the husband but also in many households the wife and older children, had enough to keep them healthy and able to work. They might feel entitled to spend some of their income on minor luxuries, such as tea, to alleviate some of the miseries of everyday existence. And they also had to consider – even when they had a garden or allotment – the costs in their widest sense of producing the foods they would then eat. All this determined what they ate.

Seen in this light, the low consumption of fruit and vegetables is easier to understand. They contain relatively few calories – which is why they are recommended today for people trying to lose weight – and they are not very filling. This was illustrated during the Second World War in 1940, when eminent nutritionists were asked to devise an 'Iron Ration' that would keep people alive. It consisted of bread, oatmeal, potatoes (and other roots, especially carrots and swedes), leafy green vegetables (cabbages and/or kale), fats and milk; but, according to the historian Derek Oddy, while 'it yielded adequate amounts of minerals such as iron and calcium, the "Iron Ration" was low in carbohydrates and fats and so offered insufficient energy for most normal activity'.[86] Some fruit and vegetables – though not potatoes and other root crops – were expensive in relation to the energy they gave, while those that were cheap to buy, such as potatoes, required the use of expensive fuel for the lengthy cooking that made them palatable but which also destroyed a great deal of their nutritional value. Finally, although middle-class observers extolled the moral benefits that would accrue to a Sunday spent tending an allotment, those rewards must have seemed less attractive to someone who had just completed six days – of between ten and twelve hours each day – of hard manual labour and who would often live at some distance from his allotment plot. Nor would he be much comforted by the thought that he could spend more time digging his ground during the winter when he had been laid off from his job.

No wonder, in the light of these factors, that the working classes as a whole consumed little fruit and vegetables and, when they did, often bought them in the market rather than growing them themselves.[87] Better, perhaps, to brighten up their houses by using their patch of land to grow the garish bedding plants that middle-class observers also derided. Like the upper classes who exhibited at the Royal Horticultural Society, working-class gardeners could show off their prowess in competition with their neighbours at innumerable village shows around the country. But that does not mean that their produce of giant marrows, beans or cucumbers formed a major part of their diet. In other words, although there were some periods – principally the two world wars – when the working-class kitchen garden helped to keep people alive, and although some allotment holders could say the same, growing fruit and vegetables in the back garden was not an efficient choice, even if it was possible at all, for much of the population.

CONCLUSION

It may seem counter-intuitive to argue that the one product of English gardens that one can eat – fruit and vegetables – is as much of a luxury as the flowers that delight the eye or the sweeping parkland dotted with statues and temples that has, since the eighteenth century, inspired thoughts of the sublime. But they are all similar in the sense that, for almost everyone, they are not necessities of life, in the form of energy-giving foods, housing or clothing, but rather things that are 'nice to have'. The 'five-a-day' mantra for the minimum number of portions of fruit and vegetables, urged on the British population today by the government for the vitamins and minerals they contain, can be bought more cheaply in the supermarket than grown in the garden; moreover, this has been true for many centuries, as the history of the market-gardening industry demonstrates.

Both the allotment of the working classes and the walled kitchen garden of the upper classes could, through intensive cultivation, the application of large quantities of manure and a great deal of labour, produce more per acre than could field horticulture, but they could not do so more economically once the costs of labour and time were taken into account. For most people, the benefits of home-grown fruit and vegetables – for their fresh taste, the physical exercise or the pleasure of seeing them grow – were not enough to overcome those costs. But for many, gardening is a hobby, even sometimes a competitive one, which they pursue because they enjoy it. Hence many people were and are, nevertheless, prepared to pay the price – whether in their own time or in employing an expert gardener – and it is those men and women, ready to pay whatever is needed, who urge themselves or their gardeners to create the beautiful rows of peas, the swelling globes of artichokes or a juicy and succulent pineapple. Revealing the costs that underpinned English kitchen gardens in no way diminishes the splendour of what they produced.

IO

Conclusion

'I pity that man who has completed every thing in his garden'[1]

The lesson of this book is that there is no danger of Alexander Pope's pity being required. At least since the seventeenth century, most of the English population have been unable to stop making, improving, extending and dreaming of gardens. Nor have they wanted to do so. In the process, they have changed the whole appearance of the country, introducing lakes where none existed before and spreading gardens great and small across hundreds of thousands of acres. They have employed millions of people, not just in their gardens but in the thousands of acres of nurseries that once surrounded London and were emulated, on a smaller scale, in many provincial cities. They have bought millions of books and magazines as they sought advice on what to do; now, again in their millions, they watch television programmes, surf the internet and visit other people's gardens in search of inspiration and beauty. They change their gardens far more often than they change their houses or even, in some cases, their wardrobes.

Why, then, do the government, politicians and economists discount it as 'only a hobby' of little economic significance? Why do they fail to recognize that gardening has been, for centuries, one of England's greatest industries?

One reason may sound technical, relating to arcane economics, but its effects are wide-ranging. It is the fact that very little gardening is included as part of the output of our economy, usually measured as gross domestic product, a statistic that is regularly quoted but little understood. GDP is defined, as the concept has been since it was first invented in the United States in the 1930s, to be the monetary value

285

of all the final goods and services produced by a country, normally in a year.* The crucial words are 'monetary value' because they mean that only the value of goods and services that are sold are included as part of GDP.[2] Activities such as housework, looking after our children or decorating our homes, together with the gardening that we do ourselves or the value of the fruit, flowers and vegetables that we produce, are not counted.[3] When there were plenty of paid gardeners, their wages would be counted as part of GDP; today, when we largely do it ourselves, our time and effort are not.

So the role of gardening in the economy is much understated because it is largely excluded from GDP. Perhaps equally serious is the fact that few people realize how much land is taken up by gardens, nor the value of that land if it could be used for other purposes. The country devotes a very large amount of land, some of it very valuable, to gardens. Current estimates by the Horticultural Trades Association are that gardens occupy 2 per cent of the UK land surface, or 1,481 square miles.† Most of that is in towns, where 18–20 per cent of the land area is taken up by gardens. If that land were to be available for building, its value would be at least £750 billion.[4]

A final reason why we do not appreciate the size of the gardening industry, past and present, has been apparent throughout this book. It is the result of the failure of most historians, and the gardening authors and TV producers who make use of their work, to translate properly the money values of the past into those of today. As indicated in Chapter 1, they use the retail price index for this purpose, but that greatly understates the money spent on gardening in the past. If you convert £1 in 1660 to modern values using the retail price index, it comes to £135.40; if, instead, you use the method employed in this book and make the conversion based on average earnings, the answer is £2,000 – a very large difference. The result of using the more accurate measure – which produces credible results for the

---

* The word 'final' reflects the fact that many goods and services – for example, engineering components or the work of recruitment consultants – become part of products that are then sold to the public. To avoid double counting, their value has to be excluded from the calculation of GDP.
† The proportion of land used for gardens in England alone would be significantly higher.

income of gardeners and the cost of gardening projects – is to demonstrate the huge size, in economic terms, of gardening and landscaping projects over several centuries.

This neglect of gardening and its full value to the economy, as well as to the well-being of our society, has consequences. It means that successive governments have been able to neglect, and have failed properly to fund, the training of young horticulturists, so that the industry today is suffering from shortages of skilled labour. Nurseries have been undervalued because they are not seen to be as important as other forms of manufacturing – yet making things is exactly what they do. Land continues to be converted from nurseries to housing developments because its value is then so much greater. The funding of public parks has greatly suffered under the impact of competitive tendering for their maintenance, which has torn out the heart of the parks departments of local authorities – one of the major training grounds for gardening apprentices – while recent cuts to local authority resources have meant that budgets for parks, which councils do not have a statutory duty to provide or maintain, have fallen sharply.

More broadly, there has been a failure to understand the long history of gardening's intimate links with many other sectors of the economy and its pivotal role in a number of innovations and new technologies. The garden industry is an excellent – but not previously noted – example of a phenomenon that has been recently analysed by economists, the importance of public spending as a catalyst to economic development. As we saw in Chapter 2, in England, the 'entrepreneurial state', to borrow the title of Mariana Mazzucato's book,[5] played a central part in the development of gardening, and thus of the industry devoted to it, initially through its spending on the royal parks and gardens and then through the lavish subsidies to the aristocracy that went under the name of the 'Old Corruption'. Direct public spending continued, to produce the public parks of the nineteenth and twentieth centuries, which were again responsible for innovations in garden design as well as in the training of the labour force. Even more important were the decisions of public authorities in the twentieth and twenty-first centuries both to build and to regulate housing so as to provide ample gardens for millions of people.

Meanwhile, as this book has shown for the first time, gardening spawned or influenced a range of other technologies that have been crucial to the growth of our economy. Chapter 7 demonstrated that the techniques used in building the canal network – crucial to the Industrial Revolution – were anticipated in the 'canals' and lakes of seventeenth- and eighteenth-century gardens. Central heating was developed in the service of plants long before it was applied to people. Steam engines were used in gardens years before their use in manufacturing industry. The application of iron and glass to construction – which has trans- formed buildings the world over – emerged first from the building of greenhouses and the need for plants to have the maximum light and air. The search for new plants and new varieties, the breeding and hybrid- izing, the technologies that enable us to grow fruit, flowers and vegetables throughout the year, have revolutionized not only the appear- ance of our gardens but also botanical science.

Gardening has changed the entire English landscape. Huge lakes, rolling parkland, expansive vistas and eye-catching buildings can be found throughout the country, all created – mostly since 1660 – in the service of gardens. As we saw in Chapter 8, the collective decision to build houses at low densities, specifically in order to provide gardens, rather than to meet housing need through flats or apartments in the manner of most other European countries, has equally transformed our townscapes.

On top of all this, through the centuries, English gardening has created beauty and brought pleasure. It certainly deserves to be seen – although it is never included in official definitions – as one of the earliest and largest of the creative industries that play an increasing role in England's economy and society.[6] In the eighteenth century, gardening was seen as equivalent among the arts to poetry, literature or painting. Good taste in gardening was as important to the culti- vated citizen as his or her taste in the other arts. In the nineteenth century, Prince Albert unsuccessfully tried to revive the public appre- ciation of its value and since then its contribution to our culture has continued to be undervalued, despite gardening's status as England's greatest leisure pursuit.

So gardening has transformed our country, occupied large amounts of our time and money and become a major topic of our conversa- tion. One reason for this is that nearly all of us are, in some sense, gardeners, as our parents have been before us. There can be few households in England today that do not contain, at the very least, some pot plants or a hanging basket, while the majority have some kind of garden and have told surveys that they enjoy gardens and gardening. They echo the millions who first delighted in the gardens of suburban homes, so different from the back yards of earlier generations, but even those earlier homes were decorated with window boxes and the aspidistras that could withstand the coal dust of the industrialized cities. Meanwhile, in the countryside there were cottage gardens – even if we should be cautious about their extent – and the thousands of country estates. So most of us have, for a very long time, been gardeners and we have chosen to spend substantial amounts of public and private money on this pleasurable pastime. That has itself been made possible by the great increase since 1660 in the income and wealth of Great Britain as a whole and of millions of individuals within it.

The temperate English climate certainly helps. It makes the country – with the other parts of the British Isles – particularly suitable as a place in which plants from every other part of the world can, with some effort, be grown. The effort of doing so spawned much of the garden industry as well as the specialist botanic gardens established from the seventeenth century onwards and led later by the Royal Botanic Gardens at Kew, which itself inspired and helped similar gardens around the world. This reminds us that we should be wary of claiming British, or English, exceptionalism in a love of gardening: many other parts of the world have shared the passion, even if the history of the garden industries of Babylon, Persia, India, China, Japan and many European countries remains to be written. But there was certainly a period in the eighteenth and nineteenth centuries when the coincidence of the strength of the economy, the amount of public and private spending on gardens, the design skills of men such as Capability Brown, the efforts of the plant hunters and the skills of hundreds of British nurserymen, produced *in toto* a garden industry unrivalled anywhere else in the world.

There are, of course, many other ways of enjoying gardens apart from doing the gardening, many other ways of spending one's time and money. There is the pleasure of visiting other people's gardens and comparing them with one's own. There are the websites, television and radio programmes, advice columns in newspapers and all the other means of finding out the mistakes one has made and the opportunities that lie ahead. There are the visits to garden centres, where one is willingly seduced into buying yet more plants that have to be nurtured. There are the advice books and the coffee table books. And there is the enjoyment that comes from being in the garden, from seeing and hearing children and grandchildren playing games and laying the foundations for their own gardening days. The Victorians, great moralists, saw gardening as good for the soul and, more prosaically, as a diversion for men from the appeal of the pub or, for women, from the dangers of reading novels. But one does not need to be a moralist to appreciate the sheer pleasure of sitting in the garden, listening to the birds and enjoying the play of light on trees and shrubs. That is what Kipling meant when he wrote of 'The Glory of the Garden' and it is what being a gardener – in all its many senses – means.

Gardening is creative and also competitive. There is a drive to purchase paradise, to create and display beauty, to produce the perfect rose or carnation. It can become an obsession. It drives men and women to scour the world for plants, nurseries to breed the best hybrid, head gardeners to grow and display tens of thousands of bedding plants, designers to fight for gold medals at the Chelsea Flower Show. It is not the only force that impels gardeners – there is the sheer excitement of seeing shoots pushing up from the ground, the sound of bees buzzing around the wisteria, or the excited shouts of grandchildren turning cartwheels on the lawn – but competition is never far away. There are flower shows, open gardens for the National Garden Scheme, the grand gardens of the National Trust, all competing for attention, while every visitor measures your garden against his or her own. After all, if even More's utopians competed, street by street, to produce the best gardens, can we lesser mortals be far behind?

This book began, like all good projects, as a question: how much did it cost to make and to maintain the gardens of England? The answer is, to my surprise and – I expect – to that of everyone else: much more than any of us could have imagined. But what has surprised me even more is that the expenditure of money is only the tip of the iceberg. What lies underneath are the connections of our gardens to the re-establishment of the English monarchy, to the expansion of the national debt, to the transport network that enabled the Industrial Revolution, to the industries of steam, glass and iron, to the suburbs and new housing schemes that are all around us. It is only a slight exaggeration to conclude that the history of English gardens is, in many senses, the history of England. But that will not come as a surprise to the millions of people who now, like their predecessors over many centuries, toil, curse, enjoy and celebrate the glory of the garden.

List of Gardens

Properties that appear in this book are listed below. If a garden is open to the public its website is given; please refer to that for details of opening times. Some of the gardens are now attached to hotels and wedding venues and are only open to guests. Other gardens may occasionally be open as part of the National Garden Scheme (www.ngs.org.uk). Virtually all the existing British gardens that appear here may be found on the Parks & Gardens website (www.parksandgardens.org), as well as being listed with either Historic England (historicengland.org.uk/listing/the-list) or Historic Environment Scotland (www.historicenvironment.scot). Any existing British gardens that are not listed with Historic England or Historic Environment Scotland are marked with an asterisk; any not listed with Parks & Gardens are marked with a cross.

Aiglemont, Picardie, 08090, France
*+ **Albert Park,** Middlesbrough, North Yorkshire TS1 3LB (middles brough.gov.uk/leisure-events-libraries-and-hubs/parks/albert-park)
Alnwick Castle, Northumberland NE66 1NQ (www.alnwickcastle.org)
* **Alnwick Garden,** Northumberland NE66 1NQ (www.alnwickgarden.com)
Althorp, Northamptonshire NN7 4HQ (www.spencerofalthorp.com)
Apothecaries' Garden, Chelsea (Chelsea Physic Garden), London SW3 4HS (www.chelseaphysicgarden.co.uk)
Appuldurcombe, Isle of Wight PO38 3EP (www.english-heritage.org.uk/visit/places/appuldurcombe-house)
Ashburnham Place, Sussex TN33 9NF (www.ashburnham.org.uk/visiting-ashburnham)
Ashridge, Hertfordshire HP4 1LX (www.nationaltrust.org.uk/ashridge-estate)
Audley End, Essex CB11 4JF (www.english-heritage.org.uk/visit/places/audley-end-house-and-gardens)

Aynhoe Park, Northamptonshire OX17 3BQ (aynhoepark.co.uk)

Badminton, Gloucestershire GL9 1DB (www.badmintonestate.com)

Balmoral Castle, Aberdeenshire, Scotland AB35 5TB (www.balmoralcastle.com/admissions)

Basildon Park, Berkshire RG8 9NR (www.nationaltrust.org/Basildon-park)

Battersea Park, London SW11 4NJ (www.batterseapark.org/info)

Beaufort House, Chelsea, London – demolished in 1739

Bedford House, Strand, London – demolished in 1800

Bemerton Lodge, Salisbury, Wiltshire SP2 7EN – now a residential care home

Birkenhead Park, Merseyside CH41 4HY (www.wirral.gov.uk/leisure-parks-and-events/parks-and-open-spaces/birkenhead-park)

Blandings Castle – fictional in the novels of P. G. Wodehouse

Blenheim Palace, Oxfordshire OX20 1PX (www.blenheimpalace.com)

Bowood House and Gardens, Wiltshire SN11 0LZ (www.bowood.org/bowood-house-gardens)

Bretby Hall, Derbyshire DE15 0QQ

Bretton Hall, West Bretton, West Yorkshire WF4 4LG (ysp.co.uk) – the Yorkshire Sculpture Park

Broadlands, Hampshire SO51 9ZD (broadlandsestates.co.uk)

Broadmoor Hospital, Berkshire RG45 7EG

Brocket Hall, Hertfordshire AL8 7XG (brocket-hall.co.uk)

Bulstrode Park, Buckinghamshire SL9 8SZ

*+ Burford House, Royal Mews, Windsor, Berkshire SL4 1NG

Burghley House, Lincolnshire PE9 3JY (burghley.co.uk)

Bushy Park, London TW12 2EJ (www.royalparks.org.uk/parks/bushy-park)

Cadland House, Hampshire SO45 1AA (visits arranged by written appointment)

Canons Park (Cannons), Middlesex HA6 8QT (www.harrow.gov.uk/info/200259/harrows_parks/33/canons_park)

Capesthorne Hall, Cheshire SK11 9JY (www.capesthorne.com)

Carlton House, London – demolished in 1826

Cassiobury Park, Hertfordshire WD18 7LD (cassioburypark.info)

Castle Howard, North Yorkshire YO60 7DA (www.castlehoward.co.uk)

Caversham Park, Berkshire RG4 6PF

Center Parcs Woburn Forest, Millbrook Road, Bedfordshire MK45 2HZ (www.centerparcs.co.uk/discover-center-parcs/holiday-locations/woburn-forest.html)

Central Park, New York, NY 10022, USA (www.centralparknyc.org)

Charleston Kedding – fictional country house in the work of the historian Susan Campbell

Château de Ferrières (École Ferrières, University of Paris), 77164 Ferrières-en-Brie, France

Chatsworth, Derbyshire DE45 1PP (www.chatsworth.org)

Chicheley Hall, Buckinghamshire MK16 9JJ (www.chicheleymilton keynes.co.uk)

Chiswick House, London W4 2RP (chiswickhouseandgardens.org.uk)

Cirencester Park, Gloucestershire GL7 2BU (cirencesterpark.co.uk)

Claremont, Surrey KT10 9JG (www.nationaltrust.org.uk/ claremont-landscape-garden)

Cliveden House, Buckinghamshire SL6 0JA (www.nationaltrust.org.uk/ Cliveden)

Constable Burton Hall, North Yorkshire DL8 5LJ (constableburton.com)

Coombe Abbey, Warwickshire CV3 2AB (www.coventry.gov.uk/info/136/ coombe-country-park)

Cragside, Northumberland NE65 7PX (www.nationaltrust.org.uk/ cragside)

Crewe Hall, Cheshire CW1 6UZ (www.crewehallcheshire.co.uk)

Croome Court, Worcestershire WR8 9DW (www.nationaltrust.org.uk/ croome)

Crystal Palace, Sydenham, London SE20 8DT (crystalpalacepark.org.uk)

Cusworth Hall, South Yorkshire DN5 7TU (www.doncaster.gov.uk/ services/culture-leisure-tourism/cusworth-hall-museum-and-park)

Daylesford House, Gloucestershire GL56 0YH

Derby Arboretum, Derbyshire DE23 8FN (www.inderby.org.uk/parks/ derbys-parks-and-open-spaces/derby-arboretum)

Dodington House, Bristol BS37 6SL

Dunham Massey, Cheshire WA14 4SJ (www.nationaltrust.org.uk/ Dunham-Massey)

East Park, Hull, East Yorkshire HU8 8JU (www.visithullandeastyorkshire. com/Hull-east-park-Hull)

Eastnor Castle, Herefordshire HR8 1RL (eastnorcastle.com)

Eaton Hall, Cheshire CH4 9JD (www.eatonestate.co.uk)

*+ Elveden, Suffolk IP24 3TQ (www.elveden.com)

Endsleigh Cottage, Devon PL19 0PG (hotelendsleigh.com)

Euston Park, Suffolk IP24 2QW (www.eustonhall.co.uk)

Exbury House, Hampshire SO45 1AZ (www.exbury.co.uk)

Exton Park, Rutland LE15 8AN (www.extonpark.co.uk)

Eythrope Park, Buckinghamshire HP18 0HT

* **Fenstanton Manor House**, Cambridgeshire PE28 9JQ
* **Fisherwick**, County Antrim, Northern Ireland BT39 0PA
Flambards (now Harrow Park and School), Harrow HA1 3NE
Fonthill House, Wiltshire SP3 5SH
Fort Belvedere, Surrey TW20 0UU (www.windsorgreatpark.co.uk)
+ **Frogmore Gardens** (including **Frogmore Kitchen Garden**), the Royal
  Estate, Windsor, Berkshire SL4 2JG
Frogmore House, Windsor, Berkshire SL4 2JG
Fulham Palace, London SW6 6EA (www.fulhampalace.org)
Gatton Park, Surrey RH2 0TW (gattonpark.com)
Glen Tanar, Aberdeenshire, Scotland AB34 5EU (glentanar.co.uk)
Goldney Hall, Bristol BS8 1BH
Gopsall Park, Leicestershire – demolished by 1952
Gower Street (13 Upper Gower Street), London – garden no longer
  survives
Gravetye Manor, West Sussex RH19 4LJ (www.gravetyemanor.co.uk)
Greenwich Palace, London SE10 8QY (www.royalparks.org.uk/parks/
  greenwich-park)
Grimsthorpe, Lincolnshire PE10 0LY (www.grimsthorpe.co.uk)
The Haining, Selkirk, Scotland TD7 5LR
Hall Barn, Buckinghamshire HP9 2SG
*+ **Hall Place**, Hampshire PO8 0RP
Halton House, Buckinghamshire HP22 5NN (www.haltonhouse.org.uk)
Hampton Court, Herefordshire HR6 0PN (www.hamptoncourt.org.uk)
Hampton Court Palace, London KT8 9AU (www.hrp.org.uk/Hampton-
  court-palace)
Harewood House, West Yorkshire LS17 9LG (harewood.org)
Hartwell House, Buckinghamshire HP17 8NR (www.hartwell-house.com)
Heathfield Park, Worcester – in *Life in the Gardeners' Bothy* by Arthur
  Hooper
Heligan (The Lost Gardens of Heligan), Cornwall PL26 6EN (www.
  heligan.com)
Herrenhausen, 30419 Hanover, Germany (www.hannover.de/en/Tourism-
  Culture/Sightseeing-City-Tours/Tourist-Highlights/
  Royal-Gardens-of-Herrenhausen)
Het Loo Palace, Koninklijk Park 1, 7315 JA Apeldoorn, Netherlands
  (www.paleishetloo.nl)
Hever Castle, Kent TN8 7NG (www.hevercastle.co.uk)
Highbury Hall, Birmingham B13 8QG (www.birmingham.gov.uk/
  directory_record/3607/highbury_hall)

Highbury Hall, Cambridge – in *Life in the Gardeners' Bothy* by Arthur Hooper

*+ Highgrove, Gloucestershire GL8 8TQ (www.highgrovegardens.com)

The Hill, Hampstead, London NW11 7EX (www.cityoflondon.gov.uk/things-to-do/green-spaces/hampstead-heath/heritage/Pages/the-pergola.aspx)

Hill Park (Valence School), Kent TN16 1QN

Hillier Gardens and Arboretum (Sir Harold Hillier Gardens), Hampshire SO51 0QA (www.hants.gov.uk/thingstodo/hilliergardens)

Holkham, Norfolk NR23 1AB (www.holkham.co.uk)

Huis Honselaarsdijk, Netherlands – demolished in 1815

Horseheath Hall, Cambridgeshire – demolished in 1777

* Horwood House, Buckinghamshire MK17 0PH (www.devere.co.uk/Horwood-estate)

* Humberstone Park, Leicester LE5 4DG (www.leicester.gov.uk/leisure-and-culture/parks-and-open-spaces/our-parks/humberstone-park)

Hyde Park, London W2 2UH (www.royalparks.org.uk/parks/Hyde-park)

Kedleston Hall, Derbyshire DE22 5JH (www.nationaltrust.org.uk/kedleston-hall)

Kelston Park, Somerset BA1 9AE

* Kennington Park, Lambeth, London SE11 4BE (www.lambeth.gov.uk/places/kennington-park)

Kensal Green Cemetery, London W10 4RA (www.kensalgreencemetery.com)

Kensington Gardens, London W2 2UH (www.royalparks.org.uk/parks/Kensington-gardens)

Kenwood House, Hampstead, London NW3 7JR (www.english-heritage.org.uk/visit/places/kenwood)

Killerton, Devon EX5 3LE (www.nationaltrust.org.uk/killerton)

Knightshayes Court, Devon EX16 7RQ (www.nationaltrust.org.uk/knightshayes-court)

Knowsley Hall, Merseyside L34 4AG (knowsley.com)

*+ The Laines, Plumpton, Sussex BN7 3AJ

Lambeth Palace, London SE1 7JU (www.archbishopofcanterbury.org)

The Leas, Folkestone, Kent CT20 2DJ (www.visitkent.co.uk/attractions/the-leas-2698)

The Leasowes, Halesowen, West Midlands B62 8QF (https://www.dudley.gov.uk/residents/environment/countryside-in-dudley/nature-reserves/the-leasowes)

Leeds Abbey, Kent – demolished at the end of the eighteenth century

*+ **Lincoln's Inn Fields**, London WC21 3BP (www.camden.gov.uk/ parks-in-camden#nfbx)

* **Little Aston Hall**, Staffordshire B74 3BJ

Llewenny Palace, Denbighshire, Wales – demolished in the early nineteenth century

* **Lockeridge House**, Wiltshire SN8 4EL

Londesborough Park, East Yorkshire YO43 3LF

Longleat, Wiltshire BA12 7NW (www.longleat.co.uk)

Luton Hoo, Bedfordshire LU1 3TQ (www.lutonhooestate.co.uk)

Lyons Demesne, Lyons Hill, County Kildare, Eire

*+ **Madeira Walk**, Ramsgate, Kent CT11 8HQ

Madeley Court, Madeley, Telford, Shropshire TF7 5DW (www.accor hotels.com/gb/hotel-8202-mercure-telford-madeley-court-hotel/index. shtml#origin=mercure)

*+ **Le Manoir aux Quat'Saisons**, Oxfordshire OX44 7PD (www.belmond. com/LeManoir/Oxfordshire)

Mentmore Towers, Buckinghamshire LU7 0QH

Moggerhanger Park, Bedfordshire MK44 3RW (www.moggerhanger park.com)

*+ **The Mount**, Shrewsbury, Shropshire SY3 8PP

Munstead Wood, Surrey GU7 1UN (munsteadwood.org.uk)

* **Navestock**, Essex RM4 1HA

New Park, Surrey – demolished *c*.1835

*+ **Newmarket Palace**, Suffolk – partly demolished in the nineteenth century, the remains now incorporated in Palace House, Newmarket, Suffolk CB8 8EP (www.palacehousenewmarket.co.uk)

* **Norman Court**, Hampshire SP5 1NF

* **Number 4, the Circus**, Bath BA1 2EW (visitbath.co.uk/listings/single/ the-circus)

Old Warden, Bedfordshire SG18 9EP (www.shuttleworth.org/ swissgarden)

*+ **Olympic Park**, London E20 2ST (www.queenelizabetholympicpark.co.uk)

Osborne, Isle of Wight PO32 6JY (www.english-heritage.org.uk/visit/ places/osborne)

Painshill, Surrey KT11 1JE (www.painshill.co.uk)

Palace of Peace, The Hague, Carnegieplein 2, 2517 KJ Den Haag, Netherlands (www.vredespaleis.nl)

Patshull Hall, Staffordshire WV6 7HR (patshull-park.co.uk)

* **Peasholm Park**, Scarborough, North Yorkshire YO12 6AG (www. peasholmpark.com)

*+ **Peel Park**, Bradford, West Yorkshire BD2 4BX (www.visitbradford.com/thedms.aspx?dms=3&venue=2180200)

Petworth House and Park, West Sussex GU28 0AE (www.nationaltrust.org.uk/petworth-house-and-park)

Pierrepont House, Nottingham – demolished in the early nineteenth century

*+ **The Plantation Garden**, Norwich NR2 3RA (www.plantationgarden.co.uk)

Pontypool Park, Monmouthshire, Wales NP4 8AT (www.torfaen.gov.uk/en/LeisureParksEvents/ParksandOpenSpaces/Pontypool-Park/Pontypool-Park.aspx)

+ **3–5 Porchester Terrace**, Bayswater, London W2 3TH

Powderham Castle, Devon EX6 8JQ (www.powderham.co.uk)

Primrose Hill, London NW3 2NA (www.royalparks.org.uk/parks/the-regents-park)

Prince's Park, Toxteth, Liverpool L8 3TZ (liverpool.gov.uk/leisure-parks-and-events/parks-and-greenspaces/princes-park)

*+ **Queen Mary's Gardens**, Regent's Park, London NW1 4NR (www.royalparks.org.uk/parks/the-regents-park/things-to-see-and-do/gardens-and-landscapes/queen-marys-gardens)

Regent's Park, London NW1 4NR (www.royalparks.org.uk/parks/the-regents-park)

Richmond Park, London TW10 5HS (www.royalparks.org.uk/parks/Richmond-park)

Rivington Terraced Gardens, Bolton, Lancashire BL6 7SB (www.visit-northwest.com/sights/lever-park)

Rolleston Hall, Leicestershire – demolished in 1928

Rousham, Oxfordshire OX25 4QX (rousham.org)

Royal Botanic Garden, Edinburgh EH3 5LR (www.rbge.org.uk)

Royal Botanic Gardens, Kew TW9 3AB (www.kew.org)

Royal Botanic Gardens, Regent's Park, London – closed in 1932

*+ **Royal Victoria Park**, Bath BA1 2NQ (visitbath.co.uk/listings/single/royal-Victoria-park)

* **Rycote**, Oxfordshire OX9 2PE

*+ **Sainsbury Arts Centre**, University of East Anglia, Norwich NR4 7TJ (scva.ac.uk)

St James's Park, London SW1A 2BJ (www.royalparks.org.uk/parks/st-Jamess-park)

Saltram House, Devon PL7 1UH (www.nationaltrust.org.uk/saltram)

Sandringham House, Norfolk PE35 6EN (sandringham-estate.co.uk)

Shardeloes, Buckinghamshire HP7 0RL

Sherborne Castle, Dorset DT9 3PY (www.sherbornecastle.com)

Shotover Park, Oxfordshire OX33 1QS

Sissinghurst Castle Garden, Kent TN17 2AB (www.nationaltrust.org.uk/
sissinghurst-castle-garden)

Sotherton Court – fictional in *Mansfield Park* by Jane Austen

Stansted Park, Hampshire PO9 6DX (www.stanstedpark.co.uk)

Staunton Harold Hall, Leicestershire LE65 1RT (www.stauntonharold
estate.co.uk)

Stourhead, Wiltshire BA12 6QD (www.nationaltrust.org.uk/stourhead)

Stow Hall Gardens, West Norfolk PE34 3HT (www.churchfarmstowbar
dolph.co.uk)

Stowe, Buckinghamshire MK18 5EH (www.nationaltrust.org.uk/Stowe)

Studley Royal (Fountains Abbey and Studley Royal Water Garden), North
Yorkshire HG4 3DY (www.nationaltrust.org.uk/fountains-abbey-and-
studley-royal-water-garden)

Syon Park, London TW8 8JF (www.syonpark.co.uk)

Tatton Park, Cheshire WA16 6QN (www.tattonpark.org.uk/home.aspx)

*+ **Thenford,** Northamptonshire OX17 2BX (thenfordarboretum.com)

Thornton Manor, Cheshire CH63 1JB (thorntonmanor.co.uk)

Tottenham Park, Wiltshire SN8 3BD

Trentham, Staffordshire ST4 8JG (www.trentham.co.uk)

Tyntesfield, Somerset BS48 1NX (www.nationaltrust.org/tyntesfield)

Vauxhall Garden, Dudston Hall, Birmingham – closed in 1850

Vauxhall Pleasure Gardens, London – closed in 1859

+ **Vernon Park,** Stockport SK1 4AR (www.stockport.gov.uk/vernon-park)

Versailles, Palace of, 78000 Versailles, France (www.chateauversailles.fr)

*+ **Victoria Park,** London E3 5TB (www.towerhamlets.gov.uk/lgnl/
leisure_and_culture/parks_and_open_spaces/victoria_park/victoria_
park.aspx)

Villa d'Este, 00019 Tivoli RM, Italy (www.villadestetivoli.info/storiae.htm)

Villa Victoria, Grasse, France – sold in 1922; just the tea house remains
and the garden no longer survives

Virginia Water, Surrey TW20 0UU (www.windsorgreatpark.co.uk)

Waddesdon Manor, Buckinghamshire HP18 0JH (www.nationaltrust.org.
uk/waddesdon-manor)

Wakefield Lodge, Northamptonshire NN12 7QX

Wanstead, Essex E11 2LT (www.cityoflondon.gov.uk/things-to-do/
green-spaces/epping-forest/visitor-information/wheretogoineppingforest/
Pages/wanstead-park.aspx) and (www.wansteadpark.org.uk)

Warwick Castle, Warwickshire CV34 4QU (www.warwick-castle.com)

*+ **Warwick Priory**, Warwickshire CV3 4 4XW (www.warwickdc.gov.uk)

Welbeck Abbey, Nottinghamshire S80 3LZ (www.welbeck.co.uk)

Wentworth Castle Gardens, South Yorkshire S75 3EW (www.wentworth
castlegardens@nationaltrust.org.uk)

West Wycombe Park, Buckinghamshire HP14 3AJ (www.nationaltrust.
org.uk/westwycombe)

Westbury Court Garden, Gloucestershire GL14 1PD (www.nationaltrust.
org.uk/westbury-court-garden)

Weston Park, Staffordshire TF11 8LE (www.weston-park.com)

Westwood Park, Worcestershire WR9 0AD

*+ **Whetham House**, Calne, Wiltshire SN11 OPT

Whitehall Palace, Westminster, Middlesex – demolished in 1698

Wilton House, Wiltshire SP2 0BJ (www.wiltonhouse.co.uk)

Wimbledon House, Surrey – garden no longer survives

Wimbledon Park, London SW19 7HX (www.merton.gov.uk/leisure-
recreation-and-culture/parks-and-open-spaces/
parks-and-recreation-grounds/wimbledon/wimbledon-park)

Windsor Great Park, Berkshire SL4 2HT (www.windsorgreatpark.co.uk)

Woburn Abbey, Bedfordshire MK17 9WA (www.woburnabbey.co.uk)

* **Worksop Manor**, Nottinghamshire S80 3DG

Wotton House, Wotton Underwood, Buckinghamshire HP18 0SB

Wrest Park, Bedfordshire MK45 4HS (www.englishheritage.org.uk/visit/
places/wrest-park)

Notes

For a list of abbreviations used in bibliographical references, see Selected Bibliography, p. 347.

INTRODUCTION

1. One partial exception is the work of the late John Harvey, who collected and published an enormous amount of information about early nurseries, but even he refers only occasionally to overall economic conditions.
2. One example among many is David Jacques's monumental and beautiful study *Gardens of Court and Country, 1630–1730* (New Haven: Yale University Press, 2017), which 'touches on politics, religion, taste, men's fashion, cuisine, transport, "undertakers", attitudes to flowers, husbandry, plant introductions and many other topics' (p. 15) but not, except very occasionally and in passing, money.
3. This term was first used by the Norwegian-American economist Thorstein Veblen in *The Theory of the Leisure Class* (1899) to describe expenditure that demonstrated wealth and social status.
4. When I first told a distinguished garden historian that I wanted to write this book, he said it was impossible – that the records don't exist. In fact, the bulk of records in the National Archives, in every county record office and in the muniment rooms of numerous stately homes, is so great that there is scope for much more work on the economic history of gardening.

I. THE ENGLISH GARDEN IN 1660 AND 2020

1. Josuah Sylvester (trans.), 'Hortus', in *Du Bartas: His Divine Weekes and Workes* (London: Robert Young, 1641).

2. The order is reprinted in full in Catherine Parsons, 'Horseheath Hall and its Owners', *Proceedings of the Cambridge Antiquarian Society*, 41 (1943–7), pp. 1–51, at pp. 16–17, which is also available on the Horseheath village website (www.horseheath.info).

3. William Shakespeare, *A Midsummer Night's Dream*, Act III, scene i, lines 173–4, in W. J. Craig (ed.), *The Complete Works of William Shakespeare* (London: Oxford University Press, 1957), quoted in Gerit Quealy, *Botanical Shakespeare: An Illustrated Compendium* (New York: HarperCollins, 2017), p. 200.

4. www.parksandgardens.org, entry on Horseheath Hall. The house was sold in 1700 to John Bromley, a Barbados sugar importer, and demolished in 1777 when one of Bromley's descendants gambled away his inheritance.

5. John Evelyn, *The Diary of John Evelyn* (1818; reprinted Woodbridge: Boydell Press, 1995), entry for 20 July 1670.

6. David Jacques, *Gardens of Court and Country: English Design, 1630–1730* (New Haven: Yale University Press, 2017), pp. 64, 73.

7. The methods used in compiling the data are described in Gregory Clark, 'What were the British Earnings and Prices Then? (New Series)', MeasuringWorth, 2019 (www.measuringworth.com/ukearncpi/).

8. *Guardian*, 11 July 2018.

9. John H. Harvey, *Early Nurserymen* (London and Chichester: Phillimore, 1974), p. 5.

10. Linda Farrar, *Ancient Roman Gardens* (Stroud: History Press, 1998); Wilhelmina F. Jashemski, *The Gardens of Pompeii, Herculaneum and the Villas Destroyed by Vesuvius* (New Rochelle, NY: Caratzas Brothers, 1979).

11. Harvey, *Early Nurserymen*, p. 25.

12. Ibid., p. 28.

13. C. Paul Christianson, *The Riverside Gardens of Sir Thomas More's London* (New Haven: Yale University Press, 2005), pp. 109–12; Harvey, *Early Nurserymen*, pp. 29–30.

14. Quoted in Christianson, *Riverside Gardens of Sir Thomas More's London*, pp. 4–5.

15. Harvey, *Early Nurserymen*, p. 32.

16. Stephen Broadberry, Bruce M. S. Campbell, Alexander Klein, Mark Overton and Bas van Leeuwen, *British Economic Growth, 1270–1870* (Cambridge: Cambridge University Press, 2015), pp. 210–11 and table 5.07.

17. Harvey, *Early Nurserymen*, ch. 3.

18. An excellent account of our knowledge of gardens and gardening before 1660 can be found in Jill Francis, *Gardens and Gardening in Early Modern England and Wales* (New Haven and London: Yale University Press, 2018).

19. Eurostat, 'Harmonised European Time Use Survey' (Brussels: Eurostat, 2014) – found at the HETUS website (www.tus.scb.se).

20. Agriculture and Horticulture Development Board, *Import Substitution and Export Opportunities for UK Ornamentals Growers* (London: Agriculture and Horticulture Development Board, 2014), p. 9.

21. Alan B. Krueger (ed.), *Measuring the Subjective Well-Being of Nations: National Accounts of Time Use and Well-Being* (Chicago: Chicago University Press, 2009), table 1.16.

22. Horticultural Trades Association, *The Great British Gardener: A Profile of Gardeners in 2006* (Didcot: Horticultural Trades Association, 2006), pp. 2–35.

23. Oxford Economics, *The Economic Impact of Ornamental Horticulture and Landscaping in the UK: A Report for the Ornamental Horticulture Round Table Group* (London: Oxford Economics, 2018), pp. 12–21. This study also quantifies the 'indirect and induced impacts' of ornamental horticulture, including the wages of workers in the industry and the tax that they paid, together with tourist expenditure attributable to visiting parks and gardens. All this adds up to a contribution of £24.2 billion annually to the GDP of the United Kingdom (see p. 3 of the study). This does not include the wider benefits of gardens and green spaces, such as the 'aggregate boost' of £131 billion to Britain's house prices attributable to the presence and proximity of the natural environment (see p. 4).

24. John Claudius Loudon, *An Encyclopaedia of Gardening* (London: Longman, Hurst, Rees, Orme and Brown, 1822), p. 1202.

2. GARDENS AND THE STATE

1. Edmund Waller, *On St James's Park, as Lately Improved by His Majesty* (London: Gabriel Bedel and Thomas Collins, 1661).

2. John Evelyn quoted in Caroline Grigson, *Menagerie: The History of Exotic Animals in England* (Oxford: Oxford University Press, 2000), pp. 30–33.

3. Ibid., pp. 30–35.

4. Evelyn, *Diary*, entry for 1 December 1662.

5. *The Diary of Samuel Pepys* (www.pepysdiary.com).

6. The Office of Works was set up in the English royal household in 1378 to oversee the building and maintenance of the royal castles and residences.
7. NA Work 6/2, f. 162.
8. NA Work 5/141 and 5/142, the Paymaster's accounts, record annual expenditure on both houses and gardens from 1706 to 1780.
9. Grigson, *Menagerie*, p. 76.
10. Edward Impey, *Kensington Palace* (London: Merrell Publishers, 2012), pp. 39–40, 49–53.
11. Jacques, *Gardens of Court and Country*, pp. 20, 262.
12. Department of Culture, Media and Sport, *The Royal Parks: Annual Report and Accounts 2015–2016* (London: Department of Culture, Media and Sport, 2016), laid before Parliament on 19 July 2016. The parks comprise: Bushy Park, Green Park, Greenwich Park, Hyde Park, Kensington Gardens, Regent's Park and Primrose Hill, Richmond Park, and St James's Park. Maintenance of the Royal Parks also embraces a number of other spaces in London, including Brompton Cemetery, Victoria Tower Gardens, the gardens of 10, 11 and 12 Downing Street, and Grosvenor Square Garden. The grounds of Buckingham Palace, Kensington Palace and Windsor Castle are not included.
13. Jacques, *Gardens of Court and Country*, p. 20. Versailles still covers nearly 2,000 acres even after parcels of land were sold off during the French Revolution. For comparison, the cultivated area of garden at Hampton Court extended for only about 60 acres even after the alterations by William and then Anne, although Bushy Park next to it is 1,100 acres; Richmond Park, the largest of the London Royal Parks, created by Charles I for hunting deer, is 2,360 acres, while another hunting park, Windsor Great Park, is nearly 5,000.
14. The gardens in the sixteenth century are described, for example, by Roy Strong, who comments that 'Nothing quite like them had ever been seen before' – *The Renaissance Garden in England* (London: Thames and Hudson, 1979), p. 25.
15. Simon Thurley, *Hampton Court: A Social and Architectural History* (New Haven: Yale University Press, 2003), p. 209.
16. David Coombs, in 'The Garden at Carlton House of Frederick Prince of Wales and Augusta Princess Dowager of Wales: Bills in their Household Accounts 1728 to 1772', *Garden History*, 25 (1997), pp. 153–7, transcribes the list of plants. The remainder of the items that make up the total project are recorded in Frederick's household accounts, held in the archives of the Prince of Wales and available on microfilm in the British Library.

17. Alistair Roach, 'Miniature Ships in Designed Landscapes', *The Mariner's Mirror*, 98 (2012), pp. 43–54, at p. 43.

18. Grigson, *Menagerie*, pp. 165–72.

19. Susan Campbell, 'The Genesis of Queen Victoria's Great New Kitchen Garden', *Garden History*, 12 (1984), pp. 100–119, at p. 117.

20. Roderick Floud, Kenneth Wachter and Annabel Gregory, *Height, Health and History: Nutritional Status in the United Kingdom, 1750– 1980* (Cambridge: Cambridge University Press, 1990); Roderick Floud, Robert W. Fogel, Bernard Harris and Sok Chul Hong, *The Changing Body: Health, Nutrition and Human Development in the Western World since 1700* (Cambridge: Cambridge University Press, 2011).

21. PP 1833: xxi, *First Report from Commissioners Appointed to Collect Information in the Manufacturing Districts, Relative to the Employment of Children in Factories*; Michael Flinn (ed.), *Report on the Sanitary Condition of the Labouring Population, by Edwin Chadwick* (Edinburgh: Edinburgh University Press, 1965), p. 268.

22. Sanitary condition report of 1842, quoted in William Ashworth, *The Genesis of Modern British Town Planning* (London: Routledge & Kegan Paul, 1954), p. 51.

23. PP 1833: xv, 337, *Report of a Select Committee Appointed to Consider the Best Means of Securing Open Spaces in the Vicinity of Populous Towns, as Public Walks and Places of Exercise, Calculated to Promote the Health and Comfort of the Inhabitants.*

24. www.historyofparliamentonline.org, entry on Slaney.

25. PP 1833: xv, 337, p. 5.

26. Ibid., p. 8.

27. PP 1857–8: xlviii, 347, *Return, Showing the Amount of Public Money Expended in the Purchase and Formation of Public Parks, Public Walks and Recreation Grounds in Large Towns and Populous Places in Great Britain and Ireland since the Year 1840.* The largest sum went to Manchester, which got £2.2 million *(£3,000)*, while Bradford received £1.1 million *(£1,500)* for Peel Park. But London received much more: Primrose Hill in north London got £17 million *(£23,442 2s 11d)* and Victoria Park in east London £94 million *(£129,718 2s)*. South London, which the committee found had no public parks at all, received the most – £4.3 million *(£5,919 19s 7d)* for Kennington Park, which was converted from scrub and common pasture, and a whopping £223.9 million *(£308,842 1s 8d)* for Battersea Park, previously a duelling ground.

28. From Tudor times, the only legal source of bodies for dissection was executed criminals. Far more bodies were needed by anatomists and

medical schools and, from about 1730 until the passing of the Anatomy Act in 1832, which allowed corpses of paupers to be used, 'grave-robbers had been active in any area where corpses were required'. 'Sir Astley Cooper, the leading surgeon-anatomist of Regency London, boasted that he could obtain *anyone's* body, no matter whose, and from whatever station in life. Ready money was always the key.' The demand was sufficient to induce some, like Burke and Hare in Edinburgh, to murder rather than wait for people to die. James Stevens Curl (ed.), *Kensal Green Cemetery: The Origins and Development of the General Cemetery of All Souls, Kensal Green, London, 1824–2001* (Chichester: Phillimore, 2001), p. 23.

29. Quoted in Ruth Richardson and James Stevens Curl, 'George Frederick Carden and the Genesis of the General Cemetery Company', in ibid., pp. 21–48, at pp. 33–4.

30. Ibid.

31. Curl (ed.), *Kensal Green Cemetery*, pp. 14–18.

32. Brent Elliott, 'The Landscape of Kensal Green Cemetery', in ibid., pp. 287–96, at p. 288; Beverley F. Ronalds, 'Ronalds Nurserymen in Brentford and Beyond', *Garden History*, 45 (2017), pp. 82–100.

33. Sarah Rutherford, *The Victorian Cemetery* (Bodley: Shire Publications, 2010), p. 31.

34. Ibid., pp. 32–5.

35. Ibid., p. 55.

36. Tristram Hunt, *Building Jerusalem: The Rise and Fall of the Victorian City* (London: Weidenfeld & Nicolson, 2004), pp. 220–25.

37. Elaine Mitchell, 'Duddeston's "shady walks and arbours": The Provincial Pleasure Garden in the Eighteenth Century', in Malcolm Dick and Elaine Mitchell (eds), *Gardens and Green Spaces in the West Midlands since 1700* (Hatfield: West Midlands Publications, 2018), pp. 76–101, at pp. 81–93.

38. Travis Elborough, *A Walk in the Park: The Life and Times of a People's Institution* (London: Jonathan Cape, 2016), p. 61.

39. G. F. Chadwick, *The Park and the Town: Public Landscape in the 19th and 20th Centuries* (New York and Washington: Frederick A. Praeger, 1966), p. 68.

40. Vernon Park, Stockport, opened in 1858 and Albert Park, Middlesbrough, opened in 1868.

41. Much infrastructure spending was enabled by the Public Works Loan Board, although expenditure on public parks and gardens is not mentioned in Ian Webster, 'The Public Works Loan Board and the Growth

of the State in Nineteenth-Century England', *Economic History Review*, 71 (2018), pp. 887–908.

42. Claude Hitching, *Rock Landscapes: The Pulham Legacy* (Woodbridge: Antique Collectors' Club, 2012). Made to a secret recipe by James Pulham (1820–98), Pulhamite consisted of rubble and crushed brick with a mixture of sand and cement sculpted over the top, and was so realistic that it fooled even geologists at the time.

43. Local government statistics no longer give separate figures for expenditure on parks and gardens, but it seems clear that it has fallen substantially. Local government expenditure on 'open spaces' in 2016–17 was budgeted at £663 million, far less than in the 1970s.

44. Heritage Lottery Fund, *The State of UK Public Parks* (London: Heritage Lottery Fund, 2016), p. 3.

45. However, the ring-fenced fund for this purpose was ended in December 2017, leaving parks to compete for grants with other areas of cultural expenditure.

46. Sarah Rutherford, 'Landscapes for the Mind: English Asylum Designers, 1845–1914', *Garden History*, 33 (2005), pp. 61–86.

47. Sarah Rutherford, 'The Landscapes of Public Lunatic Asylums in England, 1808–1914', PhD thesis, De Montfort University, Leicester (2003), table 6.

48. Jeremy Burchardt, *The Allotment Movement in England, 1793–1873* (Woodbridge: Boydell & Brewer, 2002), p. 52.

49. Ibid., pp. 220–28. By the 1840s there were at least 100,000 plots, by the 1870s nearly 250,000 and by 1914 between 450,000 and 600,000.

50. Anthony Alexander, *Britain's New Towns: Garden Cities to Sustainable Communities* (London and New York: Routledge, 2009), p. 168.

51. Aiton left £4,880,000 *(£7,000)*, excluding the value of property, at his death in 1849.

3. THE GREAT GARDENS

1. Anon., *The Rise and Progress of the Present Taste in Planning Parks, Pleasure Grounds, Gardens etc. in a Poetic Epistle to the Right Honorable Charles, Lord Viscount Irwin* (London: C. Moran, 1767).

2. Richard Wilson and Alan Mackley, *The Building of the English Country House* (London: Continuum, 2000), p. 7.

3. Thomas Carew, 'To my friend G.N. from Wrest', in *The Works of Thomas Carew: Reprinted from the Original Edition of MDCXL [1640]* (1824; reprinted Edinburgh: W. and C. Tait, 1844).

4. Wrest was sold by the de Greys in 1917, used for a variety of purposes and then as the National Institute of Agricultural Engineering, before being taken over by English Heritage in 2006.

5. Bedfordshire Record Office L31/288, records of the de Grey family of Wrest Park.

6. *ODNB*, entry on Grey.

7. Queen Anne may not have shared the general opinion of the duke as, two years later, she made him a Knight of the Garter, the premier order of chivalry and an honour given personally by the sovereign.

8. Bedfordshire Record Office L31/319.

9. David Hume, 'Of the Standard of Taste', in *Essays Moral, Political and Literary* (1739; reprinted London: A. Millar, 1768), p. 274.

10. Horace Walpole made the connection directly: 'Poetry, Painting and Gardening, or the science of Landscape, will forever by men of Taste be deemed Three Sisters, or *the Three New Graces* who dress and adorn Nature' – quoted in Peter Willis, *Charles Bridgeman and the English Landscape Garden* (Newcastle upon Tyne: Elysium Press, 2002), p. 25.

11. J. R. Ward, *The Finance of Canal Building in Eighteenth-Century England* (Oxford: Oxford University Press, 1974), p. 7.

12. Roderick Floud, 'Capable Entrepreneur? Lancelot Brown and His Finances', *Occasional Papers from the RHS Lindley Library*, 14 (2016), pp. 19–41, at p. 29.

13. Jane Brown, *The Pursuit of Paradise: A Social History of Gardens and Gardening* (London: HarperCollins, 1999), quoted in Fiona Cowell, *Richard Woods (1715–1793): Master of the Pleasure Garden* (Woodbridge: Boydell Press, 2009), p. xix.

14. Jacques, *Gardens of Court and Country*, p. 18.

15. The *Return* was compiled by the Local Government Board at the request of the government, following a controversy about the amount of land owned by the aristocracy. It consists of a list, county by county, of the acreage and estimated yearly rental value of all landholdings of more than an acre; the name of the landowner is given but not the location of the landholding. Separate returns were published for England and Wales (excluding London) and for Scotland and Ireland. There were inaccuracies in the *Return* and these were corrected in subsequent publications by John Bateman, the last in 1883. PP 1875 (C 1097): *England and Wales (Exclusive of the Metropolis): Return of Owners of Land, 1873; Presented to both Houses of Parliament by Command of Her Majesty*; John Bateman, *The Great Landowners of Great Britain and Ireland* (1883; reprinted Leicester: Leicester University Press, 1971).

16. Wilson and Mackley, *Building of the English Country House*, p. 7. The estimate is based on returns from the second half of the nineteenth century, but probably is not too far distant from the position a century earlier.

17. T. C. G. Rich, G. Hutchinson, R. Randall and R. G. Ellis, 'List of Plants Native to the British Isles', *BSBI News*, 80 (1999), pp. 23–7.

18. Harvey, *Early Nurserymen*, p. 23.

19. Andrea Wulf, *The Brother Gardeners: Botany, Empire and the Birth of an Obsession* (London: Windmill Books, 2009).

20. Edmund Berkeley and Dorothy Smith Berkeley (eds), *The Correspondence of John Bartram, 1734–1777* (Gainesville: University Press of Florida, 1992), p. 24. A peck is a measure of volume, equivalent to 8 quarts or a quarter of a bushel.

21. Edmund Berkeley and Dorothy Smith Berkeley, *The Life and Travels of John Bartram* (Tallahassee: Florida State University Press, 1982), p. 202.

22. John Harvey, 'Prices of Shrubs and Trees in 1754', *Garden History*, 2 (1974), pp. 34–44, and *Early Nurserymen*, p. 68.

23. Cowell, *Richard Woods*, p. 59.

24. See Miles Hadfield, *A History of British Gardening*, 3rd edn (London: John Murray, 1979), pp. 99–101, for a discussion of greens in England in 1659.

25. *Guardian*, 15 August 2018, p. 1, reporting the annual review by the High Pay Centre.

26. Catherine Horwood, *Gardening Women: Their Stories from 1600 to the Present* (London: Virago Press, 2010), pp. 11–15.

27. The principal sources of information used were the *ODNB*, the websites History of Parliament Online (historyofparliamentonline.org) and Parks & Gardens (parksandgardens.org – which incorporates the English Heritage register of parks and gardens) and Wikipedia entries on individuals and their families. To avoid overburdening the text with notes, the exact sources are not always given.

28. www.parksandgardensuk.org, entry on Londesborough.

29. Grigson, *Menagerie*, pp. 225–6. The animals were later moved to Chatsworth.

30. Michael E. Turner, John V. Beckett and B. Afton, *Agricultural Rent in England, 1690–1914* (Cambridge: Cambridge University Press, 2004), p. 207.

31. *ODNB*, entry on Spencer.

32. *ODNB*, entry on Palmerston.

33. John Phibbs, *Capability Brown: Designing the English Landscape* (New York: Rizzoli International Publications, 2016), pp. 156–73; www.parksandgardensuk.org, entry on Broadlands.

34. *ODNB*, entry on Percy. Originally called Smithson, he inherited his baronetcy at the age of seventeen and by 1740, at the age of thirty-eight, successive inheritances meant that he had an annual income of £7.5 million *(£4,000)* and expectations of £5.6 million *(£3,000)* a year more. He married Elizabeth Seymour, who soon became sole heir to the immense Percy estates in Northumberland, worth £14 million *(£8,000)* a year.

35. *ODNB*, entry on Percy.

36. Ibid.

37. www.parksandgardensuk.org, entry on Staunton Harold.

38. Quoted in *ODNB*, entry on Shirley. William Hickey (1749–1830) was a lawyer who composed in the early 1800s a set of memoirs, in ten volumes, which are an excellent source on life in England in the late 1700s.

39. Information taken from *The Database of Court Officers: 1660–1837*, compiled by R. O. Bucholz (courtofficers.ctsdh.luc.edu).

40. *ODNB*, entry on Wills Hill, 1st Marquess of Downshire.

41. Leonard Knyff and Johannes Kip, *Britannia Illustrata*, ed. John Harris and Gervase Jackson-Stops (Bungay: Paradigm Press for the National Trust, 1984), p. 179.

42. Samuel Molyneux in 1712–13, quoted in Jacques, *Gardens of Court and Country*, p. 242.

43. Jacques, *Gardens of Court and Country*, p. 233.

44. Royal patronage and money came from many kinds of service. Charles Beauclerk, 1st Duke of St Albans (1670–1726), who owned a house and garden close to Windsor Castle, was an army officer and the illegitimate son of Charles II and Nell Gwyn; he inherited from her a 'considerable' estate. From 1695 he received a pension of £4.1 million *(£2,000)* a year from the crown, though by then the Stuarts had been ousted; he also had the hereditary office of Master Falconer and secured the reversion of the office of Registrar of the High Court of Chancery, which came to him in 1697 and was worth £3.1 million *(£1,500)* a year. In 1697 the king also gave him 'a sett of coach horses finely spotted like leopards' (*ODNB*, entry on Beauclerk). In 1703 a further grant of £1.5 million *(£800)* a year came from the Parliament of Ireland. So his combined offices were worth at least £8.7 million *(£4,300)* a year, but he was also Captain of the Gentlemen Pensioners on £2 million *(£1,000)* and an extra gentleman of the bedchamber, for another £2 million *(£1,000)*.

45. *ODNB*, entry on Hill.

46. His daughter was a maid of honour to the queen at £600,000 *(£300)* a year; his first son became postmaster-general in Ireland on £1.2 million *(£600)* a year; his second son accountant-general of the excise – probably on £800,000 *(£400)* a year; his third another commissioner of excise on £1.5 million *(£800)*. Finally, his own brother was a collector of excise.

47. *ODNB*, entry on Hawkins.

48. Michael Symes, *Mr Hamilton's Elysium: The Gardens of Painshill* (London: Frances Lincoln, 2010), pp. 21–49 and information from the accounts of Frederick, Prince of Wales.

49. John Williamson, *A Treatise on Military Finance; Containing the Pay of the Forces on the British and Irish Establishment; with the Allowances in Camp, Garrison and Quarters, &c.* (London: T. Everton, 1798), pp. 5–7.

50. Brown, *Pursuit of Paradise*, pp. 82–104.

51. historyofparliamentonline.org, entry on John Calcraft Senior, MP.

52. Full details of the slave compensation payments, which varied depending on the age and sex of the slave among other factors, can be found at the Centre for the Study of the Legacies of British Slave-Ownership (www.ucl.ac.uk/lbs).

53. www.parksandgardensuk.org, entry on Dodington; Roger Turner, *Capability Brown and the Eighteenth-Century English Landscape* (1985; reprinted Stroud: History Press, 2013), p. 178.

54. *ODNB*, entry on Lascelles.

55. Henry Lascelles bought the manors of Gawthorpe and Harewood for £113 million *(£63,827)*. He died in 1753, allegedly by suicide, with a probate value of £689 million *(£392,704)*, which didn't take account of the value of his landed property nor the £175 million *(£100,000)* provided to his second son.

56. T. Hinde, *Capability Brown: The Story of a Master Gardener* (London: Hutchinson, 1987), p. 161.

57. *ODNB*, entry on Lascelles.

58. Clive was given £415 million *(£237,000)* in 1757 and, in 1759, a grant of land revenues worth £46.5 million *(£27,000)* a year.

59. His estate of 13,500 acres cost him £267 million *(£162,000)* and his London house £17.3 million *(£10,500)*.

60. Grigson, *Menagerie*, p. 141.

61. Clive left an estate worth over £835 million *(£500,000)*, excluding the value of landed property.

62. www.parksandgardensuk.org, entry on Claremont; Turner, *Capability Brown and the Eighteenth-Century English Landscape*, pp. 115–18.

63. *ODNB*, entry on Sykes.

64. Royal Horticultural Society, Lindley Library, Lancelot Brown's account book, p. 112, viewable on the Royal Horticultural Society website (www. rhs.org.uk/education-learning/libraries-at-rhs/collections/library-online/ capability-brown-account-book).

65. Alan Valentine, *The British Establishment, 1760–84,* 2 vols (Norman, OK: University of Oklahoma Press), p. 846.

66. *ODNB,* entry on Barwell.

67. Ibid.

68. Quoted in an entry on Booth in *Dictionary of National Biography,* vol. 5 (London: Smith, Elder & Co., 1886).

69. www.parksandgardensuk.org, entry on Dunham Massey.

70. *ODNB,* entry on Pakington.

71. *ODNB,* entry on Smith Stanley.

72. historyofparliamentonline.org, entry on Thomas Fitzmaurice.

73. Ibid.

74. Andrea Wulf and Emma Gieben-Gamal, *This Other Eden: Seven Great Gardens and 300 Years of English History* (London: Little, Brown, 2005), p. 220.

75. Pleydell family papers, West Devon Record Office 69/M/6/112, and www.parksandgardensuk.org, entry on Saltram House.

76. Christopher Plumb, *The Georgian Menagerie: Exotic Animals in Eighteenth-Century London* (London: I. B. Tauris, 2015), pp. 144–5.

77. They can be seen in the Buckinghamshire County Museum, Aylesbury.

78. Plumb, *Georgian Menagerie,* p. 141.

79. See, for example, https://thegardenstrustblog.wordpress.com/2018/03/17/ whats-going-on-the-shrubbery-and-whats-it-got-to-do-with-mr-repton/ (posted 17 March 2018).

4. DESIGNERS

1. William Shakespeare, *The Life and Death of King John,* Act IV, scene ii, line 11, in Craig (ed.), *Complete Works of Shakespeare.*

2. Royal Horticultural Society, Lindley Library, Lanning Roper papers, LRO/1/69, letter of 4 March 1981.

3. Ibid., letter of 1 June 1981.

4. Phibbs, *Capability Brown,* p. 13.

5. Timothy Mowl, *Gentlemen Gardeners: The Men Who Created the English Landscape Garden* (Stroud: History Press, 2000), p. 162.

6. Broadberry et al., *British Economic Growth, 1270–1870,* Appendix 5.3.

7. David Brown and Tom Williamson, *Lancelot Brown and the Capability Men* (London: Reaktion Books, 2016).

8. I am grateful to the Royal Horticultural Society and the Royal Bank of Scotland for granting access to these documents, all of which have been digitized. The RBS also holds the records of Brown's executors following his death.

9. But David Brown – in Brown and Williamson, *Lancelot Brown and the Capability Men*, and David Brown, 'People, Places and Payments: Lancelot Brown's Account Book', *Occasional Papers from the RHS Lindley Library*, 14 (2016), pp. 3–18 – has succeeded in identifying many of them.

10. Cowell, *Richard Woods*, pp. 3, 146–53. Only one of Repton's account books survives, as MS 10 in the Norfolk Record Office, and it is often difficult to interpret the entries.

11. A more detailed analysis of Brown's accounts is in Floud, 'Capable Entrepreneur?'

12. Changing wages are documented, as described in Chapter 6, in estate archives and, with respect to Richard Woods, in Cowell, *Richard Woods*, pp. 146–53.

13. Hugh Ferguson and Mike Chrimes, *The Contractors* (London: ICE Publishing, 2014), ch. 2.

14. Brown and Williamson, *Lancelot Brown and the Capability Men*. The men – some of them the same – who worked with Richard Woods are described in Cowell, *Richard Woods*.

15. Brown and Williamson *Lancelot Brown and the Capability Men*, pp. 141–51; Brown, 'People, Places and Payments'.

16. Jane Brown, *Lancelot 'Capability' Brown: The Omnipotent Magician 1716–1783* (London: Pimlico, 2012), pp. 117–19.

17. Ibid., pp. 301–2.

18. David Green, *Gardener to Queen Anne: Henry Wise (1653–1738) and the Formal Garden* (Oxford: Oxford University Press, 1956), pp. 5–8.

19. Eleanor Joan Willson, *West London Nursery Gardens* (London: Fulham and Hammersmith Historical Society, 1982), p. 9.

20. Harvey, *Early Nurserymen*, p. 55.

21. Green, *Gardener to Queen Anne*.

22. Ibid., p. 163.

23. Mowl, *Gentlemen Gardeners*, p. 85.

24. *ODNB*, entry for Switzer.

25. Hadfield, *History of British Gardening*, p. 184.

26. Willis, *Charles Bridgeman and the English Landscape Garden*, pp. 41–3 and 53.

27. Jane Austen, *Mansfield Park* (1814; reprinted London: Macmillan, 1950), p. 47.

28. Stephen Daniels, *Humphry Repton: Landscape Gardening and the Geography of Georgian England* (New Haven and London: Yale University Press, 1999), p. 37, where it is pointed out that these rates were similar to those charged by the architects with whom Repton worked.

29. Ibid., p. 154.

30. See ibid., pp. 299–302, for a list of them.

31. Sarah Rutherford (ed.), *Humphry Repton in Buckinghamshire and Beyond* (Aylesbury: Buckinghamshire Gardens Trust, 2018), p. 25.

32. Cowell, *Richard Woods*, p. 163.

33. Thomas H. Mawson, *The Life and Work of an English Landscape Architect: An Autobiography* (London: Chapman and Hall, 1927); Janet Waymark, *Thomas Mawson: Life, Gardens and Landscapes* (London: Frances Lincoln, 2009).

34. J. P. Craddock, *Paxton's Protégé: The Milner White Gardening Dynasty* (Layerthorpe, York: York Publishing Services, 2012), p. 6.

35. Ibid., p. 14.

36. Ibid., p. 36.

37. Ibid., p. 55.

38. Richard Bisgrove, *William Robinson: The Wild Gardener* (London: Frances Lincoln, 2008).

39. Ibid., p. 235.

40. Christopher Hussey, quoted in Sally Festing, *Gertrude Jekyll* (London: Viking, 1991), p. xi.

41. Michael Tooley, *Gertrude Jekyll as Landscape Gardener* (Witton-le-Wear: Michaelmas Books, 1984), p. 65; Tooley – in 'The Plant Nursery at Munstead Wood', in Michael Tooley and Primrose Alexander (eds), *Gertrude Jekyll: Essays on the Life of a Working Amateur* (Witton-le-Wear: Michaelmas Books, 1995), pp. 114–24, at p. 120 – says that her account books between 1903 and 1929 show that she designed and provided plants for 398 projects.

42. Lanning Roper papers LRO 1/130, letter of 15 December 1982.

43. Cowell, *Richard Woods*, p. 170.

44. Daniels, *Humphry Repton*, p. 42.

45. Ibid., p. 41.

46. Kate Colquhoun, *A Thing in Disguise: The Visionary Life of Joseph Paxton* (London: Harper Perennial, 2004), p. 139.

47. Mawson, *Life and Work of an English Landscape Architect*, p. 68.

48. Ibid., p. 69.

49. Jane Brown, *Lanning Roper and His Gardens* (London: Weidenfeld & Nicolson, 1987), p. 197.
50. Mawson, *Life and Work of an English Landscape Architect*, p. 46. Cunliffe-Brooks was a banker and Conservative politician, MP for East Cheshire and later Altrincham between 1869 and 1892.
51. Ibid., pp. 46–7.
52. Ibid., pp. 116–17, 190–91.
53. NA WORK 1/4:87, letter to the Board of Works by Lancelot Brown.
54. For example, Penelope J. Corfield, *Power and the Professions in Britain, 1700–1850* (London: Routledge, 1995), W. J. Reader, *Professional Men: The Rise of the Professional Classes in Nineteenth-Century England* (London: Weidenfeld & Nicolson, 1966), and F. M. L. Thompson, *Chartered Surveyors: The Growth of a Profession* (London: Routledge & Kegan Paul, 1968).
55. Anthony Trollope, *The Bertrams* (1859), ed. S. Michell (Gloucester: Sutton, 1986), p. 88, quoted in Corfield, *Power and the Professions in England*, p. 174.
56. Quoted in Brown, *Lancelot 'Capability' Brown*, p. 262.
57. Ibid.
58. Cowell, *Richard Woods*, p. 171.
59. Mowl, *Gentlemen Gardeners*, p. 177.
60. Charles Quest-Ritson, *The English Garden: A Social History* (London: Viking, 2001), p. 152.
61. Corfield, *Power and the Professions in Britain*, p. 227.
62. Mawson, *Life and Work of an English Landscape Architect*, p. 46.
63. Horwood, *Gardening Women*, p. 123.
64. Lanning Roper papers LRO/1/130, letter of 30 June 1977.
65. Private information.

5. THE NURSERY TRADE

1. Vita Sackville-West, 'Winter', in *The Garden* (London: Michael Joseph, 1946).
2. A figure of £200 million as Maria's inheritance in today's money is the estimate in www.historyofparliamentonline.org, entry on Baring. The sum that is consistent with those used in this book is £198 million.
3. *Morning Post*, 5 June 1818, quoted in Todd Longstaffe-Gowan, *The London Town Garden 1700–1840* (New Haven and London: Yale University Press, 2001), at p. 178.

4. The records of James Cochran are in the National Archives: C 111/132 and 133; C 13/2160/5 and C 13/3245/18. Samuel Harrison's are in the London Metropolitan Archives B/HRS.

5. Harvey, *Early Nurserymen*, pp. 7–8, estimates that the nurseries listed in Thomas Milne's land-use map of London of 1795–9 (see note 17 below) occupied nearly 325 hectares (800 acres), but others may have been omitted.

6. Yet they have been entirely ignored by historians; neither of the two histories of shops in the eighteenth, nineteenth and twentieth centuries – James B. Jeffreys, *Retail Trading in Britain 1850–1950* (Cambridge: Cambridge University Press, 1954), and Hoh-Cheung Mui and Lorna H. Mui, *Shops and Shopkeepers in Eighteenth-Century England* (London: Routledge, 1989) – mentions them at all.

7. R. Campbell, *The London Tradesman* (London: T. Gardner, 1747), pp. 274–5.

8. NA PROB 4/2316, Gurle inventory. Taking such an inventory was a prerequisite for proving a will. No inheritance tax was levied, so there was no incentive to undervalue.

9. *ODNB*, entry on Gurle.

10. NA PROB 32/26, Gurle inventory.

11. Ibid.

12. *Ichnographia Rustica: or, The Nobleman, Gentleman, and Gardener's Recreation* (1741–2; reprinted London: Garland, 1982), p. 78. Harvey, *Early Nurserymen*, p. 52, argues that George London was initially the junior partner and that the lead was taken by Roger Looker (Lucre).

13. Green, *Gardener to Queen Anne*, p. 7.

14. Green, ibid., pp. 214–15, quoting an indenture of December 1714; NA PROB 3/29/111, inventory of Henry Wise.

15. PROB 31/79/292, inventory of Peter Mason.

16. Ibid.

17. Harvey, *Early Nurserymen*, pp. 4–7. Thomas Milne's *Plan of the Cities of London and Westminster, Circumjacent Towns and Parishes etc. Laid Down from a Trigonometrical Survey Taken in the Years 1795–9* is reproduced, with an introduction by G. B. G. Bull, in *Thomas Milne's Land Use Map of London and Environs in 1800* (London: London Topographical Society, 1975–6).

18. Peter Mathias, *Retailing Revolution: A History of Multiple Retailing in the Food Trades Based upon the Allied Suppliers Group of Companies* (London: Longman, 1967).

19. Mui and Mui, *Shops and Shopkeepers in Eighteenth-Century England*, p. 11. Drysalters were dealers in a range of chemical products,

including glue, varnish, dye and colourings. They might supply salt or chemicals for preserving food and sometimes also sold pickles, dried meat or related items.

20. Quoted in Willson, *West London Nursery Gardens*, p. 93.
21. See, for example, Wulf, *The Brother Gardeners*, which makes extensive use of the letters of John Bartram.
22. Harvey, *Early Nurserymen*, p. 76.
23. Michael Leapman, *The Ingenious Mr Fairchild* (London: Headline, 2000), p. 25.
24. James Edwards, *A Companion from London to Brighthelmston* (1787–1801), quoted in John H. Harvey, 'An Early Garden Centre: South London in 1789', *Garden History Society Newsletter*, 13(1971), pp. 3–4, at, p. 3.
25. Dorian Gerhold, *Carriers and Coachmasters: Trade and Travel before the Turnpike* (Chichester: Phillimore, 2005), p. 3.
26. NA C 111/132, records of James Cochran.
27. William Albert, 'The Turnpike Trusts', in Derek H. Aldcroft and Michael J. Freeman (eds), *Transport in the Industrial Revolution* (Manchester: Manchester University Press, 1983), pp. 31–63, at p. 60.
28. M. E. Porter, *On Competition* (Boston: Harvard Business Review Press, 1998), p. 97, defines such clusters as 'geographic concentrations of interconnected companies, specialist suppliers, service providers, firms in related industries and associated institutions'.
29. Malcolm Thick, *The Neat House Gardens: Early Market Gardening around London* (Totnes: Prospect Books, 1998), pp. 33–7.
30. F. M. L. Thompson, '19th Century Horse Sense', *Economic History Review*, 29 (1976), pp. 60–81, at pp. 70–71. By the 1880s, there were 80,000 vans plying the streets of London, each drawn by at least one horse; there were probably about the same number of coach horses and private carriage horses, a rough total of 150,000. The London population of that period was about three times what it had been in 1800, so a rough estimate might be that there were 50,000 horses in London at the beginning of the century. Each horse produced four to five tons of droppings each year, a total of 250,000 tons.
31. The more poetic Tudor name, 'gong farmer' (a 'gong' referring to a privy and its contents), appears to have fallen into disuse.
32. Emily Cockayne, *Hubbub: Filth, Noise and Stench in England* (New Haven and London: Yale University Press, 2007), p. 191.
33. M. Galinou, *London's Pride: The Glorious History of the Capital's Gardens* (London: Museum of London, 1990), p. 90.
34. The tradition continued: in 1923, Suttons Seeds, of Reading, were selling manure made with peat and charcoal saturated with London sewage.

35. Henry Hunter, *The History of London and its Environs*, 2 vols (London: John Stockdale, 1811), vol. 2 p. 3.
36. Harvey, *Early Nurserymen*, p. 124.
37. Ibid., p. 114.
38. Joy Uings, Barbara Moth and Moira Stevenson, *Caldwells: Nurserymen of Knutsford for Two Centuries* (Chester: Cheshire Gardens Trust, 2016), p. 47.
39. The Princess of Wales Conservatory at Kew, named after Princess Augusta but opened, in 1987, by Diana, Princess of Wales, was to incorporate the same technology nearly two centuries later.
40. David Solman, *Loddiges of Hackney: The Largest Hothouse in the World* (London: Hackney Society, 1995), p. 37.
41. Quoted in ibid., p. 46.
42. Sue Shephard, *Seeds of Fortune: A Gardening Dynasty* (London: Bloomsbury, 2003), pp. 121–2.
43. Alison M. Benton, *Cheals of Crawley* (Uckfield: Moira, 2002), pp. 122–7.
44. Society of Antiquaries MS 872/3, records of Loddiges nursery.
45. Roy W. Briggs, *'Chinese' Wilson: A Life of Ernest H. Wilson, 1876–1930* (London: HMSO, 1993), pp. 12–43. Wilson arrived in China only to discover that the single tree that had been identified by a previous plant hunter had been felled, though he then located a grove of trees overhanging a sheer drop. The *Davidia* samples were almost lost on Wilson's return to England, when his ship was wrecked, but he managed to save them.
46. Benton, *Cheals of Crawley*, p. 148. The money conversion is at earnings of 1910.
47. Ibid., pp. 148, 155 (quoting a guidebook) and 156.
48. www.historyofparliamentonline.org, entry for Codrington.
49. Longstaffe-Gowan, 'James Cochran: Florist and Plant Contractor', p. 59.
50. NA C 111/132.
51. Ibid.
52. NA C 111/133.
53. Department for Business, Innovation and Skills, *Business Population Estimates for the UK and Regions* (London: Office of National Statistics, 2015), p. 3, table A.
54. Shirley Heriz-Smith: 'The Veitch Nurseries of Killerton and Exeter c.1780 to 1883', *Garden History*, 16 (1988), pp. 41–57 and 174–88; 'James Veitch and Sons of Exeter and Chelsea', *Garden History*, 17 (1989), pp. 135–53; 'James Veitch and Sons, Chelsea: Harry Veitch's Reign, 1870–1890', *Garden History*, 20 (1992), pp. 57–70; 'James

Veitch and Sons of Chelsea and Robert Veitch & Son of Exeter, 1880–1969', *Garden History*, 21 (1993), pp. 91–109. See also Shephard, *Seeds of Fortune*. Historians of the Veitch family have been hampered by the destruction of all the archives of the firms by the last family member.

55. Quoted in Shephard, *Seeds of Fortune*, p. 80.

56. Quoted in ibid., p. 81.

57. The other farmer, Thomas, also finally returned to Exeter, opened a competitive seed business and, 'wholly given up to drunkenness', went bankrupt. Quoted in ibid., p. 126.

58. Ibid., p. 154.

59. Ibid., p. 258.

60. *ODNB*, entries for John Sutton and Martin Hope Sutton.

61. Earley Local History Group, *Suttons Seeds: A History 1806–2006* (Reading: Earley Local History Group, 2006), p. 224.

62. Uings et al., *Caldwells*, p. 54.

63. Ibid., p. 7.

64. Benton, *Cheals of Crawley*, p. 305.

65. Jean Hillier, *Hillier: The Plants, the People, the Passion* (Winchester: Outhouse Publishing, 2014), p. 175.

66. Ibid.

67. Diane Barre, 'Enterprising Women: Shaping the Business of Gardening in the Midlands, 1780–1830', in Dick and Mitchell (eds), *Gardens and Green Spaces*, pp. 102–19, at p. 114.

68. Ibid., p. 112.

69. Hillier, *Hillier*, p. 127.

70. Shephard, *Seeds of Fortune*, p. 159.

71. *ODNB*, entry for Martin Hope Sutton.

72. Benton, *Cheals of Crawley*, p. 215.

73. Office of National Statistics, *UK Personal Wealth Statistics 2011–2013* (London: Office of National Statistics, 30 September 2016), p. 4.

74. Loudon, *Encyclopaedia of Gardening*, quoted in John H. Harvey, *Early Horticultural Catalogues* (Bath: University of Bath Library, 1973), p. 189.

75. Callender's probate value is actually given as 'under £300' but this normally seems to have meant 'about £300'.

76. It is tempting to add up the three Veitch probate values – about £58 million – as an indication of the value of the firm, but we do not know how much of James Senior's assets had been left to James Junior and how much then to John Gould. This problem applies to the other family businesses mentioned here; we do not know how much of a given person's estate had been inherited or earned.

77. Arthur, James Junior's third son, left £13.8 million in 1880, at the age of only thirty-six, while James Junior's brother, Robert Tosswill, left £4.2 million in 1885. Harry (Sir Henry) left £13 million in 1924, although his two unfortunate nephews did not do so well: James Herbert left £1 million in 1907 and John Gould Junior in 1914 left £480,000. Of the other, Exeter, branch of the family, Peter Christian Veitch left £2.6 million in 1929.

78. Martin Hope Sutton left £34 million in 1901, Herbert Sutton £16 million in 1924, Arthur Warwick Sutton £20 million in 1925, Martin Hubert Fouquet Sutton £26 million in 1930, Leonard Goodhart Sutton £24 million in 1932, Martin Audley Fouquet Sutton £2 million in 1963, Leonard Noel Sutton £3.5 million in 1965 and Ernest Phillips Fouquet Sutton £2.5 million in 1972. Even if this suggests that the glory days of the business were over by the time of the Second World War, this is still a record of major financial achievement.

79. William George Caldwell left £3.5 million in 1873, William Caldwell IV £1.5 million in 1918, Arthur Caldwell £6 million in 1939 and William Caldwell V £2.6 million in 1953.

80. Alexander Cheal left £600,000 (£3,346) in 1933 and Joseph Cheal £560,000 (£3,288) in 1935, and the last Cheal whose probate is recorded, Ernest, left only £200,000 (£65,368) in 1961.

81. Those in between did better – Edwin left £2.5 million (£16,152) in 1926 and his eldest son, Edwin Laurence, £2.6 million (£22,155) in 1944, although another son, Arthur, had left only £430,000 (£11,063) in 1963.

82. Longstaffe-Gowan, *London Town Garden*, p. 160.

83. Todd Longstaffe-Gowan, 'James Cochran: Florist and Plant Contractor to Regency London', *Garden History*, 15 (1987), pp. 55–63, at p. 62. The account books are in NA C 111/132 and 133 and the court proceedings in C 13/2160/5 and C 13/3145/18.

84. Burial record, St John the Baptist Church, Egham, Surrey.

85. Longstaffe-Gowan, 'James Cochran: Florist and Plant Contractor', p. 56.

86. NA C 111/132, the Day Book of James Cochran.

87. Cochran's first wife was Elizabeth Hayward, whom he married at St George's, Hanover Square, on 18 April 1801; her date of death is not known. Their daughters were Elizabeth, born 28 July 1801, and Sarah, born 18 February 1804. His second wife, whom he married at St James's Piccadilly on 1 September 1815, was Anna (Maria) Louisa Harris, born 26 April 1784, and probably daughter of Stephen Harris, Gentleman of Lambeth.

88. NA C 13/2160/5. Unfortunately, the outcome of the case is not recorded.

89. NA PROB 11/1632/290, will of James Cochran.

6. THE WORKING GARDENER

1. Rudyard Kipling, 'The Glory of the Garden', in C. R. L. Fletcher and Rudyard Kipling, *A School History of England* (Oxford: Clarendon Press, 1911), p. 249.

2. Ibid.

3. 'British Gardeners – XXVIII', *Gardeners' Chronicle*, 23 October 1875. Details in the next two paragraphs of Chapter 6 all derive from this biography of Coleman, one of twenty-eight biographies of living British gardeners, most of a page or more, that were printed at different times in the *Gardeners' Chronicle*. A much larger number of obituaries of British gardeners was also published in the *Chronicle*, but these are usually shorter and less informative.

4. I am grateful to Jonathan Denby for this information.

5. Loudon, *Encyclopaedia of Gardening*, pp. 1199–202.

6. It is commonly said that the First World War destroyed the garden labour force of the great estates; the census data suggest otherwise.

7. David Mitch, 'Education and Skill of the British Labour Force', in Roderick Floud and Paul Johnson (eds), *The Cambridge Economic History of Modern Britain*, vol. 1 (Cambridge: Cambridge University Press, 2004), pp. 332–56, at p. 336.

8. William Cresswell, *Diary of a Victorian Gardener: William Cresswell and Audley End* (Swindon: English Heritage, 2006), p. 12.

9. Mitch, 'Education and Skill of the British Labour Force', p. 352.

10. E. A. Wrigley and R. S. Schofield, *The Population History of England 1541–1871* (Cambridge, MA: Harvard University Press, 1981), p. 119.

11. Although compulsory education and the state school system did not come until after 1870, male literacy in the sense of being able to sign your name at marriage (which implied an ability to read) was as high as 60 per cent in 1750 and it improved gradually thereafter, the outcome of a range of 'dame schools', private schools and Church and charitable provision.

12. The mechanics' institutes, some founded before 1850, did provide instruction in the principles underlying various trades, but not vocational training as such.

13. Percy Thrower (with Ronald Webber), *My Lifetime of Gardening* (London: Hamlyn, 1977), p. 11.

14. Ted Humphris, *Garden Glory: From Garden Boy to Head Gardener at Aynhoe Park* (London: Collins, 1969), p. 34.

15. Loudon, *Encyclopaedia of Gardening*, pp. 1199–200.

16. Alfred Wilcox, 'Owen Thomas', *Garden Life*, 23 (1913), p. 43.

17. *Gardeners' Chronicle*, 19 June 1875, p. 785.

18. Arthur Hooper, *Life in the Gardeners' Bothy* (Suffolk: Malthouse Press, 2000), p. 7.

19. CBS, D/RO/10/1, Mentmore Estate and Garden Labour Book, January 1883 to March 1884.

20. Thrower, *My Lifetime of Gardening*, p. 34.

21. Hooper, *Life in the Gardeners' Bothy*, p. 132. His then employer, Sir Jeremiah Colman at Gatton Park, Surrey, was reacting to the fact that a holiday with pay was becoming a custom in industry at the time.

22. Ibid., p. 80.

23. Ibid., p. 72.

24. *Gardeners' Chronicle*, 1874, p. 329, and private information from Sue Dickinson.

25. Thrower, *My Lifetime of Gardening*, p. 50.

26. Ibid., p. 34.

27. Hooper, *Life in the Gardeners' Bothy*, p. 93.

28. E. A. Wrigley, R. S. Davies, J. E. Oeppen and R. S. Schofield, *English Population History from Family Reconstitution, 1580–1837* (Cambridge: Cambridge University Press, 1997), p. 134.

29. CBS D/RO/10/1, records of the Rothschild family of Mentmore. At the time, the garden labour force was over forty strong.

30. Basil and Jessie Harley, *A Gardener at Chatsworth: Three Years in the Life of Robert Aughtie, 1848–1850* (Hanley Swan, Worcestershire: Self Publishing Association, 1992), p. 17.

31. Hooper, *Life in the Gardeners' Bothy*, pp. 69–70.

32. *Gardener's Magazine*, first edition (1826), p. 9.

33. An extensive search of the literature of garden history, together with research in the archives of a number of English record offices and stately homes, has produced only 188 wage rates across the entire period from 1516 to the present day. This constitutes, however, far more evidence than has ever before been available. Additional wages, for the period before 1660, can be found in Jill Francis, *Gardens and Gardening in Early Modern England and Wales* (New Haven and London: Yale University Press, 2018), pp. 127–30, but these largely duplicate those quoted here. There is no doubt that further research in local archives would produce more wage rates, which might make it possible to discover

whether there were significant regional variations, but this is beyond the scope of this book.

34. PP 1843 (C 510): xii, 302–6, *Reports of the Poor Law Commissioners on the Employment of Women and Children in Agriculture, Report by Francis Doyle*. The data have been analysed in David Meredith and Deborah Oxley, 'Food and Fodder: Feeding England, 1700–1900', *Past & Present*, 222 (2014), pp. 163–214, in the section entitled 'The Industrious Allen Family'.

35. Robert C. Allen – *The British Industrial Revolution in Global Perspective* (Cambridge: Cambridge University Press, 2009), p. 29, and 'The High Wage Economy and the Industrial Revolution: A Restatement', *Economic History Review*, 68 (2015), pp. 1–22, at p. 3 – gives a complete account of the gardener's budget. He earned £42,250 *(£37 12s)* a year. The original budget was described in Sir Frederick Eden, *The State of the Poor*, 3 vols (London: B. & J. White, G. G. & J. Robinson, T. Payne et al., 1797), vol. 2, pp. 433–5.

36. David Jacques and Arend Jan van der Horst, *The Gardens of William and Mary* (London: Christopher Helm, 1988), p. 73.

37. Chatsworth House, records of the Dukes of Devonshire and of Sir Joseph Paxton, CH14/7/2/4 Labour and Wages in Pleasure Garden, 20 January to 16 February 1918.

38. Peter King, *Women Rule the Plot: The Story of the 100 Year Fight to Establish Women's Place in Farm and Garden* (London: Duckworth, 1999), p. 7.

39. Anne Meredith, 'Horticultural Education in England, 1900–40: Middle-Class Women and Private Gardening Schools', *Garden History*, 31 (2003), pp. 67–79.

40. Private information from Sue Dickinson.

41. *Gardener's Magazine*, 1826, pp. 24–6.

42. Thrower, *My Lifetime of Gardening*, p. 57.

43. Quoted in Kate Colquhoun, *A Thing in Disguise: The Visionary Life of Joseph Paxton* (London: Harper Perennial, 2004), p. 31. Chatsworth was famous for its fountains, its cascade and its artificial willow that spouted water.

44. Hooper, *Life in the Gardeners' Bothy*, p. 154.

45. Ibid., p. 172.

46. Jonathan Denby, 'As the Houses, so the People. Gardeners: Their Accommodation and Remuneration 1800–1914', MA thesis, University of Buckingham (2013), pp. 67–8.

47. Thrower, *My Lifetime of Gardening*, p. 12.

48. Ibid., p. 47.

49. Hooper, *Life in the Gardeners' Bothy*, p. 76.

50. Ibid., p. 92.

51. *Gardeners' Chronicle*, 1874, p. 329.

52. Quest-Ritson, *The English Garden*, p. 107.

53. Grigson, *Menagerie*, p. 72.

54. CBS D/DR/5/121, records of the Drake family of Shardeloes.

55. CBS D/C/4/40, records of the Chester family of Chicheley.

56. Devon Record Office 158/M, records of the Courtenay family of Powderham.

57. Quest-Ritson, *The English Garden*, p. 107.

58. West Devon Record Office 69/M/6/154, records of the Pleydell family of Saltram. 'Board wages' was the term for a separate wage in lieu of food and other perquisites, normally paid when the family was not in residence.

59. Heriz-Smith, 'The Veitch Nurseries of Killerton and Exeter', p. 41.

60. Barre, 'Enterprising Women', p. 109.

61. I'm grateful to Jonathan Denby for drawing my attention to this correspondence.

62. Hillier of Winchester, records held at the headquarters of the company, August 1895.

63. Rothschild Bank Archives, family current account ledgers. I am grateful to Sarah Rutherford for providing these data.

7. TECHNOLOGY

1. Anon., *The Rise and Progress of the Present Taste in Planting Parks, Pleasure Grounds, Gardens, Etc.* (London: C. Moran, 1767), quoted in Guy Cooper, Gordon Taylor and Clive Boursnell, *English Water Gardens* (London: Weidenfeld & Nicholson, 1987), p. 9.

2. Thomas Savery, *The Miner's Friend; or, An Engine to Raise Water by Fire* (1702; reprinted London: J. McCormick, 1827).

3. Stephen Switzer, *An Introduction to a General System of Hydrostaticks and Hydraulicks, Philosophical and Practical*, 2 vols (London: T. Astley, S. Austen and L. Gilliver, 1729), vol. 2, pp. 326, 334. It is unclear which Sion house is being referred to. Syon House was at the time the seat of the Duke of Somerset, while the Duke of Chandos built the great house and garden at Cannons. Cambden House is currently unidentified.

4. Steffie Shields, *Moving Heaven and Earth: Capability Brown's Gift of Landscape* (London: Unicorn Publishing Group, 2016), p. 36.

5. Of innumerable discussions of technological invention and innovation, see, for example, Nathan Rosenberg, *Exploring the Black Box: Technology, Economics and History* (Cambridge: Cambridge University Press, 1994).

6. Elborough, *A Walk in the Park*, p. 16. There are no data on average earnings in 1166, which would be needed to translate this sum into a modern value. That sum of £26 9s 4d in 1270, when the data series begins, would equate to £398,000 today.

7. Baron Duckham, 'Canals and River Navigation', in Aldcroft and Freeman (eds), *Transport in the Industrial Revolution*, pp. 100–141, at p. 101; Dan Bogart, 'The Transport Revolution in Industrialising Britain', in Roderick Floud, Jane Humphries and Paul Johnson (eds), *The Cambridge Economic History of Modern Britain*, vol. 1, revised edn (Cambridge: Cambridge University Press, 2014), pp. 368–91.

8. William Alvis Brogden, *Ichnographia Rustica: Stephen Switzer and the Designed Landscape* (London: Routledge, 2017), 305. The river was semicircular or parabolic, 10 feet broad and 7 feet deep. Switzer, *Hydrostaticks and Hydraulicks*, vol. 1, pp. 103–4, emphasizes the importance of the fall, or incline, in a river or canal: too steep and the water will cause damage; too shallow and it will not flow properly. He thought that 'the general Allowance is four Inches and a half by all Ingineers' to the mile.

9. Brogden, *Switzer and the Designed Landscape*, 221.1, mentions that Switzer designed, for Sir John Ernle's garden at Whetham, Wiltshire, a cascade 500 feet long with a 60-foot drop.

10. John Hemingway, 'Exploring a Landscape Garden: William Shenstone at The Leasowes', in Dick and Mitchell (eds), *Gardens and Green Spaces*, pp. 40–55, at pp. 46–8. The term *ferme ornée*, invented by Switzer, referred to a country estate laid out partly according to aesthetic principles and partly for farming.

11. Ian Thompson, *The Sun King's Garden: Louis XIV, André Le Nôtre and the Creation of the Gardens of Versailles* (London: Bloomsbury, 2006), pp. 229–61.

12. Switzer, *Hydrostaticks and Hydraulicks*, vol. 1, p. 7. One estimate (Thompson, *The Sun King's Garden*, p. 243) puts the cost of the Versailles garden as a whole at 80 million livres; the modern equivalent is approximately £8.3 billion, although this seems an extraordinarily large sum. We now know, of course, that the lead pipework would have been slowly poisoning the French court. Charles Paulet, 1st Duke of Bolton, who held a variety of government posts and sinecures, supported the accession of William III and Queen Mary and was created duke in 1689.

13. *ODNB*, entry on Isaac de Caus. See also Strong, *Renaissance Garden in England*, ch. 6.

14. Hazelle Jackson, *Shell Houses and Grottoes* (Bodley: Shire Publications, 2001).

15. Hitching, *Rock Landscapes*.

16. Brogden, *Switzer and the Designed Landscape*, 108.9, says that Switzer was in charge of forming the canal that powered the pump, by diverting the River Glyme into 'sections of canal-like water-way to deliver the greatest possible weight of water to its end'.

17. Savery, *The Miner's Friend* pp. 34–5.

18. Switzer, *Hydrostaticks and Hydraulicks*, vol. 2, p. 328.

19. Richard L. Hills, *Machines, Mills and Uncountable Costly Necessities: A Short History of the Drainage of the Fens* (Norwich: Goose and Son, 1967), p. 47.

20. Switzer, *Hydrostaticks and Hydraulicks*, vol. 2, pp. 314, 317.

21. www.hinckleypastpresent.org, entry on Gopsall.

22. Judith Roberts, 'Cusworth Park: The Making of an Eighteenth Century Designed Landscape', *Landscape History*, 21 (1999), pp. 77–93, at p. 87.

23. Judith Roberts and Martin Hargreaves, 'Stephen Switzer: *Hydrostaticks* and Technology in the Country House Landscape', *Transactions of the Newcomen Society*, 73 (2003), pp. 163–78, at pp. 173–5. The poem is entitled 'Stourton Gardens'.

24. Symes, *Mr Hamilton's Elysium*, pp. 65, 70–74. The waterwheel was built to the design of Joseph Bramah, who had died in 1814, and installed at Painshill after 1832.

25. Brogden, *Switzer and the Designed Landscape*, 628.2.

26. Aislabie was accused of corruptly promoting the South Sea Company, having been given £20,000 (in original values) of its stock while he was Chancellor of the Exchequer. Although there was more than a whiff of political score-settling in the judgement, he was found by an investigation of the House of Commons to be guilty of 'most notorious, dangerous, and infamous corruption', expelled from the House and temporarily imprisoned in the Tower of London. Part of his estate was confiscated in order to compensate ordinary stock-holders of the company, who had suffered in the collapse, but he was still allowed to retire to Studley Royal with £233 million *(£119,000)*, part of which must have derived from his earlier tenure of the lucrative post of Treasurer of the Navy. See *ODNB*, entry on Aislabie.

27. Mark Newman, *The Wonder of the North: Fountains Abbey and Studley Royal* (Woodbridge: Boydell Press for the National Trust, 2015); *ODNB*, entry on Aislabie.

28. Steffie Shields, ' "Mr Brown Engineer": Lancelot Brown's Early Work at Grimsthorpe Castle and Stowe', *Garden History*, 34 (2006), pp. 174–91, argues that Brown was employed in Lincolnshire, but as a journeyman overseer at Grimsthorpe Castle constructing two dams, rather than in fen work.

29. John Phibbs, *Place-Making: The Art of Capability Brown* (Swindon: Historic England, 2017), p. 44, quoting E. Hall, ' "Mr Brown's directions": Capability Brown's Landscaping at Burton Constable (1767–82)', *Garden History*, 23 (1995), pp. 145–74, reporting Brown's visit to Constable Burton Hall on 1 September 1778.

30. Shields, *Moving Heaven and Earth*, p. 74.

31. Ibid., p. 106. However, the distinguished garden historian Mavis Batey believed that the poem was based on Nuneham Courtenay.

32. Hinde, *Capability Brown*, p. 141.

33. Judith Roberts, ' "Well Temper'd Clay": Constructing Water Features in the Landscape Park', *Garden History*, 29 (2001), pp. 12–28, at p. 15.

34. Hinde, *Capability Brown*, p. 51.

35. Jeri Bapasola, *The Finest View in England: The Landscape and Gardens of Blenheim Palace* (Woodstock: Blenheim Palace, 2009), p. 64. Roberts, 'Cusworth Park', p. 82, describes in detail the construction by Richard Woods, one of Brown's competitors, of the lakes and dams at Cusworth Hall, South Yorkshire. Roberts, ' "Well Temper'd Clay" ', pp. 15–19, explains the methods used, including puddling, by John Grundy, another of Brown's contemporaries, in creating the Great Lake at Grimsthorpe, Lincolnshire.

36. Bapasola, *Finest View in England*, p. 65.

37. Shields, *Moving Heaven and Earth*, p. 216.

38. G. M. Binnie, *Early Dam Builders in Britain* (London: Thomas Telford, 1987), pp. 61–2.

39. Quoted in Cowell, *Richard Woods*, p. 117.

40. Switzer, *Hydrostaticks and Hydraulicks*, vol. 1, p. ii.

41. Information from Center Parcs.

42. Switzer, *Hydrostaticks and Hydraulicks*, vol. 1, p. 10. Roberts and Hargreaves, 'Stephen Switzer: *Hydrostaticks* and Technology in the Country House Landscape', p. 163, emphasizes that Switzer was concerned with waterworks as an aspect of the management of the landed estate as a whole.

43. Shields, *Moving Heaven and Earth*, p. 40.
44. Hinde, *Capability Brown*, p. 52.
45. Ibid., p. 42.
46. Ibid., pp. 62, 110.
47. Roberts, ' "Well Temper'd Clay" ', p. 25.
48. Hinde, *Capability Brown*, p. 161.
49. Shields, *Moving Heaven and Earth*, p. 73.
50. Private information from the current owner.
51. L. T. C. Rolt, *Navigable Waterways* (Harlow: Longman, 1969).
52. Information from Wotton's owner, David Gladstone, and his gardener, Michael Harrison.
53. Grigson, *Menagerie*, p. 229.
54. Michael and Anne Heseltine, *Thenford: The Creation of an English Garden* (London: Head of Zeus, 2016).
55. Philip Miller, *The Gardeners Dictionary: Containing the Methods of Cultivating and Improving the Kitchen, Fruit and Flower Garden* (1731; 2nd edn London: C. Rivington, 1733), described tan or tanner's bark as 'the Bark of the Oak-tree, chopped and ground into coarse Powder, to be used in Tanning or Dressing of Skins; after which it is of great Use in Gardening: First, by its Fermentation . . . which is always moderate, and of a long Duration, which renders it of great Service to Hot-beds: and secondly, after it is well rotted, it becomes excellent Manure for all Sorts of cold stiff Land.' See also May Woods and Arete Warren, *Glass Houses: A History of Greenhouses, Orangeries and Conservatories* (London: Aurum Press, 1988), p. 61.
56. John Hix, *The Glasshouse* (London: Phaidon, 2005), p. 46.
57. Richard Bradley, *New Improvements of Planting and Gardening, both Philosophical and Practical* (Dublin: George Grierson, 1710), p. 34.
58. Ibid., pp. 128–9, 64. John Theophilus Desaguliers (1683–1744) was a French-born natural philosopher and engineer, member of the Royal Society and experimental assistant to Sir Isaac Newton, who developed an expertise in ventilation/fireplace efficiency.
59. Woods and Warren, *Glass Houses*, p. 35.
60. Robert Bruegmann, 'Central Heating and Forced Ventilation: Origins and Effects on Architectural Design', *Journal of the Society of Architectural Historians*, 37 (1978), pp. 143–60, at p. 144.
61. Jennifer Tann, *The Development of the Factory* (London: Cornmarket, 1970), p. 109.

62. Mark Girouard, *The Victorian Country House* (New Haven and London: Yale University Press, 1979), p. 23.

63. Ibid.

64. Duchess of Devonshire, *The Garden at Chatsworth* (London: Frances Lincoln, 1999), p. 67. The coal reached the boilers by tram through a tunnel, whose entrance can still be seen but was carefully concealed from visitors.

65. *Gardeners' Chronicle*, editions for 1841, 1860 and 1870.

66. www.beeston-notts.co.uk/pearson.shtml (accessed 13 September 2015).

67. Quoted in Lyndon F. Cave, *The Smaller English House: Its History and Development* (London: Robert Hale, 1981), p. 223. The fear of running out of coal was a major preoccupation of late Victorian society.

68. Duchess of Devonshire, *The Garden at Chatsworth*, p. 67.

69. Ibid., pp. 95–6.

70. Bruegmann, 'Central Heating and Forced Ventilation', pp. 149–52. The scheme was abandoned in 1846, with local turrets being adopted instead – Henrik Schoenefeldt, 'How the Palace of Westminster Gets Rid of All that Hot Air in the House of Commons' (theconversation.com/how-the-palace-of-westminster-gets-rid-of-all-that-hot-air-in-the-house-of-commons-39135).

71. Stefan Koppelkamm, *Glass Houses and Winter Gardens of the Nineteenth Century* (London: Granada Publishing, 1982), p. 11.

72. Bradley, *New Improvements*, p. 69.

73. Theo C. Barker, *The Glassmakers: Pilkington. The Rise of an International Company 1826–1976* (London: Weidenfeld & Nicolson, 1977), pp. 76–8. It is sometimes suggested that the tax on windows, also introduced in the 1690s, not temporarily repealed like the glass tax, but in fact substantially increased over the eighteenth century, inhibited greenhouse construction. However, the tax was levied by the time of repeal (1851) only on domestic houses with more than seven windows and was paid by only 380,000 out of 2,750,000 houses in England, Wales and Scotland, so it is unlikely to have had much effect on the price or supply of glass.

74. Hix, *The Glasshouse*, pp. 32–3, 36; the inventions were described in John Claudius Loudon, *Remarks on the Construction of Hothouses* (London: J. Taylor, 1817).

75. Koppelkamm, *Glass Houses and Winter Gardens*, p. 25, points out that prefabrication of buildings predated the Crystal Palace by at least fifty years. E. T. Bellhouse of Manchester was making prefabricated

iron buildings in the 1790s for shipping to Australia to house emigrants. By 1882, the catalogue of Walter MacFarlane of Glasgow ran to 700 pages of cast-iron parts of 'every imaginable type'. It was possible to buy complete kits for iron churches to equip missionaries around the British Empire.

76. C. M. Yonge, *The Daisy Chain* (London: John W. Parker, 1856), ch. 15, quoted in Girouard, *The Victorian Country House*, p. 38.

77. J. C. Loudon, 'Botanical Garden (Kew)', *Gardener's Magazine*, 16 (1840), pp. 365–6, discussed in Dustin Valen, 'On the Horticultural Origins of Victorian Glasshouse Design', *Journal of the Society of Architectural Historians*, 75 (2016), pp. 403–23, at p. 12.

78. Valen, 'On the Horticultural Origins', p. 415.

79. Woods and Warren, *Glass Houses*, p. 64.

80. Ibid., p. 98.

81. Loudon, *Encyclopaedia of Gardening*, p. 339.

82. Girouard, *The Victorian Country House*, p. 28.

83. Hix, *The Glasshouse*, p. 87.

84. As it was built with economy in mind, it required extensive and expensive repairs in the 1930s and 1970s – Ray Desmond, *Kew: The History of the Royal Botanic Gardens* (London: Harvill Press, 1995), p. 335. The latest restoration, completed in 2018, cost £41 million – *Guardian*, 16 January 2018.

85. Birmingham Archives MS 1056, records of Jones & Clark, later Henry Hope & Sons Ltd.

86. Ibid., MS 1056/249, Order Book, Messrs Clark and Hope, 1818–58.

87. *Gardeners' Chronicle*, 3 March 1860.

88. Anon., *The Leaf and the Tree: The Story of Boulton and Paul Ltd., 1797–1947* (Norwich: Boulton and Paul, 1947), p. 27.

89. Quoted in Kay N. Sanecki, *Old Garden Tools* (Botley: Shire Publications, 1987), p. 4.

90. NA T1/81.37, contract of Henry Wise.

91. Loudon, *Encyclopaedia of Gardening*, pp. 336–7.

92. I am grateful to Sean Bottomley for giving me access to his database on patented inventions during the Industrial Revolution.

93. NA T1/81.37, contract of Henry Wise.

94. Tom Fort, *The Grass is Greener: Our Love Affair with the Lawn* (London: HarperCollins, 2000), p. 34.

95. Brogden, *Switzer and the Designed Landscape*, 43.8.

96. Quoted in Sanecki, *Old Garden Tools*, p. 21.

97. Ibid.

98. Further models and their prices are discussed in Fort, *The Grass is Greener*, a light-hearted but informative discussion of lawns and lawnmowers.

99. Ibid., p. 129.

100. Ibid., p. 172.

101. The use of mercury in many infectious diseases was justified by the humoral theory of medicine, the basis of Western medicine from the Greeks to the nineteenth century, because it induced sweating, vomiting and purging, which were thought to draw out the excess humours. Here Bradley may have been using it in an attempt to destroy aphid larvae in the tree.

102. Bradley, *New Improvements* p. 117.

103. Loudon, *Encyclopaedia of Gardening*, pp. 283, 290.

104. Ibid., p. 293.

105. Kate Minnis, '"The Electric Melon": Experiments in Electro-Horticulture at Sherwood Park, Tunbridge Wells, Kent', *Garden History*, 43 (2015), pp. 256–72. Electro-horticulture involves using electricity to boost plant growth/crop yield.

106. Loudon, *Encyclopaedia of Gardening*, p. 307.

107. The English firm of Antony Gibbs was granted a monopoly for the export of guano to Europe and North America by the Peruvian government in 1847 and, by 1862, 435,000 tonnes were being mined annually, under atrocious conditions, by Chinese labourers. The firm's profits were between £50 million *(£80,000)* and £65 million *(£100,000)* a year in the 1850s and one of its owners, William Gibbs, was said to be the richest non-noble man in England. He spent part of his wealth on the house and estate of Tyntesfield in Somerset.

108. William Williamson, *The British Gardener: A Manual of Practical Instruction in Gardening* (London: Methuen, 1901), p. 64.

109. Loudon, *Encyclopaedia of Gardening*, p. 474.

110. Williamson, *The British Gardener*, pp. 77–9.

111. C. E. Lucas Phillips, *The Small Garden* (London: Pan Books, 1952), pp. 371–82.

112. D. G. Hessayon, *The Pest and Weed Expert* (London: Expert Books, 2009), p. 3. Loudon put it like this: 'The only way to protect from diseases is by using every means to promote health and vigor' – *Encyclopaedia of Gardening*, p. 474.

113. Hessayon, *The Pest and Weed Expert*, pp. 99–102.

8. THE PEOPLE'S GARDENS

1. T. W. H. Crosland, *The Suburbans* (London: John Long, 1905), p. 144.

2. Ibid.

3. Ibid., p. 150.

4. Quoted in Elizabeth Jones, 'Keats in the Suburbs', *Keats-Shelley Journal*, 46 (1996), pp. 23–43, at p. 26.

5. In a letter of March 1821 from Byron to his publisher John Murray, quoted in ibid.

6. Cynthia Floud (ed.), *A Dysfunctional Hampstead Childhood, 1886–1911: The Memoir of Phyllis Allen Floud, née Ford* (London: Camden History Society, 2018), p. 155.

7. John Betjeman, 'Slough', in *Continual Dew* (London: John Murray, 1937).

8. The left-wing novelist George Orwell wrote in 1939 of a suburban street as a 'line of semi-detached torture chambers where the poor little five-to-ten-pound-a-weekers quake and shiver, every one of them with the boss twisting his tail and his wife riding him like a nightmare and his kids sucking his blood like leeches' – *Coming Up for Air* (London: Victor Gollancz, 1939), p. 24.

9. It is, no doubt, entirely coincidental that gardens and parks still make up 15 per cent of the land area of English towns and cities today.

10. Galinou, *London's Pride*, p. 49.

11. Leapman, *The Ingenious Mr Fairchild*. 'Sea-coal' was so named because it was brought by sea from the mines of north-east England.

12. Thomas Fairchild, *The City Gardener* (London: T. Woodward and J. Peele, 1722), pp. 1–2. The Swedish writer Pehr Kalm observed in 1748 that there were many small gardens in London, whose owners 'had commonly planted in these yards and round about them, partly in the earth and ground itself, partly in pots and boxes, several of the trees, plants and flowers which could stand the coal-smoke in London' – Longstaffe-Gowan, *London Town Garden*, p. 23.

13. Longstaffe-Gowan, *London Town Garden*, p. 4.

14. Ibid., p. x.

15. Mark Laird and John H. Harvey, 'The Garden Plan for 13 Upper Gower Street, London', *Garden History*, 25 (1997), pp. 189–211; Longstaffe-Gowan, *London Town Garden*, pp. 139–49.

16. Horticultural Trades Association, *Garden Market Analysis Report 2016* (Didcot: Horticultural Trades Association, 2016), pp. 26–7.

17. Twiss's estimate is reproduced in Longstaffe-Gowan, *London Town Garden*, p. 142, and transcribed in Laird and Harvey, 'The Garden Plan for 13 Upper Gower Street', pp. 206–8.

18. The wage at the time of a jobbing gardener is given in Toby Musgrave, *The Head Gardeners* (London: Aurum Press, 2007), p. 50.

19. Robert D. Bell, 'The Discovery of a Buried Georgian Garden in Bath', *Garden History*, 18 (1990), pp. 1–21.

20. Ann Brooks, *'A Veritable Eden'. The Manchester Botanic Garden: A History* (Oxford: Oxbow Books, 2011), p. 11.

21. Zoë Crisp, 'The Urban Back Garden in England in the Long Nineteenth Century', *Proceedings of the Annual Conference of the Economic History Society* held at the University of Cambridge (2011), pp. 135–141, at pp. 137–8. Crisp's research – fully described in her PhD thesis, 'The Urban Back Garden in England in the Nineteenth Century', University of Cambridge (2013) – involved the painstaking measurement of a random sample of plot sizes from large-scale Ordnance Survey maps of the 1880s and 1890s, followed by an identification of the date of building of the houses and the occupations and social status of their inhabitants.

22. Ruth Duthie, 'English Florists' Societies and Feasts in the Seventeenth and First Half of the Eighteenth Centuries', *Garden History*, 10 (1982), pp. 17–35, and 'Florists' Societies and Feasts after 1750', *Garden History*, 12 (1984), pp. 8–38.

23. Margaret Willes, *The Gardens of the British Working Class* (New Haven and London: Yale University Press, 2014), p. 97.

24. Ibid., pp. 96–103.

25. Brent Elliott, 'Flower Shows in Nineteenth Century England', *Garden History*, 29 (2001), pp. 171–84, at p. 171.

26. Jane Harding and Anthea Taigel, 'An Air of Detachment: Town Gardens in the Eighteenth and Nineteenth Centuries', *Garden History*, 24 (1996), pp. 237–54, at pp. 237–8.

27. Joy Uings, 'Gardens and Gardening in a Fast-Changing Urban Environment: Manchester 1750–1850', PhD thesis, Manchester Metropolitan University (2013), p. 171.

28. Willes, *Gardens of the British Working Class*, pp. 146–9.

29. James Mangles, *The Floral Calendar* (London: F. W. Calder, 1839), pp. 70–71.

30. F. M. L. [Michael] Thompson (ed.), *The Rise of Suburbia* (Leicester: Leicester University Press, 1982), pp. 15, 2.

31. P. K. Stembridge, 'The Development of Thomas Goldney's Eighteenth Century Garden', *Garden History*, 45 (2017), pp. 117–42, at p. 130.

32. Harriet Devlin, 'Coalbrookdale: More than an Eighteenth-Century Industrial Landscape', in Dick and Mitchell (eds), *Gardens and Green Spaces*, pp. 56–75, at p. 62, quoting the journal of Samuel More, entry for 17 July 1790.

33. Susan Campbell, ' "Sowed for Mr C. D.": The Darwin Family's Garden Diary for The Mount, Shrewsbury, 1838–1865', *Garden History*, 37 (2009), pp. 135–50, and ' "Its Situation … was exquisite in the extreme": Ornamental Flowers, Shrubs and Trees in the Darwin Family's Garden at The Mount, Shrewsbury, 1838–1865', *Garden History*, 40 (2012), pp. 167–98.

34. Thompson (ed.), *The Rise of Suburbia*, p. 5, states that by 1850 every town of more than 50,000 inhabitants 'thought of itself as possessing some suburbs'. He also points out that suburbanization was not the product of mass public transport, since the suburbs around larger English towns developed fifty years or so before the age of cheap mass transit from the 1890s.

35. Uings, 'Gardens and Gardening', p. 137.

36. Phillada Ballard, '*Rus in Urbe*: Joseph Chamberlain's Gardens at Highbury, Moor Green, Birmingham, 1879–1914', *Garden History*, 14 (1986), pp. 61–76.

37. C. Treen, 'The Process of Suburban Development in North Leeds, 1870–1914', in Thompson (ed.), *The Rise of Suburbia*, pp. 157–209, at p. 175.

38. John Loudon, *The Suburban Gardener and Villa Companion* (London: Longman, Orme, Brown, Green and Longmans, 1838), p. 276. Loudon was, perhaps, optimistic about the realities of a back-breaking task when (p. 3) he enthused: 'What more delightful than to see the master or the mistress of a small garden or pleasure-ground, with all the boys and girls, the maids, and, in short, all the strength of the house, carrying pots and pails of water to different parts of the garden.'

39. Ibid., p. 194. Plots of this size are about four times as large as houses on modern housing developments in England, which are built, outside London, at about thirty houses per hectare or twelve per acre.

40. Ibid., p. 195.

41. Ibid., p. 189.

42. Ibid., p. 209. The hired gardener was paid between £11,340 and £15,110 in modern values.

43. Ibid., p. 235.

44. Based on a plan of the houses and garden in Loudon, *Suburban Gardener*, p. 326. See also Oliver Wainwright, 'The Grand London "Semi" that Spawned a Housing Revolution', *Guardian*, 1 April 2015.

45. Loudon, *Suburban Gardener*, pp. 342, 347. Loudon comments (p. 349): 'Had we confined ourselves to herbaceous plants, instead of growing 2000 species at one time, we might have had 10,000 in our limited space . . . 1800 species may be grown on the tops of the boundary walls of a garden very little larger than ours.'

46. Ibid., p. 478.

47. Ibid., p. 519.

48. Ibid., pp. 628–52. Loudon's reference to 'no expense . . . plants' is from *Gardener's Magazine*, 1834, p. 337, quoted in Horwood, *Gardening Women*, p. 43.

49. Horwood, *Gardening Women*, pp. 42–4; *ODNB*, entry on Joseph Marryat.

50. Loudon, *Suburban Gardener*, p. 661.

51. *ODNB*, entry on Loudon.

52. M. Simo, *Loudon and the Landscape* (New Haven and London: Yale University Press, 1988), p. 1.

53. Loudon, *Suburban Gardener*, p. 216.

54. Jones, 'Keats in the Suburbs', p. 33.

55. Quoted in Peter Bailey, *Leisure and Class in Victorian England: Rational Recreation and the Contest for Control, 1830–1855* (London: Routledge & Kegan Paul, 1978), p. 86.

56. Ibid., p. 88.

57. T. Geoffrey W. Henslow, *Suburban Gardens* (London: Rich and Cowan, 1934), p. 56.

58. Loudon thought that the middle-class suburban woman was fortunate by comparison with the lady of a great estate, because she had fewer duties and could, in the garden, work equally with a man – Sarah Bilston, ' "They congregate – in towns and suburbs": The Shape of Middle-Class Life in John Claudius Loudon's *The Suburban Gardener*', *Victorian Review*, 37 (2011), pp. 144–59, at p. 152. Michael Waters, *The Garden in Victorian Literature* (Aldershot: Scolar Press, 1988), p. 223, concludes from his study of Victorian literature that it was 'in their gardens more than anywhere else that privileged Victorians sported, courted and conversed' and that minor novels are pervaded with scenes in gardens.

59. B. R. Mitchell, *British Historical Statistics* (Cambridge: Cambridge University Press, 1988), pp. 389–91.

60. S. Constantine, 'Amateur Gardening and Popular Recreation in the 19th and 20th Centuries', *Journal of Social History*, 14 (1981), pp. 387–406, at p. 397.

61. For example, Judith Roberts estimates that 4 million houses were built in the inter-war period and had about half a million acres of new gardens between them – 'The Gardens of Dunroamin: History and Cultural

Values with Specific Reference to the Gardens of the Inter-War Semi', *International Journal of Heritage Studies*, 1 (1996), pp. 229–37, at p. 230.

62. Mark Swenarton and Sandra Taylor, 'The Scale and Nature of the Growth of Owner-Occupation in Britain between the Wars', *Economic History Review*, 38 (1985), pp. 373–92, at pp. 376–7. However, there was major class distinction: in the 1930s about 55 per cent of the middle class, but under 20 per cent of the working class, owned their own homes (ibid., p. 391).

63. Richard Rodger, *Housing in Urban Britain, 1780–1914* (Cambridge: Cambridge University Press, 1989), p. 14.

64. Ibid., p. 29.

65. John Burnett, *A Social History of Housing, 1815–1985* (London: Routledge, 1986), p. 199.

66. Thompson (ed.), *The Rise of Suburbia*, p. 22.

67. Constantine, 'Amateur Gardening and Popular Recreation', p. 393. Crisp, 'The Urban Back Garden' (2011), pp. 137–8, shows that in Preston 80 per cent of houses built between 1870 and 1881 had a private plot – with an average size of 190 square feet – and in Northampton 100 per cent did so, with an average size of 360 square feet.

68. Burnett, *A Social History of Housing*, pp. 154–71.

69. An alternative was to develop derelict land within the town, as Henry Trevor, a prosperous upholsterer and cabinet maker, did in Norwich in 1856. He transformed a disused chalk quarry into an elaborate garden – now known as 'The Plantation Garden' – and built a house next to it.

70. Ashworth, *Genesis of Modern British Town Planning*, p. 187.

71. *Municipal Journal*, 1906, p. 595.

72. Martin Gaskell, 'Gardens for the Working Class: Victorian Practical Pleasures', *Victorian Studies*, 23 (1980), pp. 479–501, at p. 498.

73. Raymond Unwin, *Town Planning in Practice: An Introduction to the Art of Designing Cities and Suburbs* (London: T. Fisher Unwin, 1909), p. 320. Barry Parker, Unwin's partner, expressed the same sentiment when, in 1927, he designed the Wythenshawe estate for Manchester City Council: 'the object is to secure around the house the air space requisite for health, to grow fruit and vegetables for our table . . . to surround ourselves with pleasant places in which to live and work, rest and play, and to entertain friends' – quoted in Matthew Hollow, 'Suburban Ideals on England's Interwar Council Estates', *Garden History*, 39 (2011), pp. 203–17, at p. 205.

74. The report of a committee chaired by Sir John Tudor Walters, a Welsh architect and politician, set the standards for the construction of council houses – and greatly influenced private housebuilding – until the 1960s, when they were replaced by standards proposed by the Parker Morris

committee. In the 1980s, the standards ceased to be mandatory and many houses built since then have been constructed to lower specifications.

75. Alan A. Jackson, *Semi-Detached London: Suburban Development, Life and Transport, 1900–1978* (London: George Allen & Unwin, 1978), pp. 149–73.

76. Another – discussed in Unwin, *Town Planning in Practice*, pp. 302–8 – was to reduce the cost of roads. The building by-laws had required that a high proportion of the land should be used for roads, while the local authorities, who would be required to maintain them, demanded that they be built to very high – and very costly – standards and paid for by the housebuilders. Lower-density housing allowed the proportion of road space to overall land area to fall. It meant using twice as much land, but – in contrast to today – no one seemed to worry about using agricultural land for houses. Even now, when far more concern is expressed about land use, average density has hardly shifted; in 2009 new houses on greenfield sites were built at thirty-two dwellings per hectare. Department of Communities and Local Government, Planning and Statistical Release, Land Use Change Statistics, England, May 2010 (www.assets. publishing.service.gov.uk/government/uploads/system/. . ./155850).

77. Mark Swenarton, *Homes Fit for Heroes: The Politics and Architecture of Early State Housing in Britain* (London: Heinemann, 1981), p. 79. The Parliamentary Secretary to the Local Government Board said in April 1919 that 'the money that we are going to spend on housing is an insurance against Bolshevism and Revolution' – speech by W. Astor to the House of Commons, from *Hansard*, 8 April 1919, column 1956.

78. Quoted in Constantine, 'Amateur Gardening and Popular Recreation', p. 391.

79. NA MAF 156/375, records of allotments, quoted in Ursula Buchan, *A Green and Pleasant Land: How England's Gardeners Fought the Second World War* (London: Hutchinson, 2013), pp. 78–9.

80. David L. North, 'Middle-Class Suburban Lifestyles and Culture in England, 1919–1939', DPhil thesis, University of Oxford (1989), p. 92.

81. Waters, *The Garden in Victorian Literature*, pp. 185–222, demonstrates through his study of Victorian literature that the demand for a garden existed in every social class and throughout the period.

82. One reason was the abandonment by the Conservative government under Margaret Thatcher of the Parker Morris space standards (see note 74 above).

83. Burnett, *A Social History of Housing*, p. 257, and Department of Transport, 'Vehicle Licensing Statistics' (www.gov.uk/government/collections/ vehicles-statistics).

84. Gaskell, 'Gardens for the Working Class', p. 479.

85. Quoted in Swenarton, *Homes Fit for Heroes*, p. 86.

86. The institution of British Summer Time in 1916, giving an extra hour's daylight after work in the summer months, has also been credited with encouraging gardening; an overall reduction in working hours probably had the same effect. Jackson, *Semi-Detached London*, p. 149

87. Hollow, 'Suburban Ideals on England's Interwar Council Estates', p. 205.

88. Simon Gunn and Rachel Bell, *Middle Classes: Their Rise and Spread* (London: Phoenix, 2002), p. 76.

89. Fort, *The Grass is Greener*, p. 154.

90. Ibid., p. 167.

91. *Winterwatch*, February 2018, BBC Two.

92. A survey conducted by the Horticultural Trades Association in 2006 found that about 8 per cent of respondents used a paid gardener, spending on average £301 *(£249)* a year. Unusually, the highest spending was by people under the age of thirty-five, where 5 per cent of respondents spent on average £661 *(£546)* a year. Ten per cent of respondents had used a landscaper in the previous ten years.

93. Oxford Economics, *The Economic Impact of Ornamental Horticulture and Landscaping in the UK: A Report for the Ornamental Horticulture Round Table Group* (London: Oxford Economics, 2018).

94. Ebenezer Howard, *Garden Cities of Tomorrow* (London: S. Sonnenschein, 1902), p. 31.

95. Willes, *Gardens of the British Working Class*, ch. 5.

96. Blanche Henrey, *British Botanical and Horticultural Literature before 1800*, 3 vols (London: Oxford University Press, 1975).

97. Sarah Dewis, *The Loudons and the Gardening Press: A Victorian Cultural Industry* (Farnham: Ashgate, 2014), p. 33; Sarah Bilston, 'Queens of the Garden: Victorian Women Gardeners and the Rise of the Gardening Advice Text', *Victorian Literature and Culture*, 36 (2008), pp. 1–19, at p. 4.

98. Ray G. C. Desmond, 'Victorian Gardening Magazines', *Garden History*, 5 (1977), pp. 47–66.

99. Dewis, *The Loudons and the Gardening Press*, p. 73.

100. One of the newcomers, George Glenny's *The Gardeners Gazette and Weekly Journal of Science, Literature and General News*, published between 1837 and 1847, made a profit of 'perhaps £1,000 per annum' by 1839 (over £750,000 today). Will Tjaden, '"The Gardeners Gazette" 1837–1847 and Its Rivals', *Garden History*, 11 (1983), pp. 70–78, at p. 70.

101. Jane Loudon, *The Ladies Flower-Garden of Ornamental Annuals* (London: W. Smith, 1842), p. i, quoted in Dewis, *The Loudons and the Gardening Press*, p. 202.

102. Horwood, *Gardening Women*, p. 253.

103. Quoted in Sophie Seifalian, 'Gardens of Metro-Land', *Garden History*, 39 (2011), pp. 218–38, at p. 225.

104. Constantine, 'Amateur Gardening and Popular Recreation', p. 387. The figure presumably relates to the total sales each month, as most were published monthly.

105. F. M. Wells, *The Suburban Garden and What to Grow in It* (London: Sampson, Low, Marston, 1901), p. 10; Lucas Phillips, *The Small Garden*, p. 17.

106. Roberts, 'The Gardens of Dunroamin', p. 233.

107. *ODNB*, entry on Cecil Henry Middleton.

108. Parks & Gardens (www.parksandgardens.org).

109. Susan Groag Bell, 'Women Create Gardens in Male Landscapes: A Revisionist Approach to Eighteenth-Century English Garden History', *Feminist Studies*, 16 (1990), pp. 471–91, at p. 473.

9. KITCHEN GARDENS

1. Thomas Tusser, *Five Hundred Points of Good Husbandry* (London: Richard Tottel, 1573), p. 69.

2. NA LS 13/237, records of the Lord Steward's department and of the royal household.

3. Mary Keen, *Paradise and Plenty: A Rothschild Family Garden* (London: Pimpernel Press, 2015).

4. Thorstein Veblen, *The Theory of the Leisure Class* (New York: Macmillan, 1899).

5. Jules Hudson, *Walled Gardens* (London: National Trust, 2018), p. 16.

6. The late Earl of Limerick, an accountant, once told me that he had calculated that each cauliflower grown in the garden of his country house in the 1990s cost him £28. But the cost of gardening seems to be beside the point here. Hudson, *Walled Gardens*, p. 225, quotes the chef at Gravetye Manor today, George Blogg, as stating that the garden's value is immeasurable: 'In pure economic terms, the figures may not add up. But can you put a price on freshness?'

7. Susan Campbell: 'Queen Victoria's Great New Kitchen Garden'; *Cottesbrooke: An English Kitchen Garden* (London: Century, 1987); *Charleston Kedding: A History of Kitchen Gardening* (London: Ebury

Press, 1996); 'Digging, Sowing and Cultivating in the Open Ground, 1600–1900', in Anne C. Wilson (ed.), *The Country House Garden, 1600–1950* (Stroud: History Press, 1998); *A History of Kitchen Gardening* (London: Frances Lincoln, 2005); *Walled Kitchen Gardens* (Bodley: Shire Publications, 2010).

8. Hudson, *Walled Gardens*. For the website, see Walled Kitchen Gardens Network (www.walledgardens.net).

9. Quoted in Campbell, 'Queen Victoria's Great New Kitchen Garden', p. 116.

10. Ibid., p. 106.

11. The full report with appendices is in NA Work 6/297. Details in the following paragraphs of Chapter 9 are taken from this report.

12. The full story of the building of Frogmore is given in NA Work 19/33/1 and in Campbell, 'Queen Victoria's Great New Kitchen Garden'.

13. NA Work 19/33/1.

14. Ibid.

15. Ibid.

16. Ibid.

17. Private information given in 2016.

18. Oxfordshire Gardens Trust, *The Walled Kitchen Gardens of Oxfordshire* (Chipping Norton: Oxfordshire Gardens Trust, 2014). Many of these gardens would have had 'slip' gardens, as well as potato patches, outside the walls.

19. Hudson, *Walled Gardens*, p. 46.

20. Bedfordshire Record Office L31/223. The sum is understated because the extra amount paid during haytime and harvest, when men were paid 8 shillings rather than 5 shillings and 6 pence per week, cannot be calculated.

21. Bedfordshire Record Office L31/319. Later records, for the nineteenth century, show that expenditure at this level continued throughout that century.

22. CBS D/DR/5/95/2, Bundle No. 6.

23. In some cases, as at the Lost Gardens of Heligan today, bark (today replaced by oak leaves) was used together with horse manure. At Heligan, 90 tons of manure are required. Hudson, *Walled Gardens*, p. 199.

24. Quoted in Campbell, *Charleston Kedding*, p. 160.

25. CBS D/DR/4/6 and D/DR/4/21.

26. Campbell, *Charleston Kedding*, pp. 156–65.

27. Quoted in A. Wilkinson, *The Victorian Gardener: The Growth of Gardening and the Floral World* (London: Sutton, 2006), pp. 143–4.

28. Hooper, *Life in the Gardeners' Bothy*, p. 18.

29. Ibid., p. 46.

30. Mrs James de Rothschild, *The Rothschilds at Waddesdon Manor* (London: Collins, 1979), p. 70.

31. The Rothschild Archive (family.rothschildarchive.org/estates/26-villa-Victoria) (accessed March 2018).

32. Rothschild, *Rothschilds at Waddesdon Manor*, p. 86.

33. Keen, *Paradise and Plenty*, pp. 13–14, 17–28.

34. Rothschild, *Rothschilds at Waddesdon Manor*, p. 81.

35. Thrower, *My Lifetime of Gardening*, p. 45.

36. Loudon, *Suburban Gardener*, p. 209. The sums mentioned equate to £11,340 to £15,110 today, which would have been a substantial weekly expenditure of £220 to £300 on fruit and vegetables, even for a large Victorian family with its servants.

37. Anthony Huxley, 'Introduction', in William Cobbett, *The English Gardener* (1829; reprinted Oxford: Oxford University Press, 1980), pp. vii–xii, at p. ix.

38. Loudon, *Encyclopaedia of Gardening* (1825), p. 1062.

39. Thick, *Neat House Gardens*, p. 41.

40. Mervyn Barnes, *Root and Branch: A History of the Worshipful Company of Gardeners of London* (London: Worshipful Company of Gardeners, 1994), p. 27.

41. Ibid., p. 30.

42. They are often known as the 'livery companies', although strictly this term applies only to companies that, after a period for establishing themselves, are given the power to appoint liverymen, senior members of the trades who can take part in the election of the Lord Mayor of London. Barnes, *Root and Branch*, chronicles the lengthy struggle of the Company of Gardeners to obtain the status of a livery company, since the City authorities for centuries held a grudge against them for having secured their first charter from the king directly rather than through the City Corporation.

43. Barnes, *Root and Branch*, p. 36.

44. Ibid., p. 37.

45. Thick, *Neat House Gardens*, p. 44.

46. Joan Thirsk, *Alternative Agriculture: A History from the Black Death to the Present Day* (Oxford: Oxford University Press, 1997), pp. 32–6.

47. Thick, *Neat House Gardens*, pp. 33–4.

48. Stephen Switzer, *Practical Kitchen Gardener* (London: G. Nicol, 1727), p. vii, quoted in Thick, *Neat House Gardens*, p. 37.

49. Thick, *Neat House Gardens*, p. 43. According to Thick (p. 80), it was a landholding of the Abbots of Westminster in the southern part of the manor of Ebury; they 'occupied the moated house of the sub-manor of the Neat (the "Neat House") as a rural retreat'.
50. Ibid., p. 101.
51. Ibid., pp. 79–102, 127–9.
52. Quoted in ibid., pp. 53, 57.
53. Uings, 'Gardens and Gardening', pp. 53–61.
54. Thirsk, *Alternative Agriculture*, pp. 148–9.
55. Ronald Webber, *Covent Garden: Mud-Salad Market* (London: Dent, 1969).
56. Quoted in ibid., pp. 118–21.
57. Ibid., p. 150. The figure of £67 million in 1865 equates to £2.2 billion today and £10 million to £332 million.
58. Thirsk, *Alternative Agriculture*, pp. 175–6.
59. Ibid., p. 179.
60. Ibid., p. 189.
61. Elizabeth David, *French Country Cooking* (London: John Lehmann, 1951), p. 153.
62. Gregory King, 'Natural and Political Conclusions upon the State and Condition of England', in G. Chalmers (ed.), *An Estimate of the Comparative Strength of Great Britain; and of the Losses of her Trade from Every War since the Revolution, with an Introduction of Previous History* (London: J. Stockdale, 1810), pp. 48–9, 67, quoted in Thick, *Neat House Gardens*, pp. 25–6.
63. R. H. Rew, 'The Nation's Food Supply', *Journal of the Royal Statistical Society*, 76 (1912), pp. 98–105, at p. 104.
64. Department of the Environment, *Food Statistics Pocketbook* (London: Department of the Environment, 2016).
65. These budgets have been described and analysed by a number of writers; see, for example, Carole Shammas, 'The Eighteenth-Century English Diet and Economic Change', *Explorations in Economic History*, 21 (1984), pp. 254–69.
66. Ibid., p. 256.
67. Jon Stobart and Lucy Bailey, 'Retail Revolution and the Village Shop, *c.* 1660–1860', *Economic History Review*, 71 (2018), pp. 393–417, at p. 397.
68. Thirsk, *Alternative Agriculture*, p. 148, referring to PP 1881: xvii, 829, *Minutes of Evidence of the Royal Commission on the Depressed Condition of the Agricultural Interests*.
69. Thirsk, *Alternative Agriculture*, p. 173.

70. Ian Gazeley and Andrew Newell, 'Urban Working-Class Food Consumption and Nutrition in Britain in 1904', *Economic History Review*, 68 (2015), pp. 101–22, at p. 120.

71. F. W. Pavy, *A Treatise on Food and Dietetics, Physiologically and Therapeutically Considered*, 2nd edn (London: J. & A. Churchill, 1875), pp. 301–7, quoted in Derek Oddy, *From Plain Fare to Fusion Food: British Diet from the 1890s to the 1990s* (Woodbridge: Boydell Press, 2003), p. 30.

72. Jeremy Burchardt, *Paradise Lost: Rural Idyll and Social Change since 1800* (London: I. B. Tauris, 2002), p. 50.

73. Leigh Shaw-Taylor, 'Parliamentary Enclosure and the Emergence of an English Agricultural Proletariat', *Journal of Economic History*, 61 (2001), pp. 640–62, at p. 654.

74. Unwin, writing both before and after the First World War, despite the very different conditions of the two periods, reiterated in a 1923 edition of *Town Planning in Practice* (p. 320) that 'Twelve houses to the net acre of building land ... has been proved to be about the right number to give gardens of sufficient size ... and not too large to be worked by an ordinary labourer and his family'.

75. Burchardt, *Allotment Movement in England*, p. 253, Appendices 4 and 5.

76. Burchardt, *Allotment Movement in England*, p. 163; B. S. Rowntree and M. Kendall, *How the Labourer Lives: A Study of the Rural Labour Problem* (London: Thomas Nelson, 1913), p. 307.

77. Burchardt, *Allotment Movement in England*, pp. 220, 228.

78. James Davidson, *Courtesans and Fishcakes: The Consuming Passions of Classical Athens* (London: Fontana, 1998).

79. Floud, Wachter and Gregory, *Height, Health and History*, pp. 136, 167. Other authors have argued that this overestimates average height in the late eighteenth century; for a description of the debate, see Floud et al., *The Changing Body*, pp. 134–45.

80. Ibid., pp. 1–40.

81. Massimo Livi-Bacci, *Population and Nutrition: An Essay on European Demographic History* (Cambridge: Cambridge University Press, 1991), p. 27, quoted in Floud et al., *The Changing Body*, p. 160.

82. Shammas, 'Eighteenth-Century English Diet', p. 257, table 1.

83. Ian Gazeley and Sara Horrell, 'Nutrition in the English Agricultural Labourer's Household Over the Course of the Long Nineteenth Century', *Economic History Review*, 66 (2013), pp. 757–84, at p. 775; Gazeley and Newell, 'Urban Working-Class Food Consumption and Nutrition', p. 118.

84. Ibid., p. 120. Most of the consumption of starchy foods in 1904 was of bread and potatoes; the unskilled working class consumed 9.4 lb per capita of these per week and the semi-skilled 9.7 lb (p. 109).
85. Quoted in Oddy, *From Plain Fare to Fusion Food*, p. 119.
86. Ibid., p. 137.
87. Preferences also played a part, even when incomes rose. As Oddy, *From Plain Fare to Fusion Food*, p. 203, comments: 'Although fresh green vegetables had been promoted heavily by Ministry of Food propaganda, wartime levels of consumption were not maintained once peacetime conditions prevailed in the food markets. Consumption had risen to 500g per head per week during the war ... but from 1987 onwards fell ... to under 250g per head per week during the 1990s.' Consumption of roots, tubers and bulbs also fell back markedly. There was some compensation from canned and frozen vegetables, which were easier to prepare.

10. CONCLUSION

1. Alexander Pope, as quoted in Loudon, *Encyclopaedia of Gardening*, p. 1202.
2. The only exception is that the cost of public services such as the police or the army, which are not sold, is also included. For an overall review of these statistics, see Charles Bean, *Independent Review of UK Economic Statistics* (London: Cabinet Office and the Treasury, 2016).
3. In recent years, economists and statisticians – who have long been aware of this – have attempted to estimate 'shadow' or 'satellite' accounts that do incorporate the value of non-traded goods and services, partially based on time-use data. These measures have not been brought into common use nor have they entered the consciousness of the general public, journalists or politicians.
4. Land with planning permission to build houses was, in 2015, worth £1 million per acre in the south of England and over £600,000 in the north. Prices have continued to rise since then. *Farmers Weekly* (www. fwi.co.uk), 23 January 2015 (accessed 12 February 2019).
5. Mariana Mazzucato, *The Entrepreneurial State: Debunking Public vs. Private Sector Myths* (London: Penguin Books, 2015).
6. See, for example, Geoffrey Crossick and Patrycja Kaszynska, *Understanding the Value of Arts and Culture: The AHRC Cultural Value Project* (Swindon: Arts and Humanities Research Council, 2016).

Selected Bibliography

ABBREVIATIONS

CBS   Centre for Buckinghamshire Studies
NA   National Archives
ODNB   *Oxford Dictionary of National Biography*
     (www.oxforddnb.com)
PP   Parliamentary Papers
PROB   Probate (in National Archives references)

ORIGINAL DOCUMENTS

Archives of HRH Frederick, Prince of Wales, and Princess Augusta, microfilm in British Library
Bedfordshire Record Office L31, records of the de Grey family of Wrest Park
Birmingham Archives MS 1056, records of Jones and Clark, later Henry Hope & Sons Ltd
CBS D/C, records of the Chester family of Chicheley
CBS D/DR, records of the Drake family of Shardeloes
CBS D/RO, records of the Rothschild family of Mentmore
Chatsworth House, records of the Dukes of Devonshire and of Sir Joseph Paxton
Devon Record Office 158/M, records of the Courtenay family of Powderham
Hillier of Winchester, records held at the headquarters of the company
London Metropolitan Archives B/HRS, records of Samuel Harrison
NA C 111 and C 13, records of James Cochran
NA LS 13, records of the Lord Steward's department and of the royal household
NA MAF 156, records of allotments
NA PROB 3 and PROB 31, records of Henry Wise
NA PROB 4 and PROB 32, records of Leonard Gurle

NA PROB 11/1632/290, will of James Cochran
NA PROB 31, records of Peter Mason
NA T1, contract of Henry Wise
NA Work 1/4, letter to the Board of Works from Lancelot Brown
NA Work 5/141, the Paymaster's accounts
NA Work 5/142, the Paymaster's accounts
NA Work 6/2, correspondence between the Treasury and Board of Works
NA Work 6/297, the Report of the Inquiry into the Management, Superintendence and Expenditure of the Royal Gardens, 1837
NA Work 19/33/1, the building of the royal kitchen garden at Frogmore
Norfolk Record Office MS 10, records of Humphry Repton
Rothschild Bank Archives, family current account ledgers
Royal Horticultural Society, Lindley Library, Lancelot Brown's account book
Royal Horticultural Society, Lindley Library, LRO/1, Lanning Roper papers
Society of Antiquaries MS 872/3, records of Loddiges nursery
West Devon Record Office 69/M/6, records of the Pleydell family of Saltram

BOOKS, JOURNAL ARTICLES, STUDIES AND REPORTS

Agriculture and Horticulture Development Board, *Import Substitution and Export Opportunities for UK Ornamentals Growers* (London: Agriculture and Horticulture Development Board, 2014)
Albert, William, 'The Turnpike Trusts', in Aldcroft and Freeman (eds), *Transport in the Industrial Revolution*, pp. 31–63
Aldcroft, Derek H., and Freeman, Michael J. (eds), *Transport in the Industrial Revolution* (Manchester: Manchester University Press, 1983)
Alexander, Anthony, *Britain's New Towns: Garden Cities to Sustainable Communities* (London and New York: Routledge, 2009)
Allen, Robert C., *The British Industrial Revolution in Global Perspective* (Cambridge: Cambridge University Press, 2009)
——, 'The High Wage Economy and the Industrial Revolution: A Restatement', *Economic History Review*, 68 (2015), pp. 1–22
Anon., *The Rise and Progress of the Present Taste in Planning Parks, Pleasure Grounds, Gardens etc. in a Poetic Epistle to the Right Honorable Charles, Lord Viscount Irwin* (London: C. Moran, 1767)
Anon., *The Leaf and the Tree: The Story of Boulton and Paul Ltd, 1797–1947* (Norwich: Boulton and Paul Ltd, 1947)
Ashworth, William, *The Genesis of Modern British Town Planning* (London: Routledge & Kegan Paul, 1954)

Austen, Jane, *Mansfield Park* (1814; reprinted London: Macmillan, 1950)

Bailey, Peter, *Leisure and Class in Victorian England: Rational Recreation and the Contest for Control, 1830–1855* (London: Routledge & Kegan Paul, 1978)

Ballard, Phillada, '*Rus in Urbe*: Joseph Chamberlain's Gardens at Highbury, Moor Green, Birmingham, 1879–1914', *Garden History*, 14 (1986), pp. 61–76

Bapasola, Jeri, *The Finest View in England: The Landscape and Gardens of Blenheim Palace* (Woodstock: Blenheim Palace, 2009)

Barker, Theo C., *The Glassmakers: Pilkington. The Rise of an International Company 1826–1976* (London: Weidenfeld & Nicolson, 1977)

Barnes, Melvyn, *Root and Branch: A History of the Worshipful Company of Gardeners of London* (London: Worshipful Company of Gardeners, 1994)

Barre, Diane, 'Enterprising Women: Shaping the Business of Gardening in the Midlands, 1780–1830', in Dick and Mitchell (eds), *Gardens and Green Spaces*, pp. 102–19

Bateman, John, *The Great Landowners of Great Britain and Ireland* (1883; reprinted Leicester: Leicester University Press, 1971)

Bean, Charles, *Independent Review of UK Economic Statistics* (London: Cabinet Office and the Treasury, 2016)

Bell, Robert D., 'The Discovery of a Buried Georgian Garden in Bath', *Garden History*, 18 (1990), pp. 1–21

Bell, Susan Groag, 'Women Create Gardens in Male Landscapes: A Revisionist Approach to Eighteenth-Century English Garden History', *Feminist Studies*, 16 (1990), pp. 471–91

Benton, Alison M., *Cheals of Crawley* (Uckfield: Moira, 2002)

Berkeley, Edmund, and Berkeley, Dorothy Smith, *The Life and Travels of John Bartram* (Tallahassee: Florida State University Press, 1982)

—— (eds), *The Correspondence of John Bartram, 1734–1777* (Gainesville: University Press of Florida, 1992)

Betjeman, John, 'Slough', in *Continual Dew* (London: John Murray, 1937)

Bilston, Sarah, 'Queens of the Garden: Victorian Women Gardeners and the Rise of the Gardening Advice Text', *Victorian Literature and Culture*, 36 (2008), pp. 1–19

——, '"They congregate – in towns and suburbs": The Shape of Middle-Class Life in John Claudius Loudon's *The Suburban Gardener*', *Victorian Review*, 37 (2011), pp. 144–59

Binnie, G. M., *Early Dam Builders in Britain* (London: Thomas Telford, 1987)

Bisgrove, Richard, *William Robinson: The Wild Gardener* (London: Frances Lincoln, 2008)

Bogart, Dan, 'The Transport Revolution in Industrialising Britain', in Floud, Humphries and Johnson (eds), *Cambridge Economic History of Modern Britain*, vol. 1 (2014), pp. 368–91

Bradley, Richard, *New Improvements of Planting and Gardening, Both Philosophical and Practical* (Dublin: George Grierson, 1710)

——, *The Gentleman and Gardener's Kalendar . . . To which is added, The Design of a Green-House* (London: W. Mears, 1718)

Briggs, Roy W., *'Chinese' Wilson: A Life of Ernest H. Wilson, 1876–1930* (London: HMSO, 1993)

Broadberry, Stephen, Campbell, Bruce M. S., Klein, Alexander, Overton, Mark and van Leeuwen, Bas, *British Economic Growth, 1270–1870* (Cambridge: Cambridge University Press, 2015)

Brogden, William Alvis, *Ichnographia Rustica: Stephen Switzer and the Designed Landscape* (London: Routledge, 2017)

Brooks, Ann, *'A Veritable Eden'. The Manchester Botanic Garden: A History* (Oxford: Oxbow Books, 2011)

Brown, David, 'People, Places and Payments: Lancelot Brown's Account Book', *Occasional Papers from the RHS Lindley Library*, 14 (2016), pp. 3–18

——, and Williamson, Tom, *Lancelot Brown and the Capability Men* (London: Reaktion Books, 2016)

Brown, Jane, *Lanning Roper and His Gardens* (London: Weidenfeld & Nicolson, 1987)

——, *The Pursuit of Paradise: A Social History of Gardens and Gardening* (London: Harper Collins, 1999)

——, *Lancelot 'Capability' Brown: The Omnipotent Magician 1716–1783* (London: Pimlico, 2012)

Bruegmann, Robert, 'Central Heating and Forced Ventilation: Origins and Effects on Architectural Design', *Journal of the Society of Architectural Historians*, 37 (1978), pp. 143–60

Buchan, Ursula, *A Green and Pleasant Land: How England's Gardeners Fought the Second World War* (London: Hutchinson, 2013)

Burchardt, Jeremy, *The Allotment Movement in England, 1793–1873* (Woodbridge: Boydell & Brewer, 2002)

——, *Paradise Lost: Rural Idyll and Social Change since 1800* (London: I. B. Tauris, 2002)

Burnett, John, *A Social History of Housing, 1815–1985* (London: Routledge, 1986)

Campbell, R., *The London Tradesman: Being a Compendious View of All the Trades, Professions, Arts, Both Liberal and Mechanic, Now Practised in the Cities of London and Westminster* (London: T. Gardner, 1747)

Campbell, Susan, 'The Genesis of Queen Victoria's Great New Kitchen Garden', *Garden History*, 12 (1984), pp. 100–119

——, *Cottesbrooke: An English Kitchen Garden* (London: Century, 1987)

——, *Charleston Kedding: A History of Kitchen Gardening* (London: Ebury Press, 1996)

——, 'Digging, Sowing and Cultivating in the Open Ground, 1600–1900', in Anne C. Wilson (ed.), *The Country House Garden, 1600–1950* (Stroud: History Press, 1998)

——, *A History of Kitchen Gardening* (London: Frances Lincoln, 2005)

——, '"Sowed for Mr C. D.": The Darwin Family's Garden Diary for The Mount, Shrewsbury, 1838–1865', *Garden History*, 37 (2009), pp. 135–50

——, *Walled Kitchen Gardens* (Bodley: Shire Publications, 2010)

——, '"Its Situation . . . was exquisite in the extreme": Ornamental Flowers, Shrubs and Trees in the Darwin Family's Garden at The Mount, Shrewsbury, 1838–1865', *Garden History*, 40 (2012), pp. 167–98

Carew, Thomas, 'To my friend G.N. from Wrest', in *The Works of Thomas Carew: Reprinted from the Original Edition of MDCXL [1640]* (1824; reprinted Edinburgh: W. and C. Tait, 1844)

Cave, Lyndon F., *The Smaller English House: Its History and Development* (London: Robert Hale, 1981)

Chadwick, G. F., *The Park and the Town: Public Landscape in the 19th and 20th Centuries* (New York and Washington: Frederick A. Praeger, 1966)

Christianson, C. Paul, *The Riverside Gardens of Sir Thomas More's London* (New Haven: Yale University Press, 2005)

Cobbett, William, *The American Gardener* (London: C. Clement, 1821)

——, *The English Gardener* (1829), with an introduction by Anthony Huxley (Oxford: Oxford University Press, 1980)

Cockayne, Emily, *Hubbub: Filth, Noise and Stench in England* (New Haven and London: Yale University Press, 2007)

Colquhoun, Kate, *A Thing in Disguise: The Visionary Life of Joseph Paxton* (London: Harper Perennial, 2004)

Constantine, S., 'Amateur Gardening and Popular Recreation in the 19th and 20th Centuries', *Journal of Social History*, 14 (1981), pp. 387–406

Coombs, David, 'The Garden at Carlton House of Frederick Prince of Wales and Augusta Princess Dowager of Wales: Bills in their Household Accounts 1728 to 1772', *Garden History*, 25 (1997), pp. 153–77

Cooper, Guy, Taylor, Gordon, and Boursnell, Clive, *English Water Gardens* (London: Weidenfeld & Nicholson, 1987)

Corfield, Penelope J., *Power and the Professions in Britain 1700–1850* (London: Routledge, 1995)

Cowell, Fiona, *Richard Woods (1715–1793): Master of the Pleasure Garden* (Woodbridge: Boydell Press, 2009)

Craddock, J. P., *Paxton's Protégé: The Milner White Gardening Dynasty* (Layerthorpe, York: York Publishing Services, 2012)

Cresswell, William, *Diary of a Victorian Gardener: William Cresswell and Audley End* (Swindon: English Heritage, 2006)

Crisp, Zoë, 'The Urban Back Garden in England in the Long Nineteenth Century', *Proceedings of the Annual Conference of the Economic History Society* held at the University of Cambridge (2011), pp. 135–41

——, 'The Urban Back Garden in England in the Nineteenth Century', PhD thesis, University of Cambridge (2013)

Crosland, T. W. H., *The Suburbans* (London: John Long, 1905)

Crossick, Geoffrey, and Kaszynska, Patrycja, *Understanding the Value of Arts and Culture: The AHRC Cultural Value Project* (Swindon: Arts and Humanities Research Council, 2016)

Curl, James Stevens (ed.), *Kensal Green Cemetery: The Origins and Development of the General Cemetery of All Souls, Kensal Green, London, 1824–2001* (Chichester: Phillimore, 2001)

Daniels, Stephen, *Humphry Repton: Landscape Gardening and the Geography of Georgian England* (New Haven and London: Yale University Press, 1999)

David, Elizabeth, *A Book of Mediterranean Food* (London: John Lehmann, 1950)

——, *French Country Cooking* (London: John Lehmann, 1951)

Davidson, James, *Courtesans and Fishcakes: The Consuming Passions of Classical Athens* (London: Fontana, 1998)

Denby, Jonathan, 'As the Houses, so the People. Gardeners: Their Accommodation and Remuneration 1800–1914', MA thesis, University of Buckingham (2013)

Department for Business, Innovation and Skills, *Business Population Estimates for the UK and Regions* (London: Office of National Statistics, 2015)

Department of Culture, Media and Sport, *The Royal Parks: Annual Report and Accounts 2015–2016* (London: Department of Culture, Media and Sport, 2016)

Department of the Environment, *Food Statistics Pocketbook* (London: Department of the Environment, 2016)

Desmond, Ray G. C., 'Victorian Gardening Magazines', *Garden History*, 5 (1977), pp. 47–66

——, *Kew: The History of the Royal Botanic Gardens* (London: Harvill Press, 1995)

Devlin, Harriet, 'Coalbrookdale: More than an Eighteenth-Century Industrial Landscape', in Dick and Mitchell (eds), *Gardens and Green Spaces*, pp. 56–75

Devonshire, Duchess of, *The Garden at Chatsworth* (London: Frances Lincoln, 1999)

Dewis, Sarah, *The Loudons and the Gardening Press: A Victorian Cultural Industry* (Farnham: Ashgate, 2014)

Dick, Malcolm, and Mitchell, Elaine (eds), *Gardens and Green Spaces in the West Midlands since 1700* (Hatfield: West Midlands Publications, 2018)

Dictionary of National Biography, vol. 5 (London: Smith, Elder & Co., 1885)

Duckham, Baron, 'Canals and River Navigation', in Aldcroft and Freeman (eds), *Transport in the Industrial Revolution*, pp. 100–141

Duthie, Ruth, 'English Florists' Societies and Feasts in the Seventeenth and First Half of the Eighteenth Centuries', *Garden History*, 10 (1982), pp. 17–35

——, 'Florists' Societies and Feasts after 1750', *Garden History*, 12 (1984), pp. 8–38

Earley Local History Group, *Suttons Seeds: A History 1806–2006* (Reading: Earley Local History Group, 2006)

Eden, Sir Frederick, *The State of the Poor*, 3 vols (London: London: B. & J. White, G. G. & J. Robinson, T. Payne et al., 1797)

Elborough, Travis, *A Walk in the Park: The Life and Times of a People's Institution* (London: Jonathan Cape, 2016)

Elliott, Brent, 'Flower Shows in Nineteenth Century England', *Garden History*, 29 (2001), pp. 171–84

——, 'The Landscape of Kensal Green Cemetery', in Curl (ed.), *Kensal Green Cemetery*, pp. 287–96

Evelyn, John, *The Diary of John Evelyn* (1818; reprinted Woodbridge: Boydell Press, 1995)

——, *Elysium Britannicum, or The Royal Gardens* (1659), ed. John E. Ingram (Philadelphia: University of Pennsylvania Press, 2000)

——, *Sylva, or a Discourse of Forest-Trees and the Propagation of Timber in His Majesty's Dominions* (London: John Martyn for the Royal Society, 1664)

Fairchild, Thomas, *The City Gardener* (London: T. Woodward and J. Peele, 1722)

Farrar, Linda, *Ancient Roman Gardens* (Stroud: History Press, 1998)

Ferguson, Hugh, and Chrimes, Mike, *The Contractors* (London: ICE Publishing, 2014)

Festing, Sally, *Gertrude Jekyll* (London: Viking, 1991)

Flinn, Michael W. (ed.), *Report on the Sanitary Condition of the Labouring Population, by Edwin Chadwick* (Edinburgh: Edinburgh University Press, 1965)

Floud, Cynthia (ed.), *A Dysfunctional Hampstead Childhood, 1886–1911: The Memoir of Phyllis Allen Floud, née Ford* (London: Camden History Society, 2018)

Floud, Roderick, 'Capable Entrepreneur? Lancelot Brown and His Finances', *Occasional Papers from the RHS Lindley Library*, 14 (2016), pp. 19–41

——, Fogel, Robert W., Harris, Bernard, and Hong, Sok Chul, *The Changing Body: Health, Nutrition and Human Development in the Western World since 1700* (Cambridge: Cambridge University Press, 2011)

——, Humphries, Jane, and Johnson, Paul (eds), *The Cambridge Economic History of Modern Britain*, vol. 1, revised edn (Cambridge: Cambridge University Press, 2014)

——, and Johnson, Paul (eds), *The Cambridge Economic History of Modern Britain*, vol. 1 (Cambridge: Cambridge University Press, 2004)

——, Wachter, Kenneth, and Gregory, Annabel, *Height, Health and History: Nutritional Status in the United Kingdom, 1750–1980* (Cambridge: Cambridge University Press, 1990)

Fort, Tom, *The Grass is Greener: Our Love Affair with the Lawn* (London: HarperCollins, 2000)

Francis, Jill, *Gardens and Gardening in Early Modern England and Wales* (New Haven and London: Yale University Press, 2018)

Galinou, M., *London's Pride: The Glorious History of the Capital's Gardens* (London: Museum of London, 1990)

Gaskell, Martin, 'Gardens for the Working Class: Victorian Practical Pleasures', *Victorian Studies*, 23 (1980), pp. 479–501

Gazeley, Ian, and Horrell, Sara, 'Nutrition in the English Agricultural Labourer's Household Over the Course of the Long Nineteenth Century', *Economic History Review*, 66 (2013), pp. 757–84

Gazeley, Ian, and Newell, Andrew, 'Urban Working-Class Food Consumption and Nutrition in Britain in 1904', *Economic History Review*, 68 (2015), pp. 101–22

Gerhold, Dorian, *Carriers and Coachmasters: Trade and Travel before the Turnpike* (Chichester: Phillimore, 2005)

Girouard, Mark, *The Victorian Country House* (New Haven and London: Yale University Press, 1979)

Green, David, *Gardener to Queen Anne: Henry Wise (1653–1738) and the Formal Garden* (Oxford: Oxford University Press, 1956)

Grigson, Caroline, *Menagerie: The History of Exotic Animals in England* (Oxford: Oxford University Press, 2000)

Gunn, Simon, and Bell, Rachel, *Middle Classes: Their Rise and Spread* (London: Phoenix, 2002)

Hadfield, Miles, *A History of British Gardening*, 3rd edn (London: John Murray, 1979)

Hall, E., '"Mr Brown's directions": Capability Brown's Landscaping at Burton Constable (1767–82)', *Garden History*, 23 (1995), pp. 145–74

Harding, Jane, and Taigel, Anthea, 'An Air of Detachment: Town Gardens in the Eighteenth and Nineteenth Centuries', *Garden History*, 24 (1996), pp. 237–54

Harley, Basil and Jessie, *A Gardener at Chatsworth: Three Years in the Life of Robert Aughtie, 1848–1850* (Hanley Swan, Worcestershire: Self Publishing Association, 1992)

Harvey, John H., 'An Early Garden Centre: South London in 1789', *Garden History Society Newsletter*, 13 (1971), pp. 3–4

——, *Early Gardening Catalogues* (Chichester: Phillimore, 1972)

——, *Early Horticultural Catalogues* (Bath: University of Bath Library, 1973)

——, *Early Nurserymen* (London and Chichester: Phillimore, 1974)

Hemingway, John, 'Exploring a Landscape Garden: William Shenstone at The Leasowes', in Dick and Mitchell (eds), *Gardens and Green Spaces*, pp. 40–55

Henrey, Blanche, *British Botanical and Horticultural Literature before 1800*, 3 vols (London: Oxford University Press, 1975)

Henslow, T. Geoffrey W., *Suburban Gardens* (London: Rich and Cowan, 1934)

Heritage Lottery Fund, *The State of UK Public Parks* (London: Heritage Lottery Fund, 2016)

Heriz-Smith, Shirley, 'The Veitch Nurseries of Killerton and Exeter *c.*1780 to 1883', *Garden History*, 16 (1988), pp. 41–57, 174–88

——, 'James Veitch and Sons of Exeter and Chelsea', *Garden History*, 17 (1989), pp. 135–53

——, 'James Veitch and Sons, Chelsea: Harry Veitch's Reign, 1870–1890', *Garden History*, 20 (1992), pp. 57–70

——, 'James Veitch and Sons of Chelsea and Robert Veitch & Son of Exeter, 1880–1969', *Garden History*, 21 (1993), pp. 91–109

Heseltine, Michael and Anne, *Thenford: The Creation of an English Garden* (London: Head of Zeus, 2016)

Hessayon, D. G., *The Pest and Weed Expert* (London: Expert Books, 2009)

Hillier, Jean, *Hillier: The Plants, the People, the Passion* (Winchester: Outhouse Publishing, 2014)

Hills, Richard L., *Machines, Mills and Uncountable Costly Necessities: A Short History of the Drainage of the Fens* (Norwich: Goose and Son, 1967)

Hinde, T., *Capability Brown: The Story of a Master Gardener* (London: Hutchinson, 1987)

Hitching, Claude, *Rock Landscapes: The Pulham Legacy* (Woodbridge: Antique Collectors' Club, 2012)

Hix, John, *The Glasshouse* (London: Phaidon, 2005)

Hollow, Matthew, 'Suburban Ideals on England's Interwar Council Estates', *Garden History*, 39 (2011), pp. 203–17

Hooper, Arthur, *Life in the Gardeners' Bothy* (Suffolk: Malthouse Press, 2000)

Horticultural Trades Association, *The Great British Gardener: A Profile of Gardeners in 2006* (Didcot: Horticultural Trades Association, 2006)

——, *Garden Market Analysis Report 2016* (Didcot: Horticultural Trades Association, 2016)

Horwood, Catherine, *Gardening Women: Their Stories from 1600 to the Present* (London: Virago Press, 2010)

Howard, Ebenezer, *Garden Cities of Tomorrow* (London: S. Sonnenschein, 1902)

Hudson, Jules, *Walled Gardens* (London: National Trust, 2018)

Hume, David, 'Of the Standard of Taste', in *Essays Moral, Political and Literary* (1739; reprinted London: A. Millar, 1768)

Humphris, Ted, *Garden Glory: From Garden Boy to Head Gardener at Aynhoe Park* (London: Collins, 1969)

Hunt, Tristram, *Building Jerusalem: The Rise and Fall of the Victorian City* (London: Weidenfeld & Nicolson, 2004)

Hunter, Henry, *The History of London and its Environs*, 2 vols (London: John Stockdale, 1811)

Impey, Edward, *Kensington Palace* (London: Merrell Publishers, 2012)

Jackson, Alan A., *Semi-Detached London: Suburban Development, Life and Transport, 1900–1978* (London: George Allen & Unwin, 1978)

Jackson, Hazelle, *Shell Houses and Grottoes* (Bodley: Shire Publications, 2001)

Jacques, David, *Gardens of Court and Country: English Design, 1630–1730* (New Haven: Yale University Press, 2017)

——, and van der Horst, Arend Jan, *The Gardens of William and Mary* (London: Christopher Helm, 1988)

Jashemski, Wilhelmina F., *The Gardens of Pompeii, Herculaneum and the Villas Destroyed by Vesuvius* (New Rochelle, NY: Caratzas Brothers, 1979)

Jeffreys, James B., *Retail Trading in Britain 1850–1950* (Cambridge: Cambridge University Press, 1954).

Jones, Elizabeth, 'Keats in the Suburbs', *Keats-Shelley Journal*, 46 (1996), pp. 23–43

Keen, Mary, *Paradise and Plenty: A Rothschild Family Garden* (London: Pimpernel Press, 2015)

King, Gregory, 'Natural and Political Conclusions upon the State and Condition of England', in G. Chalmers (ed.), *An Estimate of the Comparative Strength of Great Britain; and of the Losses of her Trade from Every War since the Revolution, with an Introduction of Previous History* (London: J. Stockdale, 1810), pp. 48–9, 67

King, Peter, *Women Rule the Plot: The Story of the 100 Year Fight to Establish Women's Place in Farm and Garden* (London: Duckworth, 1999)

Kipling, Rudyard, 'The Glory of the Garden', in C. R. L. Fletcher and Rudyard Kipling, *A School History of England* (Oxford: Clarendon Press, 1911)

Knyff, Leonard, and Kip, Johannes, *Britannia Illustrata: or Views of Several of the Queen's Palaces. As also of the Principal Seats of the Nobility and Gentry of Great Britain* (London: David Mortier, 1707), reprinted with an introduction and notes by John Harris and Gervase Jackson-Stops (Bungay: Paradigm Press for the National Trust, 1984)

Koppelkamm, Stefan, *Glass Houses and Winter Gardens of the Nineteenth Century* (London: Granada Publishing, 1982)

Krueger, Alan B. (ed.), *Measuring the Subjective Well-being of Nations: National Accounts of Time Use and Well-being* (Chicago: Chicago University Press, 2009)

Laird, Mark, and Harvey, John H., 'The Garden Plan for 13 Upper Gower Street, London', *Garden History*, 25 (1997), pp. 189–211

Leapman, Michael, *The Ingenious Mr Fairchild* (London: Headline, 2000)

Livi-Bacci, Massimo, *Population and Nutrition: An Essay on European Demographic History* (Cambridge: Cambridge University Press, 1991)

Longstaffe-Gowan, Todd, 'James Cochran: Florist and Plant Contractor to Regency London', *Garden History*, 15 (1987), pp. 55–63

——, *The London Town Garden 1700–1840* (New Haven and London: Yale University Press, 2001)

Loudon, Jane, *The Ladies' Flower-Garden of Ornamental Annuals* (London: W. Smith, 1842)

——, *The Ladies' Companion to the Flower Garden* (London: W. Smith, 1846)

Loudon, John Claudius, *Remarks on the Construction of Hothouses* (London: J. Taylor, 1817)

——, *An Encyclopaedia of Gardening* (London: Longman, Hurst, Rees, Orme and Brown, 1822; revised 1825)

——, *An Encyclopaedia of Cottage, Farm and Villa Architecture and Furniture* (London: Longman, Brown, Green and Longmans, 1834)

——, *The Suburban Gardener and Villa Companion* (London: Longman, Orme, Brown, Green and Longmans, 1838)

——, *The Suburban Horticulturist: or, An Attempt to Teach the Science and Practice of the Culture and Management of the Kitchen, Fruit & Forcing Garden to Those Who have had no Previous Knowledge or Practice in these Departments of Gardening* (London: W. Smith, 1842)

——, *Arboretum et Fruticetum Britannicum*, 8 vols (London: Henry G. Bohn, 1783–1843)

Lucas Phillips, C. E., *The Small Garden* (London: Pan Books, 1952)

Mangles, James, *The Floral Calendar* (London: F. W. Calder, 1839)

Mathias, Peter, *Retailing Revolution: A History of Multiple Retailing in the Food Trades Based upon the Allied Suppliers Group of Companies* (London: Longman, 1967)

Mawson, Thomas H., *The Life and Work of an English Landscape Architect: An Autobiography* (London: Chapman and Hall, 1927)

Mazzucato, Mariana, *The Entrepreneurial State: Debunking Public vs. Private Sector Myths* (London: Penguin Books, 2015)

Meredith, Anne, 'Horticultural Education in England, 1900–40: Middle-Class Women and Private Gardening Schools', *Garden History*, 31 (2003), pp. 67–79

Meredith, David, and Oxley, Deborah, 'Food and Fodder: Feeding England, 1700–1900', *Past & Present*, 222 (2014), pp. 163–214

Miller, Philip, *The Gardeners Dictionary: Containing the Methods of Cultivating and Improving the Kitchen, Fruit and Flower Garden* (1731; 2nd edn London: C. Rivington, 1754)

Minnis, Kate, '"The Electric Melon": Experiments in Electro-Horticulture at Sherwood Park, Tunbridge Wells, Kent', *Garden History*, 43 (2015), pp. 256–72

Mitch, David, 'Education and Skill of the British Labour Force', in Floud and Johnson (eds), *Cambridge Economic History of Modern Britain*, vol. 1 (2004), pp. 332–56

Mitchell, B. R., *British Historical Statistics* (Cambridge: Cambridge University Press, 1988)

Mitchell, Elaine, 'Duddeston's "shady walks and arbours": The Provincial Pleasure Garden in the Eighteenth Century', in Dick and Mitchell (eds), *Gardens and Green Spaces*, pp. 76–101

Mowl, Timothy, *Gentlemen Gardeners: The Men Who Created the English Landscape Garden* (Stroud: History Press, 2000)

Mui, Hoh-Cheung, and Mui, Lorna H., *Shops and Shopkeepers in Eighteenth-Century England* (London: Routledge, 1989)

Musgrave, Toby, *The Head Gardeners* (London: Aurum Press, 2007)

Newman, Mark, *The Wonder of the North: Fountains Abbey and Studley Royal* (Woodbridge: Boydell Press for the National Trust, 2015)

North, David L., 'Middle-Class Suburban Lifestyles and Culture in England, 1919–1939', DPhil thesis, University of Oxford (1989)

Oddy, Derek, *From Plain Fare to Fusion Food: British Diet from the 1890s to the 1990s* (Woodbridge: Boydell Press, 2003)

Office of National Statistics, *UK Personal Wealth Statistics 2011–2013* (London: Office of National Statistics, 30 September 2016)

Orwell, George, *Coming Up for Air* (London: Victor Gollancz, 1939)

Oxford Economics, *The Economic Impact of Ornamental Horticulture and Landscaping in the UK: A Report for the Ornamental Horticulture Round Table Group* (London: Oxford Economics, 2018)

Oxfordshire Gardens Trust, *The Walled Kitchen Gardens of Oxfordshire* (Chipping Norton: Oxfordshire Gardens Trust, 2014)

Parsons, Catherine, 'Horseheath Hall and its Owners', *Proceedings of the Cambridge Antiquarian Society*, 41 (1943–7), pp. 1–51

Pavy, F. W., *A Treatise on Food and Dietetics, Physiologically and Therapeutically Considered*, 2nd edn (London: J. & A. Churchill, 1875)

Phibbs, John, *Capability Brown: Designing the English Landscape* (New York: Rizzoli International Publications, 2016)

——, *Place-Making: The Art of Capability Brown* (Swindon: Historic England, 2017)

Plumb, Christopher, *The Georgian Menagerie: Exotic Animals in Eighteenth-Century London* (London: I. B. Tauris, 2015)

Porter, M. E., *On Competition* (Boston: Harvard Business Review Press, 1998)

PP 1833: xv, 337, *Report of a Select Committee Appointed to Consider the Best Means of Securing Open Spaces in the Vicinity of Populous Towns, as Public Walks and Places of Exercise, Calculated to Promote the Health and Comfort of the Inhabitants*

PP 1833: xxi, *First Report from Commissioners Appointed to Collect Information in the Manufacturing Districts, Relative to the Employment of Children in Factories*

PP 1843 (C 510): xii, 302–6, *Reports of the Poor Law Commissioners on the Employment of Women and Children in Agriculture, Report by Francis Doyle*

PP 1857–8: xlviii, 347, *Return, Showing the Amount of Public Money Expended in the Purchase and Formation of Public Parks, Public Walks and Recreation Grounds in Large Towns and Populous Places in Great Britain and Ireland since the Year 1840*

PP 1875 (C 1097): *England and Wales (Exclusive of the Metropolis): Return of Owners of Land, 1873; Presented to both Houses of Parliament by Command of Her Majesty*

PP 1881: xvii, 829, *Minutes of Evidence of the Royal Commission on the Depressed Condition of the Agricultural Interests*

Quealy, Gerit, *Botanical Shakespeare: An Illustrated Compendium* (New York: Harper Collins, 2017)

Quest-Ritson, Charles, *The English Garden: A Social History* (London: Viking, 2001)

Reader, W. J., *Professional Men: The Rise of the Professional Classes in Nineteenth-Century England* (London: Weidenfeld & Nicolson, 1966)

Rew, R. H., 'The Nation's Food Supply', *Journal of the Royal Statistical Society*, 76 (1912), pp. 98–105

Rich, T. C. G., Hutchinson, G., Randall, R., and Ellis, R. G., 'List of Plants Native to the British Isles', *BSBI News*, 80 (1999), pp. 23–7

Richardson, Ruth, and Curl, James Stevens, 'George Frederick Carden and the Genesis of the General Cemetery Company', in Curl (ed.), *Kensal Green Cemetery*, pp. 21–48

Roach, Alistair, 'Miniature Ships in Designed Landscapes', *The Mariner's Mirror*, 98 (2012), pp. 43–54

Roberts, Judith, 'The Gardens of Dunroamin: History and Cultural Values with Specific Reference to the Gardens of the Inter-War Semi', *International Journal of Heritage Studies*, 1 (1996), pp. 229–37

——, 'Cusworth Park: The Making of an Eighteenth Century Designed Landscape', *Landscape History*, 21 (1999), pp. 77–93

——, '"Well Temper'd Clay": Constructing Water Features in the Landscape Park', *Garden History*, 29 (2001), pp. 12–28

——, and Hargreaves, Martin, 'Stephen Switzer: *Hydrostaticks* and Technology in the Country House Landscape', *Transactions of the Newcomen Society*, 73 (2003), pp. 163–78

Rodger, Richard, *Housing in Urban Britain, 1780–1914* (Cambridge: Cambridge University Press, 1989)

Rolt, L. T. C., *Navigable Waterways* (Harlow: Longman, 1969)

Ronalds, Beverley F., 'Ronalds Nurserymen in Brentford and Beyond', *Garden History*, 45 (2017), pp. 82–100

Rosenberg, Nathan, *Exploring the Black Box: Technology, Economics and History* (Cambridge: Cambridge University Press, 1994)

Rothschild, Mrs James de, *The Rothschilds at Waddesdon Manor* (London: Collins, 1979)

Rowntree, B. S., and Kendall, M., *How the Labourer Lives: A Study of the Rural Labour Problem* (London: Thomas Nelson, 1913)

Rutherford, Sarah, 'The Landscapes of Public Lunatic Asylums in England, 1808–1914', PhD thesis, De Montfort University, Leicester (2003)

——, 'Landscapes for the Mind: English Asylum Designers, 1845–1914', *Garden History*, 33 (2005), pp. 61–86

——, *The Victorian Cemetery* (Bodley: Shire Publications, 2010)

—— (ed.), *Humphry Repton in Buckinghamshire and Beyond* (Aylesbury: Buckinghamshire Gardens Trust, 2018)

Sackville-West, Vita, *The Garden* (London: Michael Joseph, 1946)

Sanecki, Kay N., *Old Garden Tools* (Botley: Shire Publications, 1987)

Savery, Thomas, *The Miner's Friend; or, An Engine to Raise Water by Fire* (1702; reprinted London: J. McCormick, 1827)

Seifalian, Sophie, 'Gardens of Metro-Land', *Garden History*, 39 (2011), pp. 218–38

Sellar, W. C., and Yeatman, R. J., *1066 and All That* (London: Methuen, 1930)

——, *Garden Rubbish and Other Country Bumps* (London: Methuen, 1936; reprinted 1999)

Shakespeare, William, *The Complete Works of William Shakespeare*, ed. W. J. Craig (London: Oxford University Press, 1957)

Shammas, Carole, 'The Eighteenth-Century English Diet and Economic Change', *Explorations in Economic History*, 21 (1984), pp. 254–69

Shaw-Taylor, Leigh, 'Parliamentary Enclosure and the Emergence of an English Agricultural Proletariat', *Journal of Economic History*, 61 (2001), pp. 640–62

Shephard, Sue, *Seeds of Fortune: A Gardening Dynasty* (London: Bloomsbury, 2003)

Shields, Steffie, '"Mr Brown Engineer": Lancelot Brown's Early Work at Grimsthorpe Castle and Stowe', *Garden History*, 34 (2006), pp. 174–91

——, *Moving Heaven and Earth: Capability Brown's Gift of Landscape* (London: Unicorn Publishing Group, 2016)

Simo, M., *Loudon and the Landscape* (New Haven and London: Yale University Press, 1988)

Solman, David, *Loddiges of Hackney: The Largest Hothouse in the World* (London: Hackney Society, 1995)

Stembridge, P. K., 'The Development of Thomas Goldney's Eighteenth Century Garden', *Garden History*, 45 (2017), pp. 117–42

Stobart, Jon, and Bailey, Lucy, 'Retail Revolution and the Village Shop, *c.*1660–1860', *Economic History Review*, 71 (2018), pp. 393–417

Strong, Roy, *The Renaissance Garden in England* (London: Thames and Hudson, 1979)

Swenarton, Mark, *Homes Fit for Heroes: The Politics and Architecture of Early State Housing in Britain* (London: Heinemann, 1981)

——, and Taylor, Sandra, 'The Scale and Nature of the Growth of Owner-Occupation in Britain between the Wars', *Economic History Review*, 38 (1985), pp. 373–92

Switzer, Stephen, *Practical Kitchen Gardener* (London: G. Nicol, 1727)

——, *An Introduction to a General System of Hydrostaticks and Hydraulicks, Philosophical and Practical*, 2 vols (London: T. Astley, S. Austen and L. Gilliver, 1729)

——, *Ichnographia Rustica: or, The Nobleman, Gentleman, and Gardener's Recreation* (1741–2; reprinted London: Garland, 1982)

Sylvester, Joshua (trans.), 'Hortus', in *Du Bartas: His Divine Weekes and Workes* (London: Robert Young, 1641), p. 605

Symes, Michael, *Mr Hamilton's Elysium: The Gardens of Painshill* (London: Frances Lincoln, 2010)

Tann, Jennifer, *The Development of the Factory* (London: Cornmarket, 1970)

Thick, Malcolm, *The Neat House Gardens: Early Market Gardening around London* (Totnes: Prospect Books, 1998)

Thirsk, Joan, *Alternative Agriculture: A History from the Black Death to the Present Day* (Oxford: Oxford University Press, 1997)

Thompson, F. M. L., *Chartered Surveyors: The Growth of a Profession* (London: Routledge & Kegan Paul, 1968)

——, '19th Century Horse Sense', *Economic History Review*, 29 (1976), pp. 60–81

—— (ed.), *The Rise of Suburbia* (Leicester: Leicester University Press, 1982)

Thompson, Ian, *The Sun King's Garden: Louis XIV, André Le Nôtre and the Creation of the Gardens of Versailles* (London: Bloomsbury, 2006)

Thrower, Percy (with Webber, Ronald), *My Lifetime of Gardening* (London: Hamlyn, 1977)

Thurley, Simon, *Hampton Court: A Social and Architectural History* (New Haven: Yale University Press, 2003)

Tjaden, Will, '"The Gardeners Gazette" 1837–1847 and Its Rivals', *Garden History*, 11 (1983), pp. 70–78

Tooley, Michael, *Gertrude Jekyll as Landscape Gardener* (Witton-le-Wear: Michaelmas Books, 1984)

——, 'The Plant Nursery at Munstead Wood', in Michael Tooley and Primrose Alexander (eds), *Gertrude Jekyll: Essays on the Life of a Working Amateur* (Witton-le-Wear: Michaelmas Books, 1995), pp. 114–24

Treen, C., 'The Process of Suburban Development in North Leeds, 1870–1914', in Thompson (ed.), *The Rise of Suburbia*, pp. 157–209

Trollope, Anthony, *The Bertrams* (1859), ed. S. Michell (Gloucester: Sutton, 1986)

Turner, Michael E., Beckett, John V., and Afton, B., *Agricultural Rent in England, 1690–1914* (Cambridge: Cambridge University Press, 2004)

Turner, Roger, *Capability Brown and the Eighteenth-Century English Landscape* (1985; reprinted Stroud: History Press, 2013)

Tusser, Thomas, *A Hundred Good Points of Husbandry* (London: Richard Tottel, 1573)

Uings, Joy, 'Gardens and Gardening in a Fast-Changing Urban Environment: Manchester 1750–1850', PhD thesis, Manchester Metropolitan University (2013)

——, Moth, Barbara, and Stevenson, Moira, *Caldwells: Nurserymen of Knutsford for Two Centuries* (Chester: Cheshire Gardens Trust, 2016)

Unwin, Raymond, *Town Planning in Practice: An Introduction to the Art of Designing Cities and Suburbs* (London: T. Fisher Unwin, 1909; revised 1923)

Valen, Dustin, 'On the Horticultural Origins of Victorian Glasshouse Design', *Journal of the Society of Architectural Historians*, 75 (2016), pp. 403–23

Valentine, Alan, *The British Establishment, 1760–84*, 2 vols (Norman, OK: University of Oklahoma Press)

Veblen, Thorstein, *The Theory of the Leisure Class* (New York: Macmillan, 1899)

Waller, Edmund, *On St James's Park, as Lately Improved by His Majesty* (London: Gabriel Bedel and Thomas Collins, 1661)

Ward, J. R., *The Finance of Canal Building in Eighteenth-Century England* (Oxford: Oxford University Press, 1974)

Waters, Michael, *The Garden in Victorian Literature* (Aldershot: Scolar Press, 1988)

Waymark, Janet, *Thomas Mawson: Life, Gardens and Landscapes* (London: Frances Lincoln, 2009)

Webber, Ronald, *Covent Garden: Mud-Salad Market* (London: Dent, 1969)

Webster, Ian, 'The Public Works Loan Board and the Growth of the State in Nineteenth-Century England', *Economic History Review*, 71 (2018), pp. 887–908

Wells, F. M., *The Suburban Garden and What to Grow in It* (London: Sampson, Low, Marston, 1901)

Wilcox, Alfred, 'Owen Thomas', *Garden Life*, 23 (1913), p. 43

Wilkinson, A., *The Victorian Gardener: The Growth of Gardening and the Floral World* (London: Sutton, 2006)

Willes, Margaret, *The Gardens of the British Working Class* (New Haven and London: Yale University Press, 2014)

Williamson, John, *A Treatise on Military Finance; Containing the Pay of the Forces on the British and Irish Establishment; with the Allowances in Camp, Garrison and Quarters, &c.* (London: T. Everton, 1798)

Williamson, William, *The British Gardener: A Manual of Practical Instruction in Gardening* (London: Methuen, 1901)

Willis, Peter, *Charles Bridgeman and the English Landscape Garden* (Newcastle upon Tyne: Elysium Press, 2002)

Willson, Eleanor Joan, *West London Nursery Gardens* (London: Fulham and Hammersmith Historical Society, 1982)

Wilmot, John, Earl of Rochester, 'A Ramble in St James's Park', in *Poems on Several Occasions* (London: A. Thorncome, 1685)

Wilson, Richard, and Mackley, Alan, *The Building of the English Country House* (London: Continuum, 2000)

Woods, May, and Warren, Arete, *Glass Houses: A History of Greenhouses, Orangeries and Conservatories* (London: Aurum Press, 1988)

Wrigley, E. A., and Schofield, R. S., *The Population History of England 1541–1871* (Cambridge, MA: Harvard University Press, 1981)

——, Davies, R. S., Oeppen, J. E., and Schofield, R. S., *English Population History from Family Reconstitution, 1580–1837* (Cambridge: Cambridge University Press, 1997)

Wulf, Andrea, *The Brother Gardeners: Botany, Empire and the Birth of an Obsession* (London: Windmill Books, 2009)

——, and Gieben-Gamal, Emma, *This Other Eden: Seven Great Gardens and 300 Years of English History* (London: Little, Brown, 2005)

SELECTED WEBSITES

The following list includes websites cited in this book but not those for specific gardens/properties.

Assets: How They Work (www.assets.publishing.service.gov.uk/government/uploads/system/ . . . /155850)

Centre for the Study of the Legacies of British Slave-Ownership (www.ucl.ac.uk/lbs)

The Conversation (theconversation.com)

The Database of Court Officers: 1660–1837 (courtofficers.ctsdh.luc.edu)

Department of Transport, 'Vehicle Licensing Statistics' (www.gov.uk/government/collections/vehicles-statistics)

The Diary of Samuel Pepys (www.www.pepysdiary.com)

Farmers Weekly (www.fwi.co.uk)

The Garden's Trust (thegardenstrust.org)

HETUS – Harmonised European Time Use Survey (www.h6.scb.se/tus/tus)

Hinckley Past & Present (www.hinckleypastpresent.org)

Historic England (historicengland.org.uk)

Historic Environment Scotland (www.historicenvironment.scot)

History of Parliament Online (www.historyofparliamentonline.org)

Industry in Beeston, Nottinghamshire – The Pearson Family (www.beeston-notts.co.uk/pearson.shtml)

MeasuringWorth (www.measuringworth.com)

National Garden Scheme (www.ngs.org.uk)

Oxford Dictionary of National Biography (www.oxforddnb.com)

Parks & Gardens (www.parksandgardens.org)

The Rothschild Archive (www.rothschildarchive.org)

Royal Horticultural Society (www.rhs.org.uk)

Walled Kitchen Gardens Network (www.walledgardens.net)

Acknowledgements

Writing history, even in a new way, is always incremental; I have built on many books and journal articles that have instructed me and given me pointers to where information may be found. As always, librarians, archivists and the staff of record offices have been universally helpful. I am particularly grateful to the staff of the Bodleian Library, the British Library, the National Archives, the record offices of Buckinghamshire, Bedfordshire, Norfolk, Devon and Staffordshire, the Society of Antiquaries and the London Metropolitan Archives, the archivists of Chatsworth, Longleat and Waddesdon, and, above all, the Lindley Library of the Royal Horticultural Society. Many gardeners and garden owners have been immensely helpful: Robert and Jean Hillier (Hillier), Philip Oldham (Holkham Hall), Sonia Blackmore (Kitley), David Gladstone and Michael Harrison (Wotton), Richard Watson and Ian Buck (Center Parcs), Sir Beville Stanier (Shotover Park), Corinne Price (Old Warden), Robert Adams and Mark Thompson (Bledlow Manor), Janet Flinn (St Bridget's Nursery), Sue Dickinson and Tim Hicks (Eythrope), Lord and Lady Heseltine (Thenford) and Sir Roy Strong (The Laskett). Historians who have given much help and advice include Jonathan Denby, Jane Brown, David Brown, Brent Elliott, Jane Humphries, June Ellis, Claire de Carle, Martin Gaskell, Richard Bisgrove, Bernard Harris, Kevin Schürer, Joy Uings and Sam Williamson, while I have benefited from the comments of seminar audiences at the University of Oxford, the Royal Overseas League, the Bucks Local History Network, Wotton House, the Institute of Historical Research, the London School of Economics and the University of Cambridge. I began this book when I was Provost of Gresham College, London, and the council, academic board, staff and audiences at that unique institution have supported me throughout, not least by agreeing that a grant from the European Science Foundation could be used to finance the research. I am particularly grateful, however, to Professor Stanley Engerman, Professor Peter Solar and Dr Sarah Rutherford, who have each read the text and saved me from many errors of fact, interpretation and style; my agent, Peter Robinson, and my editors at Penguin, Cecilia Stein and Stuart Proffitt, and at Knopf, Vicky Wilson, have been very helpful and constructively critical. The book is much better than it would have been without their involvement and encouragement. My picture editor, Amanda Russell, has guided me very effectively, as have the editorial staff at Penguin led by Anna Hervé, and in particular my copy-editor, Kate Parker. None of them are responsible for any errors that remain.

Finally, my greatest debt is owed to my wife, Cynthia. She has supported me throughout the research, thought and effort that has gone into writing this book, as she has done during our marriage of more than fifty-five years. She has been a fount of ideas, questions and constructive criticism of my research and writing since we first met, but on this occasion she has also brought a love and deep knowledge of gardening; she has worked with me in record offices and together we have explored many gardens, from the most immaculate to the most overgrown, always searching for clues to the ways that gardens work and the care and expense that has been lavished upon them. She has discussed every topic, has read all that I have written, often several times, and has improved the text immeasurably. The book could not have been written without her.

Index

Properties included in the List of Gardens on pp. 293–301 are shown in bold.